Intuitive Biostatistics

Intuitive
Biostatistics

A Nonmathematical Guide to Statistical Thinking

HARVEY MOTULSKY, M.D.
GraphPad Software, Inc.

FOURTH EDITION

New York   Oxford

OXFORD UNIVERSITY PRESS

Oxford University Press is a department of the University of Oxford.
It furthers the University's objective of excellence in research,
scholarship, and education by publishing worldwide.

Oxford   New York
Auckland   Cape Town   Dar es Salaam   Hong Kong   Karachi
Kuala Lumpur   Madrid   Melbourne   Mexico City   Nairobi
New Delhi   Shanghai   Taipei   Toronto

With offices in
Argentina   Austria   Brazil   Chile   Czech Republic   France   Greece
Guatemala   Hungary   Italy   Japan   Poland   Portugal   Singapore
South Korea   Switzerland   Thailand   Turkey   Ukraine   Vietnam

For titles covered by Section 112 of the US Higher Education Opportunity
Act, please visit www.oup.com/us/he for the latest information about pricing
and alternate formats.

Published by Oxford University Press
198 Madison Avenue, New York, New York 10016
www.oup.com

Oxford is a registered trademark of Oxford University Press.

CIP data is on file at the Library of Congress
978-0-19-064356-0

987654321
Printed by LSC Communications, Inc. United States of America

I dedicate this book to my wife, Lisa, to my kids (Wendy, Nat, Joey, and Ruby), to readers who encouraged me to continue with a fourth edition, and to future scientists who I hope will avoid common mistakes in biostatistics.

PRAISE FOR INTUITIVE BIOSTATISTICS

Intuitive Biostatistics is a beautiful book that has much to teach experimental biologists of all stripes. Unlike other statistics texts I have seen, it includes extensive and carefully crafted discussions of the perils of multiple comparisons, warnings about common and avoidable mistakes in data analysis, a review of the assumptions that apply to various tests, an emphasis on confidence intervals rather than P values, explanations as to why the concept of statistical significance is rarely needed in scientific work, and a clear explanation of nonlinear regression (commonly used in labs; rarely explained in statistics books).

In fact, I am so pleased with *Intuitive Biostatistics* that I decided to make it the reference of choice for my postdoctoral associates and graduate students, all of whom depend on statistics and most of whom need a closer awareness of precisely why. Motulsky has written thoughtfully, with compelling logic and wit. He teaches by example what one may expect of statistical methods and, perhaps just as important, what one may not expect of them. He is to be congratulated for this work, which will surely be valuable and perhaps even transformative for many of the scientists who read it.

—Bruce Beutler, 2011 Nobel Laureate, Physiology or Medicine
Director, Center for the Genetics of Host Defense
UT Southwestern Medical Center

GREAT FOR SCIENTISTS

This splendid book meets a major need in public health, medicine, and biomedical research training—a user-friendly biostatistics text for non-mathematicians that clearly explains how to make sense of statistical results, how to avoid common mistakes in data analysis, how to avoid being confused by statistical nonsense, and (new in this edition) how to make research more reproducible.
You may enjoy statistics for the first time!

—Gilbert S. Omenn, Professor of Medicine,
Genetics, Public Health, and Computational Medicine &
Bioinformatics, University of Michigan

I am entranced by the book. Statistics is a topic that is often difficult for many scientists to fully appreciate. The writing style and explanations of *Intuitive Biostatistics* makes the concepts accessible. I recommend this text to all researchers. Thank you for writing it.

—Tim Bushnell, Director of Shared Resource Laboratories,
University of Rochester Medical Center

GREAT FOR STUDENTS

After struggling with books that weren't right for my class, I was delighted to find *Intuitive Biostatistics*. It is the best starting point for undergraduate students seeking to learn the fundamental principles of statistics because of its unique presentation of the important concepts behind statistics. Lots of books give you the "recipe" approach, but only *Intuitive Biostatistics* explains what it all means. It meticulously goes through common mistakes and shows how to correctly choose, perform, and interpret the proper statistical test. It is accessible to new learners without being condescending.

—Beth Dawson,
The University of Texas at Austin

This textbook emphasizes the thinking needed to interpret statistical analysis in published research over knowledge of the mathematical underpinnings. The basics of choosing tests and doing simpler analyses are covered very clearly and simply. The language is easy to understand yet accurate. It brings in the higher level of intuitive understanding that we hope students will have at the end of an honors undergraduate or MSc program, skipping over the mathematical details that are now handled by software anyway. It is the prefect approach and level for undergraduates beginning research.

—Janet E. Kübler, Biology Department,
California State University at Northridge

I read many statistics textbooks and have come across very few that actually explain statistical concepts well. Yours is a stand-out exception. In particular, I think you've done an outstanding job of helping readers understand P values and confidence intervals, and yours is one of the very first introductory textbooks to discuss the crucial concept of false discovery rates. I have already recommended your text to postgraduate students and postdoctoral researchers at my own institute.

—Rob Herbert
Neuroscience Research Australia

GREAT FOR EVERYONE

I've read several statistics books but found that some concepts I was interested in were not mentioned and other concepts were hard to understand. You can ignore the "bio" in *Intuitive Biostatistics,* as it is the best applied statistics books I have come across, period. Its clear, straightforward explanations have allowed me to better understand research papers and select appropriate statistical tests. Highly recommended.

—Ariel H. Collis, Economist,
Georgetown Economic Services

BRIEF CONTENTS

CONTENTS

PART B   **Introducing Confidence Intervals**

PART E   **Challenges in Statistics**

Tips for Understanding Models
Learn More About Nonlinear Regression
Common Mistakes: Nonlinear Regression
Q & A
Chapter Summary
Terms Introduced in this Chapter

PART H   **The Rest of Statistics**

PREFACE

My approach in this book is informal and brisk (at least I hope
it is), not ceremonious and plodding (at least I hope it isn't).

JOHN ALLEN PAULOS (2008)

Intuitive Biostatistics provides a comprehensive overview of statistics without getting bogged down in the mathematical details. I've been gratified to learn that many people have found my approach refreshing and useful. Some scientists have told me that statistics had always been baffling until they read *Intuitive Biostatistics*. This enthusiasm encouraged me to write this fourth edition.

WHO IS THIS BOOK FOR?

I wrote *Intuitive Biostatistics* for three main audiences:

- Medical (and other) professionals who want to understand the statistical portions of journals they read. These readers don't need to analyze any data, but they do need to understand analyses published by others and beware of common statistical mistakes. I've tried to explain the big picture without getting bogged down in too many details.
- Undergraduate and graduate students, postdocs, and researchers who analyze data. This book explains general principles of data analysis, but it won't teach you how to do statistical calculations or how to use any particular statistical program. It makes a great companion to the more traditional statistics texts and to the documentation of statistical software.
- Scientists who consult with statisticians. Statistics often seems like a foreign language, and this text can serve as a phrase book to bridge the gap between scientists and statisticians. Sprinkled throughout the book are "Lingo" sections that explain statistical terminology and point out when ordinary words are given very specialized meanings (the source of much confusion).

I wrote *Intuitive Biostatistics* to be a guidebook, not a cookbook. The focus is on how to interpret statistical results, rather than how to analyze data. This book presents few details of statistical methods and only a few tables required to complete the calculations.

If you think this book is too long, check out my other book, *Essential Biostatistics*, which is about one-third the size and price of this one (Motulsky, 2015).

WHAT MAKES THE BOOK UNIQUE?

Nonmathematical

Statistics is a branch of mathematics, so you can't truly understand statistics without studying many equations. In fact, some aspects of statistics cannot be understood unless you first master calculus and matrix algebra. Don't despair. This book is for the many students and scientists who may find math confusing and prefer verbal explanations. I only include equations in a few places for two reasons. One reason is that in a few places seeing an equation really helps explain a concept. The other reason is that a few simple calculations can easily be done by hand, and in these cases I present the necessary equation. Compared with most statistics books, this one uses very little math.

Statistical methods are tools, and scientists use lots of tools they don't always fully understand. For example, one can effectively use a radioactively labeled compound in an experiment without really understanding the nuclear physics of radioactive decay. You can do experiments with chemical reagents that you don't know how to synthesize. You can use scientific instruments without understanding exactly how they work. Similarly, you can interpret the results calculated by statistical software without understanding how those tests were derived and without having studied mathematical proofs that they work. What you need to know is the list of assumptions the test is based upon and the list of common conceptual traps to avoid.

Statistical lingo

In Lewis Carroll's *Through the Looking Glass*, Humpty Dumpty says, "When I use a word, it means exactly what I say it means—neither more nor less" (Carroll, 1871). Lewis Carroll (the pseudonym for Charles Dodgson) was a mathematician, and it almost seems he was thinking of statisticians when he wrote that line. But that can't be true because little statistical terminology had been invented by 1871.

Statistics books can get especially confusing when they use words and phrases that have both ordinary meanings and technical meanings. The problem is that you may think the author is using the ordinary meaning of a word or phrase, when in fact the author is using that word or phrase as a technical term with a very different meaning. I try hard to point out these potential ambiguities when I use potentially confusing terms such as:

- Significant
- Error
- Hypothesis
- Model
- Power
- Variance

- Residual
- Normal
- Independent
- Sample
- Population
- Fit
- Confidence
- Distribution
- Control

Includes topics omitted from many texts

I include many topics often omitted from short introductory texts, including:

- How common sense can mislead. Chapter 1 is a fun chapter that explains how common sense can lead you astray and why we therefore need to understand statistical principles.
- Multiple comparisons. It is simply impossible to understand statistical results without a deep understanding of how to think about multiple comparisons. Chapters 22, 23, and 40 explain several approaches used to deal with multiple comparisons, including the false discovery rate (FDR).
- Nonlinear regression. In many fields of science, nonlinear regression is used more often than linear regression, but most introductory statistics books ignore nonlinear regression completely. This book gives them equal weight. Chapters 34 and 35 set the stage by explaining the concept of fitting models to data and comparing alternative models. Chapter 36 then discusses nonlinear regression.
- Bayesian logic. Bayesian thinking is briefly mentioned in Chapter 2 and is then explored in Chapter 18 as a way to interpret a finding that a comparison is statistically significant. This topic returns in Chapter 42, which compares interpreting statistical significance to interpreting the results of clinical laboratory tests. These are only brief introductions to Bayesian thinking. However, a summary chapter about Bayesian approaches to data analysis, written with Daniel Lakens (Eindhoven University of Technology), will be posted at www.intuitivebiostatistics.com
- Lognormal distributions. These are commonly found in scientific data, but are rarely found in statistics books. They are explained in Chapter 11 and are touched upon again in several examples that appear in later chapters. Logarithms and antilogarithms are reviewed in Appendix E.
- Testing for equivalence. Sometimes the goal is not to prove that two groups differ but rather to prove that they are the same. This requires a different mindset, as explained in Chapter 21.
- Normality tests. Many statistical tests assume data are sampled from a Gaussian (also called normal) distribution, and normality tests are used to test this assumption. Chapter 24 explains why these tests are less useful than many hope.

- Outliers. Values far from the other values in a set are called outliers. Chapter 25 explains how to think about outliers.
- Comparing the fit of alternative models. Statistical hypothesis testing is usually viewed as a way to test a null hypothesis. Chapter 35 explains an alternative way to view statistical hypothesis testing as a way to compare the fits of alternative models.
- Meta-analysis as a way to reach conclusions by combining data from several studies (Chapter 43).
- Detailed review of assumptions. All analyses are based on a set of assumptions, and many chapters discuss these assumptions in depth.
- Lengthy discussion of common mistakes in data analysis. Most chapters include lists (with explanations) of common mistakes and misunderstandings.

Omits topics covered in most texts

To make space for the topics listed in the prior section, I have left out many topics that are traditionally included in introductory texts:

Probability. I assume that you have at least a vague familiarity with the ideas of probability, and this book does not explain these principles in much depth. Chapter 2 explains why probability can seem confusing. But you can still understand the rest of the book even if you skip this chapter.

Equations needed to compute statistical tests. I assume that you will be either interpreting data analyzed by others or using statistical software to run statistical tests. In only a few places do I give enough details to compute the tests by hand.

Statistical tables. If you aren't going to be analyzing data by hand, there is very little need for statistical tables. I include only a few tables in places where it might be useful to do simple calculations by hand.

Statistical distributions. You can choose statistical tests and interpret the results without knowing much about z, t, and F distributions. This book mentions them but goes into very little depth.

A unique organization

The organization of the book is unique.

Part A has three introductory chapters. Chapter 1 explains how common sense can mislead us when thinking about probability and statistics. Chapter 2 briefly explains some of the complexities of dealing with probability. Chapter 3 explains the basic idea of statistics—to make general conclusions from limited data, to extrapolate from sample to population.

Part B explains confidence intervals (CIs) in three contexts. Chapter 4 introduces the concept of a CI in the context of CIs of a proportion. I think this is the simplest example of a CI, since it requires no background information. Most books would start with the CI of the mean, but this would require first explaining the Gaussian distribution, the standard deviation, and the difference between the standard deviation and the standard error of the mean. CIs of proportions are much easier to understand. Chapters 5 and 6 are short chapters that explain CIs of survival data and Poisson (counted) data. Many instructors will choose to skip these two chapters.

Part C finally gets to continuous data, the concept with which most statistics books start. The first three chapters are fairly conventional, explaining how to graph continuous data, how to quantify variability, and the Gaussian distribution. Chapter 11 is about lognormal distributions, which are common in biology but are rarely explained in statistics texts. Chapter 12 explains the CI of the mean, and Chapter 13 is an optional chapter that gives a taste of the theory behind CIs. Then comes an important chapter (Chapter 14), which explains the different kinds of error bars, emphasizing the difference between the standard deviation and the standard error of the mean (which are frequently confused).

Part D is unconventional, as it explains the ideas of P values, statistical hypothesis testing, and statistical power without explaining any statistical tests. I think it is easier to learn the concepts of a P value and statistical significance apart from the details of a particular test. This section also includes a chapter (unusual for introductory books) on testing for equivalence.

Part E explains challenges in statistics. The first two chapters of this section explain the problem of multiple comparisons. This is a huge challenge in data analysis but a topic that is not covered by most introductory statistics books. The next two chapters briefly explain the principles of normality and outlier tests, topics that most statistics texts omit. Finally, Chapter 26 is an overview of determining necessary sample size.

Part F explains the basic statistical tests, including those that compare survival curves (an issue omitted from many introductory texts).

Part G is about fitting models to data. It begins, of course, with linear regression. Later chapters in the section explain the ideas of creating models and briefly explain the ideas of nonlinear regression (a method used commonly in biological research but omitted from most introductory texts), multiple regression, logistic regression, and proportional hazards regression.

Part H contains miscellaneous chapters briefly introducing analysis of variance (ANOVA), nonparametric methods, and sensitivity and specificity. It ends with a chapter (Chapter 43) on meta-analysis, a topic covered by few introductory texts.

Part I tries to put it all together. Chapter 44 is a brief summary of the key ideas of statistics. Chapter 45 is a much longer chapter explaining common traps in data analysis, a reality check missing from most statistics texts. Chapter 46 goes through one example in detail as a review. Chapter 47 is a new (to this edition) chapter on statistical concepts you need to understand to follow the current controversies about the lack of reproducibility of scientific works. Chapter 48, also new, is a checklist of things to think about when publishing (or reviewing) statistical results.

If you don't like the order of the chapters, read (or teach) them in a different order. It is not essential that you read the chapters in order. Realistically, statistics covers a lot of topics, and there is no ideal order. Every topic will be easier to understand if you had learned something else first. There is no ideal linear path through the material, and many of my chapters refer to later chapters. Some teachers have told me that they have successfully presented the chapters in a very different order than I present them.

WHAT'S NEW?

What was new in the second and third editions?

The second edition (published in 2010, 15 years after the first edition) was a complete rewrite with new chapters, expanded coverage of some topics that were only touched upon in the first edition, and a complete reorganization.

I substantially edited every chapter of the third edition and added new chapters on probability, meta-analysis, and statistical traps to avoid. The third edition introduced new sections in almost all chapters on common mistakes to avoid, statistical terms introduced in that chapter, and a chapter summary.

Overview of the fourth edition

In this fourth edition, I edited every chapter for clarity, to introduce new material, and to improve the Q&A and Common Mistakes sections. I substantially rewrote two chapters, Chapter 26 on sample size calculations and Chapter 28 about case-control studies. I also added two new chapters. Chapter 47 discusses statistical concepts regarding the reproducibility of scientific data. Chapter 48 is a set of checklists to use when publishing or reviewing scientific papers.

List of new topics in the fourth edition

- Chapter 1. Two new sections were added to the list of ways that statistics is not intuitive. One section points out that we don't expect variability to depend on sample size. The other points out that we let our biases determine how we interpret data.
- Chapter 2. New sections on conditional probability and likelihood. Updated examples.
- Chapter 4. Begins with a new section to explain different kinds of variables. New example (basketball) to replace a dated example about premature babies. Added section on Bayesian credible intervals. Improved discussion of "95% of what?" Took out rules of five and seven. Pie and stacked bar graphs to display a proportion.
- Chapter 7. New Q&As. Violin plot.
- Chapter 9. How to interpret a SD when data are not Gaussian. Different ways to report a mean and SD. How to handle data where you collect data from both eyes (or ears, elbows, etc.) in each person.
- Chapter 11. Geometric SD factor. Mentions (in Q&As) that lognormal distributions are common (e.g., dB for sound, Richter scale for earthquakes). Transforming to logs turns lognormal into Gaussian.
- Chapter 14. Error bars with lognormal data (geometric SD; CI of geometric mean). How to abbreviate the standard error of mean (SEM and SE are both used). Error bars with n = 2.
- Chapter 15. Stopped using the term *assume* with null hypothesis and instead talk about "what if the null hypothesis were true?" Defines null versus nil hypothesis. Manhattan plot. Advantage of CI over P. Cites the 2016 report about P values from the American Statistical Association.

- Chapter 16. Type S errors. What questions are answered by P values and CIs?
- Chapter 18. Added two examples and removed an outdated one (prednisone and hepatitis). Major rewrite.
- Chapter 19. Rewrote section on very high P values. Points out that a study result can be consistent both with an effect existing and with it not existing.
- Chapter 20. Distinguishing power from beta and the false discovery rate. When it makes sense to compute power.
- Chapter 21. Fixed 90% versus 95% confidence intervals. Two one-sided t tests.
- Chapter 22. Introduces the phrase (used in physics) *look elsewhere effect.*
- Chapter 23. Two new ways to get trapped by multiple comparisons, the garden of forking paths, and dichotomizing in multiple ways.
- Chapter 24. QQ plots. Corrected the explanation of kurtosis.
- Chapter 25. Points out that *outlier* has two meanings.
- Chapter 26. This chapter on sample size calculations has been entirely rewritten to clarify many topics.
- Chapter 28. This chapter on case-control studies has been substantially rewritten to clarify core concepts.
- Chapter 29. Improved definition of hazard ratio.
- Chapter 31. Added discussion of pros and cons of adjusting for pairing or matching.
- Chapter 32. New common mistake pointed out that if you correlate a variable A with another A-B, you expect r to be 0.7 even if data are totally random. Points out that r is not a percentage.
- Chapter 33. Which variable is X, and which is Y? Misleading results if you do one regression from data collected from two groups.
- Chapter 34. Defines the terms *response variable* and *explanatory variable.* Discusses three distinct goals of regression.
- Chapter 39. Expanded discussion of two-way ANOVA with an example.
- Chapter 42. Removed discussion of LOD score. Added example for HIV testing.
- Chapter 43. Added a discussion of meta-analyses using individual participant data, enlarged the discussion of funnel plots, added more Q&As.
- Chapter 45. New statistical traps: dichotomizing, confusing FDR with significance level, finding small differences with lots of noise, overfitting, pseudoreplication.
- Chapter 47. New chapter on reproducibility.
- Chapter 48. New chapter with checklists for reporting statistical methods.

What happened to the problems and answers?
The first three editions contained a chapter of problems and another chapter with extensive discussion of the answers. These have not been updated for the fourth edition, but the problems and answers for the third edition are available online at www.oup.com/us/motulsky and at intuitivebiostatistics.com .

WHICH CHAPTERS ARE ESSENTIAL?

If you don't have time to read this entire book, read these 16 chapters to learn the essential concepts of statistics:

1. Statistics and Probability Are Not Intuitive
3. From Sample to Population
4. Confidence Interval of a Proportion
9. Quantifying Scatter
10. The Gaussian Distribution
12. Confidence Interval of a Mean
14. Error Bars
15. Introducing P Values
16. Statistical Significance and Hypothesis Testing
18. Interpreting a Result That Is Statistically Significant
19. Interpreting a Result That Is Not Statistically Significant
22. Multiple Comparisons Concepts
23. The Ubiquity of Multiple Comparisons
33. Simple Linear Regression
44. The Key Concepts of Statistics
45. Statistical Traps to Avoid

WHO HELPED?

A huge thanks to the many people listed herein who reviewed draft chapters of the fourth edition. Their contributions were huge and immensely improved this book:

Reviewers:

John D. Bonagura, The Ohio State University
Hwanseok Choi, University of Southern Mississippi
John D. Chovan, Otterbein University
Stacey S. Cofield, The University of Alabama at Birmingham
Jesse Dallery, University of Florida
Vincent A. DeBari, Seton Hall University
Heather J. Hoffman, George Washington University
Stefan Judex, Stony Brook University
Janet E. Kübler, California State University, Northridge
Huaizhen Qin, Tulane University
Zaina Qureshi, University of South Carolina
Emily Rollinson, Stony Brook University
Walter E. Schargel, The University of Texas at Arlington
Evelyn Schlenker, University of South Dakota Sanford School of Medicine
Guogen Shan, University of Nevada Las Vegas
William C. Wimley, Tulane University School of Medicine
Jun-Yen Yeh, Long Island University
Louis G. Zachos, University of Mississippi

Thanks also to the reviewers of past editions, as many of their ideas and corrections survived into this edition: Raid Amin, Timothy Bell, Arthur Berg, Patrick Breheny, Michael F. Cassidy, William M. Cook, Beth Dawson, Vincent DeBari, Kathleen Engelmann, Lisa A. Gardner, William Greco, Dennis A. Johnston, Martin Jones, Janet E. Kubler, Janet E. Kübler, Lee Limbird, Leonard C. Onyiah, Nancy Ostiguy, Carol Paronis, Ann Schwartz, Manfred Stommel, Liansheng Tang, William C. Wimley, Gary Yellen Jan Agosti, David Airey, William (Matt) Briggs, Peter Chen, Cynthia J Coffman, Jacek Dmochowski, Jim Ebersole, Gregory Fant, Joe Felsenstein, Harry Frank, Joshua French, Phillip Ganter, Cedric Garland, Steven Grambow, John Hayes, Ed Jackson, Lawrence Kamin, Eliot Krause, James Leeper, Yulan Liang, Longjian Liu, Lloyd Mancl, Sheniz Moonie, Arno Motulsky, Lawrence "Doc" Muhlbaier, Pamela Ohman-Strickland, Lynn Price, Jeanette Ruby, Soma Roychowdhury, Andrew Schaffner, Paige Searle, Christopher Sempos, Arti Shankar, Patricia A Shewokis, Jennifer Shook, Sumihiro Suzuki, Jimmy Walker, Paul Weiss, and Dustin White.

I would also like to express appreciation to everyone at Oxford University Press: Jason Noe, Senior Editor; Andrew Heaton and Nina Rodriguez-Marty, Editorial Assistants; Patrick Lynch, Editorial Director; John Challice, Publisher and Vice President; Frank Mortimer, Director of Marketing; Lisa Grzan, Manager In-House Production; Shelby Peak, Senior Production Editor; Michele Laseau, Art Director; Bonni Leon-Berman, Senior Designer; Sarah Vogelsong, Copy Editor; and Pamela Hanley, Production Editor.

WHO AM I?

After graduating from medical school and completing an internship in internal medicine, I switched to research in receptor pharmacology (and published over 50 peer-reviewed articles). While I was on the faculty of the Department of Pharmacology at the University of California, San Diego, I was given the job of teaching statistics to first-year medical students and (later) graduate students. The syllabi for those courses grew into the first edition of this book.

I hated creating graphs by hand, so I created some programs to do it for me! I also created some simple statistics programs after realizing that the existing statistical software, while great for statisticians, was overkill for most scientists. These efforts constituted the beginnings of GraphPad Software, Inc., which has been my full-time endeavor for many years (see Appendix A). In this role, I exchange emails with students and scientists almost daily, which makes me acutely aware of the many ways that statistical concepts can be confusing or misunderstood.

I have organized this book in a unique way and have chosen an unusual set of topics to include in an introductory text. However, none of the ideas are particularly original. All the statistical concepts are standard and have been discussed in many texts. I include references for some concepts that are not widely known, but I don't provide citations for methods that are in common usage.

Please email me with your comments, corrections, and suggestions for the next edition. I'll post errata at www.intuitivebiostatistics.com.

Harvey Motulsky
hmotulsky@graphpad.com
November 2016

ABBREVIATIONS

Abbreviation	Definition	Chapter where defined
α (alpha)	Significance level	16
ANOVA	Analysis of variance	39
CI	Confidence interval	4
CV	Coefficient of variation	9
df	Degrees of freedom	9
FDR	False discovery rate	18
FPR	False positive rate	18
FPRP	False positive reporting probability	18
n	Sample size	4
OR	Odds ratio	28
SD or s	Standard deviation	9
SE	Standard error	14
SEM	Standard error of the mean	14
p (lower case)	Proportion	4
P (upper case)	P value	15
r	Correlation coefficient	32
ROC	Receiver operating characteristic curve	42
RR	Relative risk	27
W	Margin of error	4

PART A

Introducing Statistics

CHAPTER 1

Statistics and Probability
Are Not Intuitive

If something has a 50% chance of happening, then 9 times out of 10 it will.

YOGI BERRA

The word *intuitive* has two meanings. One is "easy to use and understand," which is my goal for this book—hence its title. The other meaning is "instinctive, or acting on what one feels to be true even without reason." This fun (really!) chapter shows how our instincts often lead us astray when dealing with probabilities.

WE TEND TO JUMP TO CONCLUSIONS

A three-year-old girl told her male buddy, "You can't become a doctor; only girls can become doctors." To her, this statement made sense, because the three doctors she knew were all women.

When my oldest daughter was four, she "understood" that she was adopted from China, whereas her brother "came from Mommy's tummy." When we read her a book about a woman becoming pregnant and giving birth to a baby girl, her reaction was, "That's silly. Girls don't come from Mommy's tummy. Girls come from China." With only one person in each group, she made a general conclusion. Like many scientists, when new data contradicted her conclusion, she questioned the accuracy of the new data rather than the validity of the conclusion.

The ability to generalize from a sample to a population is hardwired into our brains and has even been observed in eight-month-old babies (Xu & Garcia, 2008). To avoid our natural inclination to make overly strong conclusions from limited data, scientists need to use statistics.

WE TEND TO BE OVERCONFIDENT

How good are people at judging how confident they are? You can test your own ability to quantify uncertainty using a test devised by Russo and Schoemaker (1989). Answer each of these questions with a range. Pick a range that you think has a 90% chance of containing the correct answer. Don't use Google to find the answer. Don't give up and say you don't know. Of course, you won't know the answers precisely! The goal is not to provide precise answers, but rather is to correctly quantify your uncertainty and come up with ranges of values that you think are 90% likely to include the true answer. If you have no idea, answer with a super wide interval. For example, if you truly have no idea at all about the answer to the first question, answer with the range zero to 120 years old, which you can be 100% sure includes the true answer. But try to narrow your responses to each of these questions to a range that you are 90% sure contains the right answer:

- Martin Luther King Jr.'s age at death
- Length of the Nile River, in miles or kilometers
- Number of countries in OPEC
- Number of books in the Old Testament
- Diameter of the moon, in miles or kilometers
- Weight of an empty Boeing 747, in pounds or kilograms
- Year Mozart was born
- Gestation period of an Asian elephant, in days
- Distance from London to Tokyo, in miles or kilometers
- Deepest known point in the ocean, in miles or kilometers

Compare your answers with the correct answers listed at the end of this chapter. If you were 90% sure of each answer, you would expect about nine intervals to include the correct answer and one to exclude it.

Russo and Schoemaker (1989) tested more than 1,000 people and reported that 99% of them were overconfident. The goal was to create ranges that included the correct answer 90% of the time, but most people created narrow ranges that included only 30% to 60% of the correct answers. Similar studies have been done with experts estimating facts in their areas of expertise, and the results have been similar.

Since we tend to be too sure of ourselves, scientists must use statistical methods to quantify confidence properly.

WE SEE PATTERNS IN RANDOM DATA

Table 1.1 presents simulated data from 10 basketball players (1 per row) shooting 30 baskets each. An "X" represents a successful shot and a "–" represents a miss. Is this pattern random? Or does it show signs of nonrandom streaks? Look at Table 1.1 before continuing.

Most people see streaks of successful shots and conclude this is not random. Yet Table 1.1 was generated randomly. Each spot had a 50% chance of being "X" (a successful shot) and a 50% chance of being "–" (an unsuccessful shot), without taking into consideration previous shots. We see clusters perhaps because our brains have evolved to find patterns and do so very well. This ability may have served our ancestors well to avoid predators and poisonous plants, but it is

```
- - X - X - X X X - - - - X X X - X X - X X - - - X X - - - - X X
X - - X - X X - - X X - - X - X - X - - - X X X X - - X X - - - -
X X X X - X X - X - X - X X X - - - - - - X - X - X X X - - - - X
- X - X - - X X - X X - X X - - X X X X - - - - - X X - X - X - -
- X - X - X X - - - - X X - - - - - - - X - X - X - - X - - X - X X
- - X X X - X - X - - - X X X X - X X X X - - - - - - - X X - X X X
X - - X X - - X X X X - X X X - - - X - - X X X X X - X X X - - - -
X - X - - - X X X X X - - X X - X X - X X X - X X - X X - X - - X - X
X X X - - X X X X X - X - X - X X - X - X X X X - X X - X X X X
- - - X X X - - X X X - X X X - - X - - X - X X X X X - - - X -
```

Table 1.1. Random patterns don't seem random.

Table 1.1 represents 10 basketball players (1 per row) shooting 32 baskets each. An "X" represents a successful shot, and a "–" represents a miss.

important that we recognize this built-in mental bias. Statistical rigor is essential to avoid being fooled by seeing apparent patterns among random data.

WE DON'T REALIZE THAT COINCIDENCES ARE COMMON

In November 2008, I attended a dinner for the group Conservation International. The actor Harrison Ford is on its board, and I happened to notice that he wore an ear stud. The next day, I watched an episode of the TV show *Private Practice* and one character pointed out that another character had an ear stud that looked just like Harrison Ford's. The day after that, I happened to read (in a book on serendipity!) that the Nobel Prize–winning scientist Baruch Blumberg looks like Indiana Jones, a movie character played by Harrison Ford (Meyers, 2007).

What is the chance that I would encounter Harrison Ford, or a mention to him, three times in three days? Tiny, but that doesn't mean much. While it is highly unlikely that any particular coincidence will occur, it is almost certain that some seemingly astonishing set of unspecified events will happen often, since we notice so many things each day. Remarkable coincidences are always noted in hindsight and never predicted with foresight.

WE DON'T EXPECT VARIABILITY TO DEPEND ON SAMPLE SIZE

Gelman (1998) looked at the relationship between the populations of counties and the age-adjusted, per-capita incidence of kidney cancer (a fairly rare cancer, with an incidence of about 15 cases per 100,000 adults in the United States). First, he focused on counties with the lowest per-capita incidence of kidney cancer. Most of these counties had small populations. Why? One might imagine that something about the environment in these rural counties leads to lower rates of kidney cancer. Then, he focused on counties with the highest incidence of kidney cancer. These also tended to be the smallest counties. Why? One might imagine that lack of medical care in these tiny counties leads to higher rates of kidney cancer. But it seems pretty strange that both the highest and lowest incidences of kidney cancer be in counties with small populations?

The reason is simple, once you think about it. In large counties, there is little variation around the average rate. Among small counties, however, there is much more variability. Consider an extreme example of a tiny county with only 1,000 residents. If no one in that county had kidney cancer, that county would be among those with the lowest (zero) incidence of kidney cancer. But if only one of those people had kidney cancer, that county would then be among those with the highest rate of kidney cancer. In a really tiny county, it only takes one case of kidney cancer to flip from having one of the lowest rates to having one of the highest rates. In general, just by chance, the incidence rates will vary much more among counties with tiny populations than among counties with large populations. Therefore, counties with both the highest and the lowest incidences of kidney cancer tend to have smaller populations than counties with average incidences of kidney cancer.

Random variation can have a bigger effect on averages within small groups than within large groups. This simple principle is logical, yet is not intuitive to many people.

WE HAVE INCORRECT INTUITIVE FEELINGS ABOUT PROBABILITY

Imagine that you can choose between two bowls of jelly beans. The small bowl has nine white and one red jelly bean. The large bowl has 93 white beans and 7 red beans. Both bowls are well mixed, and you can't see the beans. Your job is to pick one bean. You win a prize if your bean is red. Should you pick from the small bowl or the large one?

When you choose from the small bowl, you have a 10% chance of picking a red jelly bean. When you pick from the large bowl, the chance of picking a red one is only 7%. So your chances of winning are higher if you choose from the small bowl. Yet about two-thirds of people prefer to pick from the larger bowl (Denes-Raj & Epstein, 1994). Many of these people do the math and know that the chance of winning is higher with the small bowl, but they feel better about choosing from the large bowl because it has more red beans and offers more chances to win. Of course, the large bowl also has more white beans and more chances to lose. Our brains have simply not evolved to deal sensibly with probability, and most people make the illogical choice.

WE FIND IT HARD TO COMBINE PROBABILITIES

Here is a classic brainteaser called the Monty Hall problem, named after the host of the game show *Let's Make a Deal*: You are a contestant on a game show and are presented with three doors. Behind one is a fancy new car. You must choose one door, and you get to keep whatever is behind it. You pick a door. At this point, the host chooses one of the other two doors to open and shows you that there is no car behind it. He now offers you the chance to change your mind and choose the other door (the one he has not opened).

Should you switch?

Before reading on, you should think about the problem and decide whether you should switch. There are no tricks or traps. Exactly one door has the prize, all doors appear identical, and the host—who knows which door leads to the new car—has a perfect poker face and gives you no clues. There is never a car behind the door the host chooses to open. Don't cheat. Think it through before continuing.

When you first choose, there are three doors and each is equally likely to have the car behind it. So your chance of picking the winning door is one-third. Let's separately think through the two cases: originally picking a winning door or originally picking a losing door.

If you originally picked the winning door, then neither of the other doors has a car behind it. The host opens one of these two doors. If you now switch doors, you will have switched to the other losing door.

What happens if you originally picked a losing door? In this case, one of the remaining doors has a car behind it and the other one doesn't. The host knows which is which. He opens the door without the car. If you now switch, you will win the car.

Let's recap. If you originally chose the correct door (an event that has a one-third chance of occurring), then switching will make you lose. If you originally picked either of the two losing doors (an event that has a two-thirds chance of occurring), then switching will definitely make you win. Switching from one losing door to the other losing door is impossible, because the host will have opened the other losing door.

Your best choice is to switch! Of course, you can't be absolutely sure that switching doors will help. One-third of the time you will be switching away from the prize. But the other two-thirds of the time you will be switching to the prize. If you repeat the game many times, you will win twice as often by switching doors every time. If you only get to play once, you have twice the chance of winning by switching doors.

Almost everyone (including mathematicians and statisticians) intuitively reaches the wrong conclusion and thinks that switching won't be helpful (Vos Savant, 1997).

WE DON'T DO BAYESIAN CALCULATIONS INTUITIVELY

Imagine this scenario: You are screening blood donors for the presence of human immunodeficiency virus (HIV). Only a tiny fraction (0.1%) of the blood donors has HIV. The antibody test is highly accurate but not quite perfect. It correctly identifies 99% of infected blood samples but also incorrectly concludes that 1% of noninfected samples have HIV. When this test identifies a blood sample as having HIV present, what is the chance that the donor does, in fact, have HIV, and what is the chance the test result is an error (a false positive)?

Try to come up with the answer before reading on.

Let's imagine that 100,000 people are tested. Of these, 100 (0.1%) will have HIV, and the test will be positive for 99 (99%) of them. The other 99,900 people will not have HIV, but the test will incorrectly return a positive result in 1% of cases. So there will be 999 false positive tests. Altogether, there will be 99 + 999 = 1,098 positive tests, of which only 99/1,098 = 9% will be true positives. The other 91% of the positive tests will be false positives. So if a test is positive, there is only a 9% chance that there is actually HIV in that sample.

Most people, including most physicians, intuitively think that a positive test almost certainly means that HIV is present. Our brains are not adept at combining what we already know (the prevalence of HIV) with new knowledge (the test is positive).

Now imagine that the same test is used in a population of intravenous drug users, of which 10% have HIV. Again, let's imagine that 100,000 people are tested. Of these, 10,000 (10%) will have HIV, and the test will be positive for 9,900 (99%) of them. The other 90,000 people will not have HIV, but the test will incorrectly return a positive result in 1% of cases. So there will be 900 false positive tests. Altogether, there will be 9,900 + 900 = 10,800 positive tests, of which 9,900/10,800 = 92% will be true positives. The other 8% of the positive tests will be false positives. So if a test is positive, there is a 92% chance that there is HIV in that sample.

The interpretation of the test result depends greatly on what fraction of the population has the disease. This example gives you a taste of what is called Bayesian logic (a subject that will be discussed again in Chapters 2 and 18).

WE ARE FOOLED BY MULTIPLE COMPARISONS

Austin and colleagues (2006) sifted through a database of health statistics of 10 million residents of Ontario, Canada. They examined 223 different reasons for hospital admission and recorded the astrological sign of each patient (computed from his or her birth date). They then asked if people with certain astrological signs are more likely to be admitted to the hospital for certain conditions.

The results seem impressive. Seventy-two diseases occurred more frequently in people with one astrological sign than in people with all the other astrological signs put together, with the difference being statistically significant. Essentially, a result that is *statistically significant* would occur by chance less than 5% of the time (you'll learn more about what *statistically significant* means in Chapter 16).

Sounds impressive, doesn't it? Indeed, those data might make you think that there is a convincing relationship between astrology and health. But there is a problem. It is misleading to focus on the strong associations between one disease and one astrological sign without considering all the other combinations. Austin et al. (2006) tested the association of 223 different reasons for hospital admissions with 12 astrological signs and so tested 2,676 distinct hypotheses (223 × 12 = 2,676). Therefore, they would expect to find 134 statistically significant associations just by chance (5% of 2,676 = 134) but in fact only found 72.

Note that this study wasn't really done to ask about the association between astrological sign and disease. It was done to demonstrate the difficulty of interpreting statistical results when many comparisons are performed.

Chapters 22 and 23 explore multiple comparisons in more depth.

WE TEND TO IGNORE ALTERNATIVE EXPLANATIONS

Imagine you are doing a study on the use of acupuncture in treating osteoarthritis. Patients who come in with severe arthritis pain are treated with acupuncture. They are asked to rate their arthritis pain before and after the treatment. The pain decreases in most patients after treatment, and statistical calculations show that such consistent findings are exceedingly unlikely to happen by chance. Therefore, the acupuncture must have worked. Right?

Not necessarily. The decrease in recorded pain may not be caused by the acupuncture. Here are five alternative explanations (adapted from Bausell, 2007):

- If the patients believe in the therapist and treatment, that belief may reduce the pain considerably. The pain relief may be a placebo effect and have nothing to do with the acupuncture itself.
- The patients want to be polite and may tell the experimenter what he or she wants to hear (that the pain decreased). Thus, the decrease in reported pain may be because the patients are not accurately reporting pain after therapy.
- Before, during, and after the acupuncture treatment, the therapist talks with the patients. Perhaps he or she recommends a change in aspirin dose, a change in exercise, or the use of nutritional supplements. The decrease in reported pain might be due to these aspects of the treatment, rather than the acupuncture.
- The experimenter may have altered the data. For instance, what if three patients experience worse pain with acupuncture, whereas the others get better? The experimenter carefully reviews the records of those three patients and decides to remove them from the study because one of those people actually has a different kind of arthritis than the others, and two had to climb stairs to get to the appointment because the elevator didn't work that day. The data, then, are biased or skewed because of the omission of these three participants.
- The pain from osteoarthritis varies significantly from day to day. People tend to seek therapy when pain is at its worst. If you start keeping track of pain on the day when it is the worst, it is quite likely to get better, even with no treatment. The next section explores this idea of *regression to the mean*.

WE ARE FOOLED BY REGRESSION TO THE MEAN

Figure 1.1 illustrates simulated blood pressures before and after a treatment. Figure 1.1A includes 24 pairs of values. The "before" and "after" groups are about the same. In some cases, the value goes up after treatment, and in others it goes down. If these were real data, you'd conclude that there is no evidence at all that the treatment had any effect on the outcome (blood pressure).

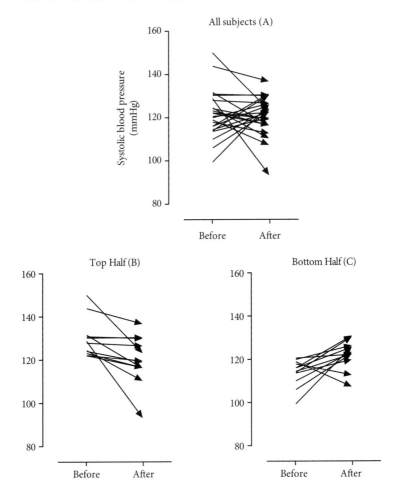

Figure 1.1. Regression to the mean.

All data in (A) were drawn from random distributions (Gaussian; mean = 120, SD = 15) without regard to the designations "before" and "after" and without regard to any pairing. (A) shows 48 random values, divided arbitrarily into 24 before–after pairs (which overlap enough that you can't count them all). (B) shows only the 12 pairs with the highest before values. In all but one case, the after values are lower than the before values. (C) shows the pairs with the lowest before measurements. In 10 of the 12 pairs, the after value is higher than the before value. If you only saw the graph in (B) or (C), you'd probably conclude that whatever treatment came between the before and after measurements had a large impact on blood pressure. In fact, these graphs simply show random values, with no systematic change between before and after. The apparent change is called *regression to the mean*.

Now imagine the study was designed differently. You've made the before measurements and want to test a treatment for high blood pressure. There is no point in treating individuals whose blood pressure is not high, so you select the people with the highest pressures to study. Figure 1.1B illustrates data for only those 12 individuals with the highest before values. In every case but one, the after values are lower. If you performed a statistical test (e.g., a paired t test; see Chapter 31), the results would convince you that the treatment decreased blood pressure. Figure 1.1C illustrates a similar phenomenon with the other 12 pairs, those with low before values. In all but two cases, these values go up after treatment. This evidence would convince you that the treatment increases blood pressure.

But these are random data! The before and after values came from the same distribution. What happened?

This is an example of *regression to the mean*: the more extreme a variable is upon its first measurement, the more likely it is to be closer to the average the second time it is measured. People who are especially lucky at picking stocks one year are likely to be less lucky the next year. People who get extremely high scores on one exam are likely to get lower scores on a repeat exam. An athlete who does extremely well in one season is likely to perform more poorly the next season. This probably explains much of the *Sports Illustrated* cover jinx—many believe that appearing on the cover of *Sports Illustrated* will bring an athlete bad luck (Wolff, 2002).

WE LET OUR BIASES DETERMINE
HOW WE INTERPRET DATA

Kahan and colleagues (2013) asked a bunch of people to analyze some simple data, similar to what you'll see in Chapter 27. While all the tables had the same values, sometimes the table was labeled to test the effectiveness of a skin cream for treating a rash (Table 1.2), and sometimes to test the effectiveness of a law prohibiting carrying concealed hand guns in public (Table 1.3).

	RASH GOT BETTER	RASH GOT WORSE	TOTAL
Patients who **did use** the cream	223	75	298
Patients who **did not use** the cream	107	21	128

Table 1.2. One of the tables that Kahan and colleagues used. After viewing this table, people were asked to determine whether the hypothetical experiment showed that the skin condition of people treated with the cream was more likely to "get better" or "get worse" compared to those who were not treated.

	DECREASE IN CRIME	INCREASE IN CRIME	TOTAL
Cities that **did ban** carrying concealed handguns in public	223	75	298
Cities that **did not ban** carrying concealed handguns in public	107	21	128

Table 1.3. The other table that Kahan and colleagues used. After viewing this table, people were asked to determine whether the made-up data show that cities that enacted a ban on carrying concealed handguns were more likely to have an increase or decrease in crime compared to cities that did not ban concealed handguns.

The experimental subjects were not asked for subtle interpretation of the data but rather were simply asked whether or not the data support a particular hypothesis. The math in Tables 1.2 and 1.3 is pretty straightforward:

- In the top row, the rash got better in 223/298 = 74.8% of the treated people, and the crime rate went down in 74.8% of the cities that banned concealed handguns.
- In the bottom row, the rash got better in 107/128 = 83.6% of the untreated people, and the crime rate when down in 83.6% of the cities that did not ban carrying concealed handguns.
- Most people had a decrease in rash, and most cities had a decrease in crime. But did the intervention matter? Since 74.8% is less than 83.6%, the data clearly show people who used the cream were less likely to have an improved rash than people who did not use the cream. Cities that passed handgun law had a smaller decrease in crime than cities who did not pass such a law.

When the data were labeled to be about the effectiveness of a skin cream, liberal Democrats and conservative Republicans (the two main political parties in the United States) did about the same. But when the data were labeled to be about the effectiveness of a gun safety policy, the results depended on political orientation. Liberal democrats tended to find that the data showed that gun safety laws reduced crime. Conservatives tended to conclude the opposite (this study was done in the United States, where conservatives tend to be against gun safety legislation.)

This study shows that when people have a preconceived notion about the conclusion, they tend to interpret the data to support that conclusion.

WE CRAVE CERTAINTY, BUT STATISTICS OFFERS PROBABILITIES

Many people expect statistical calculations to yield definite conclusions. But in fact, every statistical conclusion is stated in terms of probability. Statistics can be very difficult to learn if you keep looking for definitive conclusions. As statistician Myles Hollander reportedly said, "Statistics means never having to say you're certain!" (quoted in Samaniego, 2008).

CHAPTER SUMMARY

- Our brains do a really bad job of interpreting data. We see patterns in random data, tend to be overconfident in our conclusions, and mangle interpretations that involve combining probabilities.
- Our intuitions tend to lead us astray when interpreting probabilities and when interpreting multiple comparisons.
- Statistical (and scientific) rigor is needed to avoid reaching invalid conclusions.

TERM INTRODUCED IN THIS CHAPTER

- Regression to the mean (p. 9)

Answers to the 10 questions in the "We Tend to Be Overconfident" section.

Martin Luther King Jr.'s age at death: 39

Length of the Nile River: 4,187 miles or 6,738 kilometers

Number of countries in OPEC: 13

Number of books in the Old Testament: 39

Diameter of the moon: 2,160 miles or 3,476 kilometers

Weight of an empty Boeing 747: 390,000 pounds or 176,901 kilograms

Year Mozart was born: 1756

Gestation period of an Asian elephant: 645 days

Distance from London to Tokyo: 5,989 miles or 9,638 kilometers

Deepest known point in the ocean: 6.9 miles or 11.0 kilometers

CHAPTER 2

The Complexities of Probability

> Statistical thinking will one day be as necessary for efficient
> citizenship as the ability to read and write.
>
> H. G. WELLS

Entire books have been written on probability. It is a topic that seems simple at first but becomes more complicated as you delve into the details. This chapter provides a very broad overview that will help you understand why probability is confusing, make sense of statements about probability, and provide a solid background if you decide to learn more about the subject.

BASICS OF PROBABILITY

Probabilities range from 0.0 to 1.0 (or 100%) and are used to quantify a prediction about future events or the certainty of a belief. A probability of 0.0 means either that an event can't happen or that someone is absolutely sure that a statement is wrong. A probability of 1.0 (or 100%) means that an event is certain to happen or that someone is absolutely certain a statement is correct. A probability of 0.50 (or 50%) means that an event is equally likely to happen or not happen, or that someone believes that a statement is equally likely to be true or false.

This chapter uses the terminology of Kruschke (2011) to distinguish two uses of probability:

- Probability that is "out there," or outside your head. This is *probability as long-term frequency*. The probability that a certain event will happen has a definite value, but we rarely have enough information to know that value with certainty.
- Probability that is inside your head. This is probability as strength of subjective beliefs, so it may vary among people and even among different assessments by the same person.

PROBABILITY AS LONG-TERM FREQUENCY

Probabilities as predictions from a model
This chapter will focus on one simple example: a woman plans to get pregnant and wants to know the chance that her baby will be a boy. One way to think

14

about probabilities is as the predictions of future events that are derived by using a model. A *model* is a simplified description of a mechanism. For this example, we can create the following simple model:

- Each ovum has an X chromosome, and none have a Y chromosome.
- Half the sperm have an X chromosome (but no Y) and half have a Y chromosome (and no X).
- Only one sperm will fertilize the ovum.
- Each sperm has an equal chance of fertilizing the ovum.
- If the winning sperm has a Y chromosome, the fetus will have both an X and a Y chromosome and so will be male. If the winning sperm has an X chromosome, the fetus will have two X chromosomes and so will be female.
- Any miscarriage or abortion is equally likely to happen to male and female fetuses.

If you assume that this model is true, then the predictions are easy to figure out. Since all sperm have the same chance of being the one to fertilize the ovum and since the sperm are equally divided between those with an X chromosome and those with a Y chromosome, the chance that the fetus will have a Y chromosome is 50%. Thus, our model predicts that the chance that the fetus will be male is 0.50, or 50%. In any particular group of babies, the fraction of boys might be more or less than 50%. But in the long run, you'd expect 50% of the babies to be boys.

You can make predictions about the occurrence of future events from any model, even if the model doesn't reflect reality. Some models will prove to be useful, and some won't. The model described here is pretty close to being correct, so its predictions are pretty useful but not perfect.

Probabilities based on data

Of all the babies born in the world in 2011, 51.7% were boys (Central Intelligence Agency [CIA], 2012). We don't need to know *why* more boys than girls were born. And we don't need to know why the CIA tabulates this information (although I am curious). It is simply a fact that this sex ratio has stayed fairly consistent over many years in many countries. Based on these data, we can answer the question, what is the chance that my baby will be a boy? The answer is 51.7%. If you had a group of 1,000 pregnant women, you'd expect about 517 to have male fetuses and 483 to have female fetuses. In any particular set of 1,000 pregnant women, the number of males might be higher or lower than 517, but that is what you'd expect in the long run.

PROBABILITY AS STRENGTH OF BELIEF

Subjective probabilities

You badly want a boy. You search the Internet and read about an interesting book:

> *How to Choose the Sex of Your Baby* explains the simple, at-home, noninvasive Shettles method and presents detailed steps to take to conceive a child of a specific gender. The properly applied Shettles method gives couples a 75 percent or better chance of having a child of the desired sex. (Shettles, 1996)

The reviews of this book are glowing and convince you that its premise is correct. Therefore, you plan to follow the book's recommendations to increase your chance of having a boy.

What is the chance that you'll have a boy? If you have complete faith that the method is correct, then you believe that the probability, as stated on the book jacket, is 75%. You don't have any data supporting this number, but you believe it strongly. It is your firmly held *subjective probability*.

But what if you have doubts? You think the method probably works, and you will quantify that by saying you think there is an 85% chance that the Shettles method works (i.e., your chance of having a boy would be 75%). That leaves a 15% chance that the method is worthless and your chance of having a boy is 51.7%. What is your chance of having a boy? Calculate a weighted average of the two predictions, weighting on your subjective assessment of the chance that each theory is correct. So the chance is $(0.850 \times 0.750) + (0.150 \times 0.517) = 0.715 = 71.5\%$.

Of course, different people have different beliefs about the efficacy of the method. I didn't look into the matter carefully, but I did spend a few minutes searching and found no studies actually proving that the Shettles method works, and one (in a prestigious journal) saying it doesn't work (Wilcox et al., 1995). Given that limited amount of research, I will quantify my beliefs by saying that I believe there is only a 1% chance that the Shettles method works (and that your chance of having a boy is 75%) and thus a 99% chance that the method does not work (and so your chance of having a boy is 51.7%). The weighted average is thus $(0.010 \times 0.750) + (0.990 \times 0.517) = 0.519 = 51.9\%$.

Since you and I have different assessments of the probability of whether the Shettles method works, my answer for the chance of you having a boy (51.9%) doesn't match your answer (71.5%). The calculations depend upon a subjective assessment of the likelihood that the Shettles method works, so it isn't surprising that my answer and yours differ.

"Probabilities" used to quantify ignorance

Assume that you (or someone you know) are pregnant but haven't yet done any investigations (e.g., ultrasound, karyotype) to determine whether the fetus is male or female or undergone any interventions (see previous section) that purport to change the chances of having a boy or a girl. What is the probability that the fetus is male?

In one sense, the concept of probability is simply not involved. The random event was the race of many sperm to fertilize the egg. That already happened. One sperm won, and the fetus either is male or is female. Since the random event has already happened and the gender of the fetus is now a fact, you could argue that it makes no sense to ask about probability or chance or odds. Chance is no longer involved. The issue is ignorance, not randomness.

Another perspective: Before you became pregnant, there was a 51.7% chance that your fetus would be male. If many thousands of women were pregnant without knowing the sex of their fetus, you'd expect about 51.7% to have male fetuses. Therefore, it seems sensible to say that there is about a 51.7% chance that your fetus is male and a 48.3% chance that it is female. That was your chance before you got pregnant, so it still is your chance now.

But notice the change in perspective. Now you are no longer talking about predicting the outcome of a random future event but rather are quantifying your ignorance of an event that has already happened.

Quantitative predictions of one-time events

As I began to revise this chapter in October 2016, the US presidential election was four weeks away. Different people have different opinions and interpret the polling data differently. The *New York Times* stated on October 5, 2016, that there was an 81% chance that Clinton will be elected and a 19% chance that Trump will be. On the same day, the website fivethirtyeight.com said that there is a 76% chance that Clinton will be elected and a 24% chance that Trump will be elected.

Are these probabilities subjective? Partly. The probabilities were computed from lots of poll data. But there are many polls, and there were some subjective choices in which polls each publication decided to use and how they weighted the polls when averaging. This explains why the two estimates from reputable institutions differ.

Can you interpret the 81% value as a long-term frequency? No. It doesn't mean that the *New York Times* expected Clinton to win 81% of all the presidential elections she will ever enter or that 81% of all voters would vote for Clinton. Based on analysis of polling data, the *Times* was 81% sure that Clinton would win. That was its belief on October 5. By October 22, two weeks before the election, the *New York Times* had raised its estimate to a 93% chance of Clinton winning, and fivethirtyeight.com raised that value to 86%. The published probabilities were quantitative statements about informed beliefs. With newer polling data, they updated their beliefs.

In the end, Trump won the election. Were the pollsters wrong? Not really. They didn't say there is no chance that Trump would win, just a 24%, 29%, 7%, or 14% chance (depending on which publication you looked at and when).

CALCULATIONS WITH PROBABILITIES CAN BE EASIER IF YOU SWITCH TO CALCULATING WITH WHOLE NUMBERS

Gigerenzer (2002) showed that rephrasing probability problems can greatly impact the chance that they will be answered correctly. His example is about interpreting the results of mammograms given to asymptomatic women between the ages of 40 and 50. He asked 48 physicians to interpret a positive (abnormal) mammogram. Half of the physicians were given the question in terms of probabilities:

> The probability that one of these women has breast cancer is 0.8 percent. If a woman has breast cancer, the probability is 90 percent that she will have a positive mammogram. If a woman does not have breast cancer, the probability is 7 percent that she will have a positive mammogram. Imagine a woman who has a positive mammogram. What is the probability that she actually has breast cancer?

Go ahead. Try to figure out the answer.

The other half of the physicians were given the same question but in natural frequencies (whole numbers) rather than percentages:

> Eight out of every 1,000 women will have breast cancer. Of those 8 women with breast cancer, 7 will have a positive mammogram. Of the remaining 992 women who don't have breast cancer, about 70 will still have a positive mammogram. Imagine a woman who has a positive mammogram. What is the probability that she actually has breast cancer?

Now try to answer the question. Is it easier?

The physicians in both groups took the time to think the problem through. But those in the first group (percentages) did terribly. Only 2 out of the 24 answered the question correctly (the probability of breast cancer is about 9%). A few others were close. The majority of the physicians were very wrong, and the most common answer was 90%. The physicians in the second group (natural frequencies) did much better, with the majority answering correctly.

Why is the answer 9%? Of the 1,000 women tested, we will see 7 positive mammograms among those who actually have breast cancer and 70 among those who don't. In other words, given the setup of this problem, we expect 7 out of 77 people with positive mammograms to turn out to actually have breast cancer (about 9%).

If you get stuck thinking through problems with probabilities, try changing the problem to thinking about whole numbers. That trick often makes things much easier to think through.

COMMON MISTAKES: PROBABILITY

Mistake: Ignoring the assumptions

The first example asked, "What is the chance that a fetus will be male?" However, that question is meaningless without context. For it to be meaningful, you must accept a list of assumptions, including:

- We are asking about human babies. Sex ratios may be different in another species.
- There is only one fetus. "Will the baby be a boy or girl?" is ambiguous, or needs elaboration, if you allow for the possibility of twins or triplets.
- There is only a tiny probability (which we ignore) that the baby is neither completely female nor completely male.
- The sex ratio is the same for all countries and all ethnic groups.
- The sex ratio does not change from year to year, or between seasons.
- There will be (and have been) no sex-selective abortions or miscarriages, so the sex ratio at conception is the same as the sex ratio at birth.

Whenever you hear questions or statements about probability, remember that probabilities are *always* contingent on a set of assumptions. To think clearly about probability in any situation, you must know what those assumptions are.

Sometimes probabilities are in the form: If A is true, what is the chance that B will happen? This is called a *conditional probability*, as it is the probability of some event (B) occurring, conditional on another event (A) also occurring.

Mistake: Trying to understand a probability without clearly defining both the numerator and the denominator

The example we've used for this chapter is the fraction of babies who are boys. The numerator is the number of boy babies. The denominator is the number of babies born in a certain place in a certain time period. In this case, the fraction is simple and unambiguous.

Let's switch to another example about babies where there is some ambiguity. Infertile couples often use in vitro fertilization to increase their chance of having a baby. After the ova are fertilized in a "test tube," one or more embryos are implanted into the uterus. If you implant more than one embryo, the chance of a successful pregnancy is higher than if you implant only one, but now there is also a chance of having twins or triplets (or more).

When couples consider in vitro pregnancy, they want to know the success rate. The denominator is clearly the number of women who had one or more embryo implanted. But what is the numerator? To make the answer to be a frequency that can be interpreted as a probability, the numerator and denominator must count the same thing. So the numerator should be number of women who completed a successful pregnancy. Each woman who underwent the procedure either completed a successful pregnancy or didn't. Since there are two alternative outcomes, you can compute the percentage of women who completed a pregnancy, and use this as a frequency to predict the results of other women in the future.

Decades ago, when my wife underwent in vitro fertilization, the clinic used another way to compute the "success rate." They divided the number of babies born by the number of women who received the treatment. This ratio is not a probability and can't really be interpreted, because the numerator and denominator count different things. The numerator counts babies; the denominator counts women. This "success rate" is higher than the probability mentioned in the prior paragraph, because some women had more than one baby (twins or triplets).

You can't understand what the success rate means (or compare between clinics) without knowing exactly how the numerator and denominator were defined.

Mistake: Reversing probability statements

With the examples we've used so far, there is no danger of accidentally reversing probability statements by mistake. The probability that a baby is a boy is obviously very different from the probability that a boy is a baby. But in many situations, it is easy to get things backwards.

- The probability that a heroin addict first used marijuana is not the same as the probability that a marijuana user will later become addicted to heroin.
- The probability that someone with abdominal pain has appendicitis is not the same as the probability that someone diagnosed with appendicitis will have had abdominal pain.
- The probability that a statistics books will be boring is not the same as the probability that a boring book is about statistics.

- The fraction of studies performed under the null hypothesis in which the P value is less than 0.05 is not the same as the fraction of studies with a P value less than 0.05 for which the null hypothesis is true. (This will make more sense once you read Chapter 15.)

Knowing how easy it is to mistakenly reverse a probability statement will help you avoid that error.

Mistake: Believing that probability has a memory

If a couple has four children, all boys, what is the chance that the next child will be a boy? A common mistake is to think that since this couple already has four boys, they are somehow due to have a girl, so the chance of having a girl is elevated. This is simply wrong.

This mistake is often made in gambling, especially when playing roulette or a lottery. Some people bet on a number that has not come up for a long while, believing that there is an elevated chance of that number coming up on the next spin. This is called the *gambler's fallacy*. Probability does not have a memory.

LINGO

Probability vs. odds

So far, this chapter has quantified chance as probabilities. But it is also possible to express the same values as *odds*. Odds and probability are two alternatives for expressing precisely the same concept. Every probability can be expressed as odds. Every odds can be expressed as a probability. Some scientific fields tend to prefer using probabilities; other fields tend to favor odds. There is no consistent advantage to using one or the other.

If you search for demographic information on the fraction of babies who are boys, what you'll find is the *sex ratio*. This term, as used by demographers, is the ratio of males to females born. Worldwide, the sex ratio at birth in many countries is about 1.07. Another way to say this is that the odds of having a boy versus a girl are 1.07 to 1.00, or 107 to 100.

It is easy to convert the odds to a probability. If there are 107 boys born for every 100 girls born, the chance that any particular baby will be male is $107/(107 + 100) = 0.517$, or 51.7%.

It is also easy to convert from a probability to odds. If the probability of having a boy is 51.7%, then you expect 517 boys to be born for every 1,000 births. Of these 1,000 births, you expect 517 boys and 483 girls (which is $1,000 - 517$) to be born. So the odds of having a boy versus a girl are $517/483 = 1.07$ to 1.00. The odds are defined as the probability that the event will occur divided by the probability that the event will not occur.

Odds can be any positive value or zero, but they cannot be negative. A probability must be between zero and 1 if expressed as a fraction, or be between zero and 100 if expressed as a percentage.

A probability of 0.5 is the same as odds of 1.0. The probability of flipping a coin to heads is 50%. The odds are 50:50, which equals 1.0. As the probability goes from 0.5 to 1.0, the odds increase from 1.0 to approach infinity. For example, if the probability is 0.75, then the odds are 75:25, 3 to 1, or 3.0.

Probability vs. statistics

The words *probability* and *statistics* are often linked together in course and book titles, but they are distinct.

This chapter has discussed probability calculations. While the details can get messy and make it easy to apply probability theory incorrectly, the concepts are pretty simple. Start with the general case, called the population or model, and predict what will happen in many samples of data. Probability calculations go from general to specific, from population to sample (as you'll see in Chapter 3), and from model to data (as you'll see in Chapter 34).

Statistical calculations work in the opposite direction (Table 2.1). You start with one set of data (the sample) and make inferences about the overall population or model. The logic goes from specific to general, from sample to population, and from data to model.

Probability vs. likelihood

In everyday language, *probability* and *likelihood* are essentially synonyms. In statistics, however, the two have a distinct meaning. This book won't use the term *likelihood*, but if you read more mathematical books, you need to know that likelihood has a technical meaning distinct from *probability*. Briefly, probability answers the questions outlined in the top half of Table 2.1, and likelihood is the answer to questions outlined in the bottom half.

PROBABILITY IN STATISTICS

Earlier in this chapter, I pointed out that probability can be "out there" or "in your head." The rest of this book mostly explains confidence intervals (CIs) and

PROBABILITY		
General	→	Specific
Population	→	Sample
Model	→	Data
STATISTICS		
General	←	Specific
Population	←	Sample
Model	←	Data

Table 2.1. The distinction between probability and statistics.
Probability theory goes from general to specific, from population to sample, from model to data.

P values, which both use "out there" probabilities. This style of analyzing data is called *frequentist statistics*. Prior beliefs, or prior data, never enter frequentist calculations. Only the data from a current set are used as inputs when calculating P values (see Chapter 15) or CIs (see Chapters 4 and 12). However, scientists often account for prior data and theory when interpreting these results, a topic to which we'll return in Chapters 18, 19, and 42.

Many statisticians prefer an alternative approach called *Bayesian statistics*, in which prior beliefs are quantified and used as part of the calculations. These prior probabilities can be subjective (based on informed opinion), objective (based on solid data or well-established theory), or uninformative (based on the belief that all possibilities are equally likely). Bayesian calculations combine these prior probabilities with the current data to compute probabilities and Bayesian CIs called *credible intervals*. This book only briefly explains Bayesian statistics.

Q & A

Can all probabilities be expressed either as fractions or percentages?
> Yes. Multiple the fraction by 100 to express the same value as a percentage.

Are all fractions probabilities?
> No. A fraction is a probability only when there are two alternative outcomes. Fractions can be used in many other situations.

Do probabilities always have values between zero and 1, or between 0% and 100%?
> Yes.

CHAPTER SUMMARY

- Probability calculations can be confusing.
- There are two meanings of probability. One meaning is long-range frequency that an event will happen. It is probability "out there." The other meaning is that probability quantifies how sure one is about the truth of a proposition. This is probability "in your head."
- All probability statements are based on a set of assumptions.
- It is impossible to understand a probability or frequency until you have clearly defined both its numerator and its denominator.
- Probability calculations go from general to specific, from population to sample, and from model to data. Statistical calculations work in the opposite direction: from specific to general, from sample to population, and from data to model.
- Frequentist statistical methods calculate probabilities from the data (P values, CIs). Your beliefs before collecting the data do not enter into the calculations, but they often are considered when interpreting the results.
- Bayesian statistical methods combine prior probabilities (which may be either subjective or based on solid data) and experimental evidence as part of the calculations. This book does not explain how these methods work.

TERMS INTRODUCED IN THIS CHAPTER

- Bayesian statistics (p. 22)
- Conditional probability (p. 18)
- Credible intervals (p. 22)
- Frequentist statistics (p. 22)
- Gambler's fallacy (p. 20)
- Likelihood (p. 21)
- Model (p. 15)
- Odds (p. 20)
- Probability (p. 14)
- Probability as long-term frequency (p. 14)
- Statistics (p. 21)
- Subjective probability (p. 16)

CHAPTER 3

From Sample to Population

There is something fascinating about science. One gets such a wholesale return of conjecture out of a trifling investment of fact.

MARK TWAIN (*LIFE ON THE MISSISSIPPI*, 1850)

Before you can learn about statistics, you need to understand the big picture. This chapter concisely explains what statistical calculations are used for and what questions they can answer.

SAMPLING FROM A POPULATION

The distinction between a sample and a population is key to understanding much of statistics. Statisticians say that you extrapolate from a sample to make conclusions about a population. Note that the terms *sample* and *population* have specialized meanings in statistics that differ from the ordinary uses of those words. As you learn statistics, distinguish the specialized definitions of statistical terms from their ordinary meanings.

Here are four of the different contexts in which the terms are used:

- Quality control. A factory makes lots of items (the population) but randomly selects a few items to test (the sample). The results obtained from the sample are used to make inferences about the entire population.
- Political polls. A random sample of voters is used to make conclusions about the entire population of voters.
- Clinical studies. Data from the sample of patients included in a study are used to make conclusions useful to a larger population of future patients.
- Laboratory experiments. The data you collect are the sample. From the sample data, you want to make reliable inferences about the ideal, true underlying situation (the population).

In quality control and in political and marketing polls, the population is usually much larger than the sample, but it is finite and known (at least approximately). In biomedical research, we usually *assume* that the population is infinite or at least very large compared with our sample. Standard statistical methods assume that the population is much larger than the sample. If the population has a defined size and you have sampled a substantial fraction of the population (>10% or so), then you must use special statistical methods designed for use when a large fraction of the population is sampled.

SAMPLING ERROR AND BIAS

Much of statistics is based on the assumption that the data you are analyzing are randomly sampled from a larger population. Therefore, the values you compute from the sample (e.g., the mean, the proportion of cases that have a particular attribute, the best-fit slope of a linear regression line) are considered to be *estimates* of the true population values.

There are several reasons why a value computed from a sample differs from the true population value:

- *Sampling error.* Just by chance, the sample you collected might have a higher (or lower) mean than that of the population. Just by chance, the proportion computed from the sample might be higher (or lower) than the true population value. Just by chance, the best-fit slope of a regression line (which you will learn about in Chapter 33) might be steeper (or shallower) than the true line that defines the population.
- *Selection bias.* The difference between the value computed from your sample and the true population value might be due to more than random sampling. It is possible that the way you collected the sample is not random, but rather preferentially samples subjects with particular values. For example, political polls are usually done by telephone and thus select for people who own a phone (and especially those with several phone lines) and for people who answer the phone and don't hang up on pollsters. These people may have opinions very different than those of a true random sample of voters.
- Other forms of *bias.* The experimental methodology may be imperfect and give results that are systematically either too high or too low. Bias is not random but rather produces errors in a consistent direction.

MODELS AND PARAMETERS

The concept of sampling from a population is an easy way to understand the idea of making generalizations from limited data, but it doesn't always perfectly describe what statistics and science are all about. Another way to think about statistics is in terms of models and parameters.

A *model* is a mathematical description of a simplified view of the world. A model consists of both a general description and *parameters* that have particular values. For example, the model might describe temperature values that follow a Gaussian bell-shaped distribution. The parameters of that distribution are its mean (average) and standard deviation (a measure of variation, see Chapter 9). Another example: After a person ingests a drug, its concentration in his or her blood may decrease according to an exponential decay with consistent half-life.

The parameter is the half-life. In one half-life, the drug concentration goes down to 50% of the starting value; in two half-lives, it goes down to 25%; in three half-lives, it goes down to 12.5%; and so on.

One goal of statistics is to analyze your data to make inferences about the values of the parameters that define the model. Another goal is to compare alternative models to see which best explains the data. Chapters 34 and 35 discuss this idea in more depth.

MULTIPLE LEVELS OF SAMPLING

Roberts (2004) published 10 different self-experiments that he performed to test ideas about sleep, mood, health, and weight. Here is one example: When Roberts looked at faces on television in the morning, his mood was lower that evening (>10 hours later) but elevated the next day (>24 hours later). In another example from that same paper, Roberts found that when he drank unflavored fructose water between meals, he lost weight and maintained the lower weight for more than a year. Statistical analysis of the sample of data he collected enabled him to make conclusions that he found helpful for managing his life. But can the conclusions be extrapolated to others? These are intriguing ideas that provide a starting point for designing broader experiments, but you can't extrapolate from a sample of data drawn from one person to make firm conclusions about others.

A political poll might first randomly choose counties and then randomly select voters in those counties to survey. Here, there are two levels of extrapolation: from the voters surveyed in a county to all the voters in that county, and from the counties actually surveyed to all the counties in the state or country. Special methods are needed to deal with this kind of *hierarchical* or *multilevel sampling*.

WHAT IF YOUR SAMPLE *IS*
THE ENTIRE POPULATION?

In some situations, your sample of data *is* the entire population, or so it seems. A political scientist may analyze all the election data for a certain election. Someone doing quality control may be able to analyze every single item being manufactured. A hospital epidemiologist will have data about every single infection that occurs in a hospital.

How does statistics work in these situations when your data seem to be the entire population? There are three ways to view this situation.

- In some cases, you truly only care about the data you have collected and have no interest in making any general conclusions. In these cases, statistical inference won't be helpful, because there is no inference needed. All you need to do is describe and plot the data.
- In many cases, you do want to make inferences beyond the data at hand. The political scientist probably wants to make conclusions about future

elections. The quality control expert will want to make inferences to improve quality in the future. The epidemiologist will want to compare the current data with past data to make inferences to help future patients. In all these cases, the current set of data can sort of be considered a sample of a larger population that includes data to be collected in the future.

• Another perspective is that you might really care about the underlying probability model that generated your data, so the inferences aren't about the data at hand. Instead, you want to make inferences about the model that generated those data (Gelman, 2009).

CHAPTER SUMMARY

• The goal of data analysis is simple: to make the strongest possible conclusions from limited amounts of data.
• Statistics help you extrapolate from a particular set of data (your sample) to make a more general conclusion (about the population). Statistical calculations go from specific to general, from sample to population, from data to model.
• Bias occurs when the experimental design is faulty, so that on average the result is too high (or too low). Results are biased when the result computed from the sample is not, in fact, the best possible estimate of the value in the population.
• The results of statistical calculations are always expressed in terms of probability.
• If your sample *is* the entire population, you only need statistical calculations to extrapolate to an even larger population, other settings, or future times.

TERMS INTRODUCED IN THIS CHAPTER

• Bias (p. 25)
• Hierarchical/multilevel sampling (p. 26)
• Model (p. 25)
• Parameter (p. 25)
• Population (p. 24)
• Sample (p. 24)
• Sampling error (p. 25)
• Selection bias (p. 25)

PART B

Introducing Confidence Intervals

CHAPTER 4

Confidence Interval of a Proportion

> The first principle is that you must not fool yourself—and you
> are the easiest person to fool.
>
> RICHARD FEYNMAN

A fundamental concept in statistics is the use of a confidence interval, abbreviated CI, to make a general conclusion from limited
data. This chapter defines CIs in the context of results that have two
possible outcomes summarized as a proportion or fraction. This chapter
introduces fundamental concepts that apply to much of statistics, so it is
worth reading even if you don't work with data expressed as proportions.
Chapters 27 and 28 will explain how to compare two proportions.

DATA EXPRESSED AS PROPORTIONS

Data can be quantified by different kinds of variables (to be reviewed in
Chapter 8). Some variables are quantitative and others are categorical. Some
categorical variables can have only two possible values such as male or female,
alive or dead, graduated or not. These variables are called *binomial variables*.
This chapter considers this kind of variable, where the results can be expressed
as a proportion. Some examples of this kind of results include the proportion of
patients who already suffered a myocardial infarction (heart attack) who then go
on to have heart failure, the proportion of students who pass a particular course,
and the proportion of coin flips that are heads. Later chapters discuss different
kinds of data.

THE BINOMIAL DISTRIBUTION:
FROM POPULATION TO SAMPLE

If you flip a coin fairly, there is a 50% probability that it will land on heads and a
50% probability that it will land on tails. That means that in the long run, it will
land on heads as often as it lands on tails. But it any particular series of tosses, you
may not see equal numbers of heads and tails, and you may even see only heads
or only tails. The *binomial distribution* calculates the likelihood of observing any
particular outcome when you know its proportion in the overall population.

The binomial distribution can answer questions like these:

- If you flip a coin fairly 10 times, what is the chance of observing exactly 7 heads?
- If, on average, 5% of the patients undergoing a particular operation get an infection, what is the chance that 10 of the next 30 patients will get an infection?
- If 40% of voters are Democrats (one of the two major political parties in the United States), what is the chance that a random sample of 600 voters will be 45% Democrat?

The *cumulative binomial distribution* answers questions like these:

- If you flip a coin fairly 10 times, what is the chance of observing 7 *or more* heads?
- If, on average, 5% of the patients undergoing a particular operation get infected, what is the chance that 10 *or more* of the next 30 patients will get infected?
- If 40% of voters are Democrats, what is the chance that a random sample of 600 voters will be *at least* 45% Democrat?

This book won't show you how to answer these questions, but it is easy to find the relevant equations or to compute the answers with Excel. A helpful Web calculator can be found at http://www.graphpad.com/quickcalcs/probability1/.

The binomial distribution is rarely useful in analyzing data because it goes the wrong way. It starts with a known probability for the population and computes the chances of observing various outcomes in *random samples*. When analyzing data, we usually need to work the opposite way. We know what the result is in a sample expressed as a single proportion and want to use that proportion to make an inference about the overall proportion in the population.

EXAMPLE: FREE THROWS IN BASKETBALL

Suppose we observe a professional basketball player make 20 free throws and he succeeds in 17 of these attempts (85%).

Is it worth doing any calculations to make more general conclusions? Only if you think it is reasonable to assume that these free throws are somewhat representative of a larger population of free throws by that player, and you wish to make inferences about that larger group of throws.

If you collected data on a much larger set of free throws, the percentage of successful shots would probably not be exactly 85%. How far away from those numbers is that percentages likely to be? If you want to create a range of values that you are certain will contain the overall percentage from the much larger hypothetical population, it would have to run all the way from 0% to 100%. That wouldn't be useful. Instead, let's create ranges of percentages that we are 95% sure will contain the true population value. Such a range is called a *confidence interval*, abbreviated CI. Before reading on, write down what you think the CI is for the example (17/20 = 85%). Really. I'll give the answer in the paragraph after the next one.

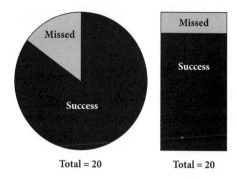

Figure 4.1. Two ways to plot data with two possible outcomes.

Figure 4.1 shows two ways of plotting these data. The pie chart on the left is conventional, but a stacked bar graph is probably easier to read.

Note that there is no uncertainty about what we observed. We are absolutely sure we counted the number of free throws and baskets correctly. Calculation of a CI will not overcome any counting mistakes. What we don't know is the overall success rate of free throws for this player.

The CI for this example ranges from 63.1% to 95.6% (calculated using the modified Wald method, explained later in this chapter). Sometimes this is written as [63.1%, 95.6%]. You might get slightly different results depending on which program you use.

EXAMPLE: POLLING VOTERS

Say you polled 100 randomly selected voters just before an election and 33 said they would vote for your candidate. What can you say about the proportion of all voters in the population who will vote for your candidate?

Again, there are two issues to deal with. First, you must think about whether your sample is really representative of the population of voters *and* whether people tell pollsters the truth about how they will vote. Statistical calculations cannot help you grapple with those issues! We'll assume that the sample is perfectly representative of the population of voters and that all people will vote as they said they would when polled. Second, you must think about *sampling error*. Just by chance, your 100-voter sample will almost certainly contain either a smaller or a larger fraction of people voting for your candidate than does the overall population.

Because we only know the proportion of favorably inclined voters in one sample, there is no way to be sure about the proportion in the population. The best we can do is to calculate a range of values that bracket the true population proportion. How wide is this range of values? In the overall population, the fraction of voters who will vote for your candidate could be almost anything. So, to create a

95% CI, we accept a 5% chance that the true population value will not be included in the range.

Before reading on, make your best guess for the 95% CI when your sample has 100 people and 33 of those people have said they would vote for your candidate (so the proportion in the sample is 0.33, or 33%).

Note that there is no uncertainty about what we observed in the voter sample. We are absolutely sure that 33.0% of the 100 people polled said they would vote for our candidate. If we weren't sure of that count, calculation of a CI could not overcome any mistakes that were made in tabulating those numbers or any ambiguity in defining "vote." What we don't know is the proportion of favorable voters in the entire population. Write down your guess before continuing.

One way to calculate the CI is explained later in this chapter. You can also use free Web-based calculators to do the math. In this case, the 95% CI extends from 25% to 43%. Sometimes this is written as [25%, 43%]. You might get slightly different results depending on which algorithm the program you use. We can be 95% confident that somewhere between 25% and 43% of the population you polled preferred your candidate on the day the poll was conducted. The phrase "95% confident" is explained in more detail in the following discussion.

Don't get confused by the two different uses of percentages. The CI is for 95% confidence. That quantifies how sure you want to be. You could ask for a 99% CI if you wanted your range to have a higher chance of including the true value. Each confidence limit (25% and 43%) represents the percentage of voters who will vote for your candidate.

How good was your guess? Many people guess that the interval is narrower than it is.

ASSUMPTIONS: CI OF A PROPORTION

The whole idea of a CI is to make a general conclusion from some specific data. This generalization is only useful if certain assumptions are true. A huge part of the task of learning statistics is to learn the assumptions upon which statistical inferences are based.

Assumption: Random (or representative) sample

All statistical conclusions are based on the assumption that the sample of data you are analyzing is drawn from a larger population of data about which you want to generalize. This assumption might be violated in the basketball example if the data were collected during a week when the player was injured or distracted by family problems. This assumption would be violated in the election example if the sample was not randomly selected from the population of voters. In fact, this mistake was made in the 1936 Roosevelt–Landon US presidential election. To find potential voters, the pollsters relied heavily on phone books and automobile registration lists. In 1936, Republicans were far more likely than Democrats to own a phone or a car, and therefore, the poll selected too many Republicans. There also was a second problem. Supporters of Landon were much more likely to return replies to the poll than were supporters of Roosevelt (Squire, 1988). The poll predicted that Landon would win by a large margin, but, in fact, Roosevelt won.

Assumption: Independent observations

The 95% CI is only valid when all subjects are sampled from the same population, and each has been selected independently of the others.

The assumption would be violated in the election example if the pollsters questioned both the husband and the wife in each family or if some voters were polled more than once.

Assumption: Accurate data

The 95% CI is only valid when the number of subjects in each category is tabulated correctly.

The data would not be accurate in the election example if the pollster deliberately recorded some of the opinions incorrectly or coerced or intimidated the respondents to answer a certain way.

Results are said to be *biased* if the experimental methods or analysis procedures ensure that the calculated results, on average, deviate from the true value.

WHAT DOES 95% CONFIDENCE REALLY MEAN?

The true population value either lies within the 95% CI you calculated or it doesn't. There is no way for you to know. If you were to calculate a 95% CI from many samples, you would expect it to include the population proportion in about 95% of the samples and to not include the population proportion in about 5% of the samples.

A simulation

To demonstrate this, let's switch to a *simulation* in which you know *exactly* the population from which the data were selected. Assume that you have a bowl of 100 balls, 25 of which are red and 75 of which are black. Mix the balls well and choose one randomly. Put it back in, mix again, and choose another one. Repeat this process 15 times, record the fraction of the sample that are red, and compute the 95% CI for that proportion. Figure 4.2 shows the simulation of 20 such samples. Each 95% CI is shown as a bar extending from the lower confidence limit to the upper confidence limit. The value of the observed proportion in each sample is shown as a line in the middle of each CI. The horizontal line shows the true population value (25% of the balls are red). In about half of the samples, the sample proportion is less than 25%, and in the other half the sample proportion is higher than 25%. In one of the samples, the population value lies outside the 95% CI. In the long run, given the assumptions previously listed, 1 of 20 (5%) of the 95% CIs will not include the population value, and that is exactly what we see in this set of 20 simulated experiments.

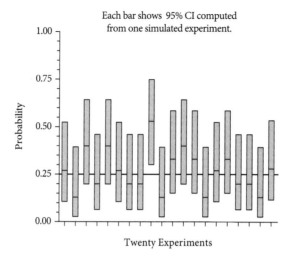

Figure 4.2. **What would happen if you collected many samples and computed a 95% CI for each?**

Each bar shows one simulated experiment, indicating the proportion of red balls chosen from a mixture of 25% red balls (the rest are black), with n = 15. The percentage success is shown as a line near the middle of each bar, which extends from one 95% confidence limit to the other. In all but one of the simulated samples, the CI includes the true population proportion of red balls (shown as a horizontal line). The 95% CI of Sample 9, however, does not include the true population value. You expect this to happen in 5% of samples. Because Figure 4.1 shows the results of simulations, we know when the CI doesn't include the true population value. When analyzing data, however, the population value is unknown, so you have no way of knowing whether the CI includes the true population value.

Figure 4.2 helps explain CIs, but you cannot create such a figure when you analyze data. When you analyze data, you don't know the actual population value. You only have results from a single experiment. In the long run, 95% of such intervals will contain the population value and 5% will not. But there is no way you can know whether or not a particular 95% CI includes the population value because you don't know that population value (except when running simulations).

95% chance of what?
When you see a 95% CI, it is easy to be confused about what the 95% refers to.

The CI you compute depends on the data you happened to collect. If you repeated the experiment, your CI would almost certainly be different, as demonstrated in Figure 4.2. If you repeated the experiment many times, you'd expect the 95% CI to include the true population value 95% of the time. That is the best way to think about the meaning of 95%.

Many simplify this concept to say there is a 95% chance that the interval computed in a particular experiment includes the population value. This is true in the sense that 95% of similar CIs will contain the true population value. But it is false in the sense that this particular interval either includes the population value or doesn't. There is no chance about it, just uncertainty. Nonetheless, I think it is clear enough to say there is a 95% chance that the interval contains the population value.

It is incorrect to interpret a CI by saying: "There is a 95% chance that the population value lies within this 95% CI." The problem is that the population proportion has one value. You don't know what it is (unless you are doing simulations), but it has one value. If you repeated the experiment, that value wouldn't change (and you still wouldn't know what it is). Therefore, it is incorrect to ask about the probability that the population proportion lies within a certain range. It either does or it doesn't. The interval is subject to random variation so it is correct to ask about the chance that the interval includes the population value. The population value is not subject to random variation, so it is not correct to ask about the chance that the population value is within the interval.

What is special about 95%?
Tradition!

CIs can be computed for any degree of confidence. By convention, 95% CIs are presented most commonly, although 90% and 99% CIs are sometimes published. With a 90% CI, there is a 10% chance that the interval does not include the population value. With a 99% CI, that chance is only 1%.

If you are willing to be less confident, then the interval will be *shorter*, so 90% CIs are narrower than 95% CIs. If you want to be more confident that your interval contains the population value, the interval must be *longer*. Thus, 99% CIs are wider than 95% CIs. If you want to be 100% confident that the interval contains the population value, the interval would need to be very wide to include every possible value, so this is not done.

ARE YOU QUANTIFYING THE EVENT YOU CARE ABOUT?

The 95% CI allows you to extrapolate from the sample to the population for the event that you tabulated. For the first example (basketball), the tabulated event (making a free throw) is exactly what you care about. But sometimes you really care about an event that is different from the one you tabulated.

In the voting example, we assessed our sample's responses to a poll conducted on a particular date, so the 95% CI lets us generalize those results to predict how the population would respond to that poll on that date. We wish to extrapolate to election results in the future but can do so only by making an additional assumption: that people will vote as they said they would. In the polling prior to the 1948 Dewey versus Truman presidential election in the United States, polls of many thousands of voters indicated that Dewey would win by a large margin. Because the CI was so narrow, the pollsters were very confident. Newspapers were so sure of the results that they prematurely printed the headline "Dewey Beats Truman." In fact, Truman won. Why was the poll inaccurate? The polls were performed in September and early October, and the election was held in November. Many voters changed their minds in the interim period. The 95% CI computed from data collected in September was inappropriately used to predict voting results two months later.

LINGO

CIs versus confidence limits

The two ends of the CI are called the *confidence limits*. The CI extends from one confidence limit to the other. The CI is a range, whereas each confidence limit is a value.

Many scientists use the terms *confidence interval* and *confidence limits* interchangeably. Fortunately, mixing up these terms does not get in the way of understanding statistical results!

Estimate

The sample proportion is said to be a *point estimate* of the true population proportion. The CI covers a range of values so is said to be an *interval estimate*.

Note that *estimate* has a special meaning in statistics. It does not mean an approximate calculation or an informed hunch, but rather is the result of a defined calculation. The term *estimate* is used because the value computed from your sample is only an estimate of the true value in the population (which you can't know).

Confidence level

A 95% CI has a confidence level of 95%. If you generate a 99% CI, the confidence level equals 99%. The term *confidence level* is used to describe the desired amount of confidence. This is also called the *coverage probability*.

Uncertainty interval

Gelman (2010) thinks that the term *confidence interval* is too confusing and too often misunderstood. He proposed using the term *uncertainty interval* instead. I think that is a good idea, but this term is rarely used.

CALCULATING THE CI OF A PROPORTION

Several methods are commonly used

Several methods have been developed for computing the CI, and there is no consensus about which is best (Brown, Cai, & DasGupta, 2001). These methods all produce results that are similar, especially with large samples. The three methods you are most likely to encounter are:

- The so-called *exact method* of Clopper and Pearson (1934) always has a confidence level of at least 95%, but the intervals are sometimes a bit wider than necessary. This method cannot be easily computed by hand.
- The *standard Wald method* (explained in the first edition of this book, and many others) is easy to compute by hand, but the modified Wald method (see next bullet point) is not much harder to compute and is much more accurate.

- The *modified Wald method,* developed by Agresti and Coull (1998) is quite accurate and is also easy to compute by hand (see next section). Simulations with many sets of data demonstrate that it works very well (Brown, Cai, & DasGupta, 2001; Ludbrook & Lew, 2009).

You can find a Web calculator that computes both the modified Wald and the Clopper–Pearson method at http://www.graphpad.com/quickcalcs/confInterval1/.

How to compute the modified Wald method by hand

If you don't want to use a calculator or a program, you can easily compute an approximate 95% CI by hand using the following steps:

1. Calculate a value called p′ from the number of successes (S) and the number of trials (n). Its value will be between the observed proportion (S/n) and 0.5:

$$p' = \frac{s+2}{n+4}.$$

2. Compute W, the margin of error (or half-width) of the CI:

$$W = 2\sqrt{\frac{p'(1-p')}{n+4}}.$$

3. Compute the 95% CI.
   From $(p' - W)$ to $(p' + W)$.

Three notes:

- The variable p′ used here is not the same as a P value (which we will discuss extensively in later chapters). This book uses an uppercase P for P values and a lowercase p for proportions, but not all books follow this convention.
- The margin of error is the direction the CI extends in each direction. The length of the interval equals twice the margin of error.
- The CI is symmetrical around p′ but is not symmetrical around the observed proportion p. The lack of symmetry is noticeable with small samples and when p is far from 0.5.

Shortcut for proportions near 50%

When the proportion is near 50%, the margin of error of the CI is approximately equal to the square root of 1/n. This leads to the rules of thumb listed in Table 4.1. Many political polls use a sample size of about 1,000, and many polls have results very near 50:50. In this case, the margin of error is 3%, a value you'll often see reported in newspapers as the margin of error of a poll.

Shortcut for proportions far from 50%

When the proportion is far from 50%, the CIs will be narrower than if the proportion is close to 50%. Here is one rough rule of thumb for large n. If the proportion is 80% or 20%, then the CI will be about 80% the length (so 20% shorter) than it would have been had the proportion been 50%.

PROPORTION	MARGIN OF ERROR (%)	APPROXIMATE 95% CI
5/10	32	18 to 82
50/100	10	40 to 60
500/1,000	3	47 to 53
5,000/10,000	1	49 to 51

Table 4.1. Approximate CI of a proportion when the observed proportion is 50%.
This table's values should not be used for formal data analysis, because they are approximations. Remember these values when you wish to rapidly evaluate published data. The margin of error is approximately equal to the reciprocal of the square root of n.

Shortcut when the numerator is zero: The rule of three

Hanley and Lippman-Hand (1983) devised a simple shortcut equation, called the *rule of three*, for determining the 95% CI of a proportion when the numerator is zero. If you observe zero events in n trials, the 95% CI extends from 0.0 to approximately 3.0/n.

You have observed no adverse drug reactions in the first 250 patients treated with a new antibiotic. The CI for the true rate of drug reactions extends from 0.0 to about 3/250, or 1.20%.

You found that zero of 29 infants born by Cesarean section were infected. Using the rule of three, the 95% CI for the proportion of infected babies extends from 0.0 to about 3/29, or 10.3%.

AMBIGUITY IF THE PROPORTION IS 0% OR 100%

In the previous example, none of the 29 babies born at 22 weeks of gestation survived. It still makes sense to compute a CI, because the true proportion in the population may not be zero. Of course, the lower confidence limit must be 0%, because there is no possibility of a proportion being negative. There are two ways to define the upper confidence limit in this situation.

As usually defined, the 95% CI allows for a 2.5% chance that the upper confidence limit doesn't go high enough to include the true population proportion and a 2.5% chance that the lower confidence limit doesn't go low enough to include the true population proportion. This leaves a 95% chance that the interval includes the true population proportion.

If you use this approach when the numerator is zero, there is a 2.5% chance that the upper interval doesn't go high enough. But there is zero chance that the lower interval doesn't go low enough, because the lower limit has to be zero. Because the uncertainty only goes in one direction, this "95%" CI really gives you 97.5% confidence. Calculated in this way, the 95% CI for the percentage of surviving babies extends from 0% to 13.9%.

The alternative approach is to compute the upper confidence limit such that there is a 5% chance it doesn't go high enough to include the true population value. This approach creates a true 95% CI but is less consistent with other intervals. Calculated this way, the 95% CI for babies surviving extends from 0% to 10.2%.

If the numerator and denominator are equal, the sample proportion is 100%, which is also the upper limit of the CI. The same logic applies, and there are two ways to define the lower limit mirroring the approach previously described to calculate the upper limit.

AN ALTERNATIVE APPROACH: BAYESIAN CREDIBLE INTERVALS

Let's return to the basketball example. Recall that a professional basketball player made 20 free throws and succeeded in 17 of these attempts (85%). What can we say about the true average fraction of free throws he makes? Using only the data at hand, our best estimate is 85% (the value we observed) and the CI ranges from 63.1% to 95.6% (calculated using the modified Wald method). This calculation comes directly from the data, without considering context.

But what if we understand the context and have some prior knowledge we want to consider? In this case, suppose we know that on average professional basketball players make about 75% of their free throws, with most players making somewhere between 50% and 90% of them. Since the observed percentage, 85%, is on the high end of this range, we might suspect that the player just had a lucky run so his true average is somewhere between 85% (what we observed) and 75% (the average from prior data).

Bayesian methods extend this reasoning formally. First, you need to state your estimate of probabilities for population proportion. This should be based only on prior data or theory, without considering the current data, and be expressed as an equation that defines the prior probability that the population percentage is any particular value. Bayesian methods have two inputs: that prior distribution of the proportion and the proportion actually observed.

For this example, it is possible to specify inputs to a mathematical equation known as the beta function to create a distribution consistent with the informal belief previously stated (centered at 75%, with more than 95% of the distribution between 50% and 90%, and with only a tiny prior probability that the proportion is less than 50% or greater than 90%). Bayesian methods combine this prior distribution with the observed value. The results are that most likely value for the proportion is 79.7% and the 95% interval ranges from 66.3% to 89.7%. (This example comes from Krushke, 2015, p. 135.)

To distinguish this from a CI, this Bayesian interval is given a different name, *credible interval*. The CIs mentioned earlier in this chapter are computed directly from the data (the number of experiments and number of times each outcome occurred). It is only when you interpret those intervals that you might also consider the context. Bayesian credible intervals are different. They are computed both from the data and from context. The latter is encoded as a prior probability distribution for the proportion. Depending on what that prior distribution is, the Bayesian credible interval can be quite similar to, or very different than, the CI.

Of course, this Bayesian approach of combining prior probabilities with data goes way beyond just thinking about using a sample proportion to make inferences about the population proportion. It can be (and is) used for any kind of data. Since this kind of Bayesian reasoning is not standard in most fields of biology, this book will only present this very brief introduction to this way of thinking.

COMMON MISTAKES: CI OF A PROPORTION

Mistake: Using 100 as the denominator when the value is a percentage

The calculation of the CI depends on the sample size. If the proportion is expressed as a percentage, it is easy to mistakenly enter 100 as the sample size. If you do this, you'll get the wrong answer unless the sample size is 100 or close to it. Look at Table 4.1. In each case, the proportion is 0.5, so the percentage is 50%. But the CIs vary enormously depending on sample size.

Mistake: Computing binomial CIs from percentage change in a continuous variable

The methods explained in this chapter apply when there are two possible outcomes (i.e., binomial outcomes), so the result is expressed as a proportion of the time that one particular outcome occurs. This value is often expressed as a percentage. But beware of percentages, which can be used to express many kinds of data. Results such as percentage change in weight, percentage change in price, or percentage change in survival time cannot be analyzed with the methods explained in this chapter. If you try to do so, the results won't make any sense.

Mistake: Computing a CI from data that look like a proportion but really is not

A fertility clinic tells you that out of 250 women given in vitro implantation, 100 got pregnant and successfully delivered a baby (or babies, if twins or triplets). The fraction 100/250 is a proportion. Each of the 250 women either had a successful delivery or they did not. You can calculate the proportion (0.40) and the 95% CI of it (0.34 to 0.46).

Another clinic tells you that their success rate is 50%. But they define this as having successfully delivered 125 babies from 250 women. But some women had twins or triplets. The ratio 125/250 looks like a proportion, but it is not. The denominator counts women; the numerator counts babies. It would be invalid to compute a CI from this pseudo ratio.

Mistake: Interpreting a Bayesian credible interval without knowing what prior probabilities (or probability distribution) were assumed for the analysis

The credible intervals computed by Bayesian methods depend both on the data and the assumption about prior probabilities. You can't interpret the interval without understanding that assumption.

Q & A

Which is wider, a 95% CI or a 99% CI?

To be more certain that an interval contains the true population value, you must generate a wider interval. A 99% CI is wider than a 95% CI. See Figure 4.3.

Is it possible to generate a 100% CI?

A 100% CI would have to include every possible value, so it would always extend from 0.0% to 100.0% and not be the least bit useful.

How do CIs change if you increase the sample size?

The width of the CI is approximately proportional to the reciprocal of the square root of the sample size. So if you increase the sample size by a factor of four, you can expect to cut the length of the CI in half. Figure 4.4 illustrates how the CI gets narrower as the sample size increases.

Can you compute a CI of a proportion if you know the proportion but not the sample size?

No. The width of the CI depends on the sample size. See Figure 4.4.

Why isn't the CI symmetrical around the observed proportion?

Because a proportion cannot go below 0.0 or above 1.0, the CI will be lopsided when the sample proportion is far from 0.50 or the sample size is small. See Figure 4.5.

You expect the population proportion to be outside your 95% CI in 5% of samples. Will you know when this happens?

No. You don't know the true value of the population proportion (except when doing simulations), so you won't know if it lies within your CI or not.

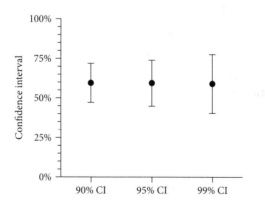

Figure 4.3. Effect of degree of confidence on the width of a CI.

Each bar represents a sample with a 60% success rate and n = 40. The graph illustrates the 90%, 95%, and 99% CIs for the percentage success in the population from which the data were drawn. When you choose to have less confidence, the CI is narrower.

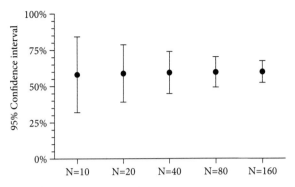

Figure 4.4. Effect of sample size on the width of a CI.
Each bar represents a sample with a 60% success rate, and the graph shows the 95% CI for the percentage success in the population from which the data were drawn. When the samples are larger, the CI is narrower.

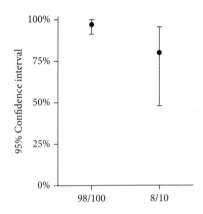

Figure 4.5. Asymmetrical CI.
If the proportion is far from 50%, the 95% CI of a proportion is noticeably asymmetrical, especially when the sample size is small.

CHAPTER SUMMARY

- This chapter discusses results that have two possible outcomes and are summarized as a proportion. The proportion is computed by dividing the number of times one outcome happened by the total number of trials.
- The binomial distribution predicts the distribution of results when you create many random samples and know the overall proportion of the two possible outcomes.
- A fundamental concept in statistics is the use of a CI to analyze a single sample of data to make conclusions about the population from which the data were sampled.

- To compute a CI of a proportion, you only need the two numbers that form its numerator and its denominator.
- Given a set of assumptions, 95% of CIs computed from a proportion will include the true population value of that proportion. You'll never know if a particular CI is part of that 95% or not.
- There is nothing magic about 95%. CIs can be created for any degree of confidence, but 95% is used most commonly.
- The width of a CI depends in part on sample size. The interval is narrower when the sample size is larger.

TERMS INTRODUCED IN THIS CHAPTER

- Bias (p. 35)
- Binomial variable (p. 31)
- Binomial distribution (p. 31)
- Confidence interval (p. 32)
- Confidence level (p. 38)
- Confidence limit (p. 38)
- Coverage probability (p. 38)
- Credible interval (p. 41)
- Cumulative binomial distribution (p. 32)
- Interval estimate (p. 38)
- Modified Wald method (p. 39)
- Point estimate (p. 38)
- Random sample (p. 32)
- Sampling error (p. 33)
- Simulation (p. 35)
- Uncertainty interval (p. 38)

CHAPTER 5

Confidence Interval of Survival Data

In the long run, we are all dead.

JOHN MAYNARD KEYNES

Outcomes that can only happen once (e.g., death) are often displayed as graphs showing percentage survival as a function of time. This chapter explains how survival curves are created and how to interpret their confidence intervals (CIs). Even if you don't use survival data, skim this chapter to review the concept of CIs with a different kind of data. This chapter explains how to analyze a single survival curve. Chapter 29 explains how to compare survival curves.

SURVIVAL DATA

The term *survival curve* is a bit limiting, because these kinds of graphs are used to plot time to any well-defined end point or event. The event is often death, but it could also be the time until occlusion of a vascular graft, first metastasis, or rejection of a transplanted kidney. The same principles are used to plot time until a lightbulb burns out, until a router needs to be rebooted, until a pipe leaks, and so on. In these cases, the term *failure time* is used instead of survival time. The event does not have to be dire. The event could be restoration of renal function, discharge from a hospital, resolution of a cold, or graduation. The event must be a one-time event. Special methods are needed when events can recur.

The methods described here (and in Chapter 29) apply when you know the survival time of each subject (or know when the data were censored, as explained in the next section). These methods are not appropriate for analyzing, for example, the survival of thousands or millions of cells, because you don't know the survival time of each individual cell. Instead, you would simply plot the percentage survival versus time and fit a curve to the data or connect the points with point-to-point lines.

CENSORED SURVIVAL DATA

If each subject's survival time were known, creating a survival curve would be trivial. But many survival times are not known.

Many studies enroll patients over a period of several years. The patients who enroll later are not followed for as many years as patients who enrolled early.

Imagine a study of a cancer treatment that enrolled patients between 1995 and 2000 and ended in 2008. If a patient enrolled in 2000 and was still alive at the end of the study, his survival time would be unknown but must exceed eight years. Although the study lasted 13 years, the fate of that patient after year 8 is unknown.

During the study, imagine that some subjects dropped out—perhaps they moved to a different city or wanted to take a medication disallowed on the protocol. If a patient moved after two years in the study, her survival time would be unknown but must exceed two years. Even if you knew how long she lived, you couldn't use the data, because she would no longer be taking the experimental drug. However, the analysis must account for the fact that she lived at least two years on the protocol.

Information about these patients is said to be censored. The word *censor* has a negative connotation. It sounds like the subject has done something bad. Not so. It's the data that have been censored, not the subject! These censored observations should not be removed from the analyses—they must just be accounted for properly. The survival analysis must take into account how long each subject is known to have been alive and following the experimental protocol, and it must not use any information gathered after that time.

Table 5.1 presents data for a survival study that only includes seven patients. That would be a very small study, but the size makes the example easy to follow. The data for three patients are censored for three different reasons. One of the censored observations is for someone who is still alive at the end of the study. We don't know how long he will live after that. Another person moved away from the area and thus left the study protocol. Even if we knew how much longer she lived, we couldn't use the information, because she was no longer following the study protocol. One person died in a car crash. Different investigators handle this kind of situation differently. Some define a death to be a death, no matter what the cause (termed *all-cause mortality*). Some investigators present the data two ways, first using all subjects, called an *intent-to-treat analysis*, and then censoring subjects who did not adhere to the full protocol, called an *according-to-protocol analysis*. Here we will define a death from a clearly unrelated cause (such as a car crash) to be a censored observation. We know he lived 3.67 years on the treatment and don't know how much longer he would have lived, because his life was cut short.

Table 5.2 demonstrates how these data are entered into a computer program when using censoring. The codes for death (1) and censored (0) are commonly used but are not completely standard.

STARTING DATE	ENDING DATE	WHAT HAPPENED?
Feb 7, 1998	Mar 2, 2002	Died
May 19, 1998	Nov 30, 2004	Moved. No longer on protocol
Nov 14, 1998	Apr 3, 2000	Died
Mar 4, 1999	May 4, 2005	Study Ended
Jun 15, 1999	May 4, 2005	Died
Dec 1, 1999	Sep 4, 2004	Died
Dec 15, 1999	Aug 15, 2003	Died in car crash

Table 5.1. Sample survival data details for the data plotted in Figure 5.1.

YEARS	CODE
4.07	1
6.54	0
1.39	1
6.17	0
5.89	1
4.76	1
3.67	0

Table 5.2. How the data of Table 5.1 are entered into a computer program.

CALCULATING PERCENTAGE SURVIVAL AT VARIOUS TIMES

There are two slightly different methods for calculating percentage survival over time. With the *actuarial method*, the X-axis is divided up into regular intervals, perhaps months or years, and survival is calculated for each interval. This method is used when the actual time of death is not known (Wormuth, 1999) or when the sample size is enormous. With the *Kaplan–Meier method*, survival time is recalculated with each patient's death. This method is preferred and is used more frequently, at least in clinical research. The term *life-table analysis* seems to be used inconsistently but usually includes both methods.

The Kaplan–Meier method is logically simple. To calculate the fraction of patients who survive on a particular day, simply divide the number alive at the end of the day by the number alive at the beginning of the day (excluding any who were censored on that day from both the numerator and the denominator). This gives you the fraction of patients who were alive at the beginning of a particular day and are still alive at the beginning of the next day. To calculate the fraction of patients who survive from day 0 until a particular day, multiply the fraction of patients who survive day 1 by the fraction of those patients who survive day 2 and then by the fraction of those patients who survive day 3, and so on, until you eventually multiply by the fraction who survive day k. This method automatically accounts for censored patients, because both the numerator and the denominator are reduced on the day a patient is censored. Because we calculate the product of many survival fractions, this method is also called the *product-limit method*. This method is part of many statistical computer programs, so there is no need to learn the tedious details.

Time zero is not some specified calendar date; rather, it is the time that each patient entered the study. In many clinical studies, time zero spans several calendar years as patients are enrolled. At time zero, by definition, all patients are alive, so survival equals 100%. On a survival graph, this is $X = 0$. Whenever a patient dies, the percentage of surviving patients decreases. If the study (and thus the X-axis) were extended far enough, survival would eventually drop to zero.

Table 5.3 includes the computed survival percentages, with 95% CIs, at each time a patient dies.

NO. OF YEARS	LOWER LIMIT	% SURVIVAL	UPPER LIMIT
0.00	100.00	100.00	100.00
1.39	33.39	85.71	97.86
3.67	33.39	85.71	97.86
4.07	21.27	68.57	91.21
4.76	11.77	51.43	81.33
5.89	4.81	34.29	68.56
6.17	4.81	34.29	68.56
6.54	4.81	34.29	68.56

Table 5.3. Kaplan–Meier survival proportions with 95% confidence limits.

GRAPHING SURVIVAL CURVES WITH CONFIDENCE BANDS

Figure 5.1 plots the example data four ways. Each time a patient dies, the curve drops. The censored subjects are shown as circles in the upper- and lower-right panels and as ticks in the lower-left panel.

An essential part of data analysis includes interpreting CIs. Survival curve programs compute a 95% CI at each time that a death occurs. These intervals can then be joined to plot 95% confidence bands, as shown in the lower-left panel of Figure 5.1. The stepped survival curve shows exactly what happened in the sample we studied. Given certain assumptions (explained later in the chapter), we can be 95% confident that the 95% confidence bands at any time include the true percentage survival in the population.

Note that the CI at any particular time is asymmetrical. This is because the percentage survival cannot possibly go below 0.0% or above 100%. The asymmetry of the CI will be especially noticeable when the sample survival percentage is far from 50% or the sample size is small. The interval is symmetrical when the survival percentage equals 50% and is always nearly symmetrical when the sample size is huge.

SUMMARIZING SURVIVAL CURVES

Median survival

It can be convenient to summarize an entire survival curve by one value, the *median survival*. The median is the middle value (the 50th percentile) of a set of numbers, so median survival is how long it takes until half the subjects have died.

It is easy to derive the median survival time from a survival curve: draw a horizontal line at 50% survival and note where it crosses the survival curve. Then look down at the X-axis to read the median survival time. The lower-right panel of Figure 5.1 indicates that the median survival in this example is 5.89 years.

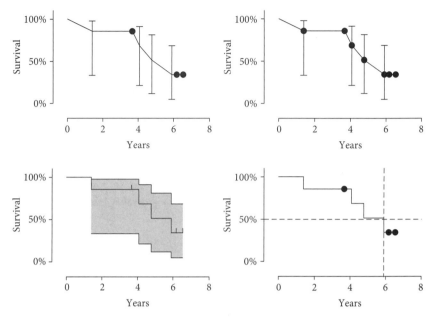

Figure 5.1. Four versions of a survival curve created from the data in Tables 5.1 through 5.3.
The X-axis plots the number of years after each subject was enrolled. Note that time zero does not have to be any particular day or year. The Y-axis plots percentage survival. The three censored patients are shown as symbols in three of the graphs and as upward blips in the fourth. Four of the subjects died. You can see each death as a downward step in the curves. The 95% CIs for the populations are shown as error bars in the top two graphs and as a shaded region (confidence bands) in the bottom-left graph. The bottom-right graph shows the median survival time, which is the time at which half of the subjects have died and half are still alive. Read across at 50% to determine median survival, which is 5.89 years in this example.

Calculations of median survival are ambiguous in two situations. Median survival is undefined when more than half the subjects are still alive at the end of the study. In this case, the survival curve never crosses the 50% survival line. The median survival is greater than the last duration of time plotted on the survival curve, but there is no way to know how much greater. The other ambiguous situation is when the survival curve is horizontal at 50% survival. One way to define median survival in this case is to average the first and last times when survival is 50%.

Mean survival?
Mean survival is rarely computed. This is because calculating a mean survival time requires that you know every single time of death. Thus, the mean survival cannot be computed if any observations are censored or if any subjects are still alive when the study ends.

In contrast, median survival can be computed when some observations are censored and when the study ends before all subjects have died. Once half of the subjects have died, the median survival is unambiguous, even without knowing how long the others will live.

Five-year survival

Survival with cancer is often quantified as *five-year survival*. A vertical line is drawn at X = 5 years, and the Y value that line intersects is the five-year percentage survival. Of course, there is nothing special about five years (rather than two or four or six years) except tradition.

ASSUMPTIONS: SURVIVAL ANALYSIS

When evaluating any statistical analysis, it is critical to review all assumptions. Analyses of survival curves depend on these assumptions.

Assumption: Random (or representative) sample

The whole idea of all statistical analyses is to generalize the data from your sample to a more general situation. If your sample is not randomly selected from a defined population, then you must *assume* that your sample is representative of that population.

Assumption: Independent subjects

The results of any statistical analyses can only be interpreted if the data from each subject offer independent information. If the study pools data from two hospitals, the subjects are not independent. It is possible that subjects from one hospital have different average survival times than subjects from another hospital, and you could alter the survival curve by choosing more subjects from one hospital and fewer from the other hospital. Because most diseases have a genetic component, another way to violate the assumption of independence would be to include two (or more) people from one family in one treatment group.

Assumption: Entry criteria are consistent

Typically, subjects are enrolled over a period of months or years. In these studies, it is important that the entry criteria do not change during the enrollment period.

Imagine a cancer survival curve starting from the date that the first metastasis was detected. What would happen if improved diagnostic technology detected metastases earlier? Patients would die at the same age they otherwise would, but now they would be diagnosed at a younger age and therefore would live longer with the diagnosis. Even with no change in therapy or in the natural history of the disease, survival time would apparently increase simply because the entry criteria changed.

Airlines used this trick to improve their reported on-time departure rates. Instead of planning to close the doors at the scheduled departure time, they now plan to close the airplane's door 10 minutes before the scheduled departure time. This means that the flight can still be recorded as having left "on time" even if they closed the doors 10 minutes later than they actually planned to.

Assumption: End point defined consistently

If the curve is plotting duration of time until death, there can be ambiguity about which deaths to count. In a cancer trial, for example, what happens to subjects who die in car crashes? Some investigators count these as deaths; others count them as censored subjects. Both approaches can be justified, but the approach should be decided before the study begins or both approaches presented in the results. If there is any ambiguity about which deaths to count, the decision should be made by someone who doesn't know which patient received which treatment.

If the curve plots duration of time until an event other than death, it is crucial that the event be assessed consistently throughout the study.

Assumption: Starting time clearly defined

The starting point should be an objective date—perhaps the date of first diagnosis or first hospital admission. You may be tempted to use an earlier starting criterion instead, such as the time when a patient remembers first observing symptoms. Don't do it. Such data are invalid because a patient's recollection of early symptoms may be altered by later events.

What happens when subjects die before they receive the treatment they were supposed to get? It is tempting to remove these subjects from the study. But this can lead to biases, especially if one treatment (such as a medication) is started immediately, but another (surgery) requires preparation or scheduling. If you remove the patients who die early from the surgical group but not from the medication group, the two groups will have different survival times, even if the treatments are equivalent. To avoid such biases, most studies follow a policy of *intention to treat*. Each subject's survival is analyzed as if he or she received the assigned treatment, even if the treatment was not actually given. The justification for using the intention-to-treat criterion is discussed in Chapter 29.

Assumption: Censoring is unrelated to survival

The survival analysis is only valid when the reasons for censoring are unrelated to survival. If data from a large fraction of subjects are censored, the validity of this assumption is critical to the integrity of the results.

Data for some patients are censored because they are alive at the end of the study. With these patients, there is no reason to doubt the assumption that censoring is unrelated to survival.

The data for other patients are censored because they drop out of the study. If the reason these patients dropped out could be related to survival, then the analysis won't be valid. Examples include patients who quit the study because they were too sick to come to the clinic, patients who felt well and stopped treatment, and patients who quit because they didn't think the treatment was working (or who were taken off the study by a physician who thought the treatment wasn't working). These reasons all relate to disease progression or response to therapy, and it is quite possible that the survival of these patients is related to their reason for dropping out. Including these subjects (and censoring their data) violates the assumption that censoring is unrelated to survival. But excluding these subjects entirely can also lead to biased results. The best plan is to analyze the data both ways. If the conclusions of the two analyses are similar, then the results are straightforward to interpret. If the conclusions of the two analyses differ substantially, then the study results are simply ambiguous.

Assumption: Average survival doesn't change during the study

Many survival studies enroll subjects over a period of several years. The analysis is only meaningful if you assume that the first few patients are sampled from the same population as the last few subjects.

   If the nature of the disease changes during the time the study accrues patients, the results will be difficult to interpret. This is quite possible when studying rapidly evolving infectious diseases. It is also important that the treatment (including supportive care) does not change over the course of the study.

Q & A

Why is percentage survival graphed as staircases, rather than as point-to-point lines?
   The tradition is to plot the actual experience in the sample. So when a subject dies, the curve drops down in a staircase fashion. Connecting the points diagonally might do a better job of demonstrating your best estimate for survival in the overall population, but this is not standard.

What if subjects enter the study on different dates?
   That's OK and is common. The X-axis shows time from entry in the study, so it does not correspond to calendar dates.

Why isn't a CI plotted at time zero?
   At time zero, 100% survival is a given. Subjects who are not alive at time zero are not included in the study! So the survival at time zero is not a value subject to sampling error. It is 100% for sure.

My confidence band is too wide to be useful. How can I make it narrower?
   Collect more data, a lot more data. The width of the CI is approximately proportional to the reciprocal of the square root of the sample size. So if you increase the sample size by a factor of four, you can expect the CI to be half as wide.

In the example, three of seven subjects survived. Why doesn't the curve end at Y = 3/7, or 43%?
   Survival curve calculations properly account for censored data. After data are censored, that subject doesn't contribute to computing the percentage survival. Data for two subjects were censored prior to the last time point, so the simple computation (2/7 = 29% still survive) is not correct. In fact, the curve ends with 34.29% survival.

Can the data be plotted as percentage death rather than percentage survival?
   Yes! See Figure 5.2.

How does one compare two survival curves?
   See Chapter 29.

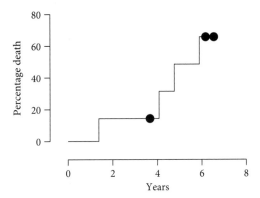

Figure 5.2. A graph of percentage death, rather than percentage survival, illustrates the same information.

When only a small percentage of subjects have died by the end of the study, this kind of graph can be more informative (because the Y-axis doesn't have to extend all the way to 100).

CHAPTER SUMMARY

- Survival analysis is used to analyze data for which the outcome is elapsed time until a one-time event (often death) occurs.
- The tricky part of survival analysis is dealing with censored observations. An observation is censored when the one-time event hasn't happened by the time the study ends, or when the patient stops following the study protocol.
- Survival curves can be plotted with confidence bands.
- A survival curve can be summarized by the median survival or the five-year survival.
- Survival curves, and their confidence bands, can only be interpreted if you accept a list of assumptions.

TERMS INTRODUCED IN THIS CHAPTER

- According-to-protocol analysis (p. 47)
- Actuarial method (p. 48)
- All-cause mortality (p. 47)
- Censored data (p. 53)
- Failure time (p. 46)
- Five-year survival (p. 51)
- Intention to treat (p. 52)
- Kaplan–Meier (product limit) method (p. 48)
- Life-table analysis (p. 48)
- Median survival (p. 49)
- Product limit method (p. 48)
- Survival curve (p. 46)

CHAPTER 6

Confidence Interval of Counted Data (Poisson Distribution)

> Not everything that can be counted counts; and not everything
> that counts can be counted.
>
> ALBERT EINSTEIN

When events occur randomly, independently of one another, and
with an average rate that doesn't change over time, the number
of events counted in a certain time interval follow a Poisson distribution.
From the number of events actually observed, a confidence interval (CI)
can be computed for the average number of events per unit of time. This
also works for computing a CI for the average number of objects in a
certain volume. Poisson distributions are common in many fields of science. In addition to teaching you about a useful distribution, this chapter
also reinforces the concept of a CI.

THE POISSON DISTRIBUTION

Some outcomes are expressed as the number of objects in some volume or the
number of events in some time interval. Examples include the number of babies
born in an obstetrics ward each day, the number of eosinophils seen in one microscope field, or the number of radioactive disintegrations detected by a scintillation
counter in one minute.

If you repeatedly sample time intervals (or objects in a volume), the number
of events (or objects) will vary. This random distribution is called a *Poisson
distribution.* Given the population's average number of occurrences in one unit
of time (or space), the Poisson distribution predicts how often you'll observe any
particular number of events or objects.

Background radiation depends on the types of radiation you measure, your
location, and the sensitivity of the detector, but it ranges from a few to a few dozen
counts per minute, on average. In any particular minute, the number of detected
radiation counts may be higher or lower than the average, but the number of disintegrations counted will always be an integer. The left half of Figure 6.1 shows
the predictions of a Poisson distribution with an average of 1.6 counts per minute.

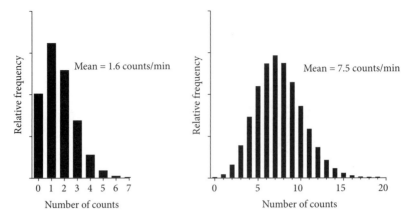

Figure 6.1. The Poisson distribution.
Assume that the average number of background radiation counts is 1.6 counts/minute (left) or 7.5 counts/minute (right). In any particular minute, the count may be higher or lower than the average, as illustrated by these Poisson distributions. The horizontal axis lists the number of radioactive disintegrations actually counted (not converted to counts per minute), and the vertical axis plots the frequency.

This Poisson distribution is notably asymmetrical. This makes sense, because the number of radioactive background counts cannot be less than zero but has no upper limit. The right half of Figure 6.1 shows the predictions of a Poisson distribution with an average of 7.5 counts per minute. This Poisson distribution is nearly symmetrical and almost looks like a Gaussian distribution (see Chapter 10).

ASSUMPTIONS: POISSON DISTRIBUTION

Calculations based on the Poisson distribution can be used for data expressed as the number of events per unit time or the number of objects in a certain volume.

Assumptions for number of events
Calculations about events assume the following:

- The event is clearly defined.
- Each event occurs randomly, independent of other events. This assumption might be violated if you tracked the number of babies born each day, because of the occurrence of twins and triplets. Instead, count the number of deliveries (counting twins and triplets as one delivery).
- The average rate doesn't change over time.
- Each event is counted only once. This assumption was violated in a study conducted to find out how often airplanes nearly collide. The investigators separately surveyed pilots and copilots about the number of times they experienced a near miss. A single near miss, therefore, could be reported four times—by the pilot and by the copilot in each plane. For this reason,

the initial analysis overestimated the number of near misses ("NASA Blows Millions on Flawed Airline Safety Survey," 2007).

Assumptions for number of objects

Calculations about objects assume the following:

- Objects are randomly dispersed.
- Each object is only counted once. If you are counting cells, any clumping would make it difficult to comply with this assumption.
- Objects are well defined, with no ambiguity about what to count (or what not to count). For example, if you are counting cells in a microscope field, there must be no ambiguity about differentiating cells from debris.

CONFIDENCE INTERVALS BASED ON POISSON DISTRIBUTIONS

The Poisson distribution can be used to compute a CI. As long as you know the actual number of objects counted in a volume, it is possible to compute a CI for the average number of objects in that volume. Similarly, if you know the number of events that occurred in a single time interval, you can compute a CI for the average number of counts in that time interval.

Raisins in bagels

Imagine that you carefully dissect a bagel and find 10 raisins. You assume that raisins are randomly distributed among bagels, that the raisins don't stick to each other to create clumps (perhaps a dubious assumption), and that the average number of raisins per bagel doesn't change over time (the recipe is constant). The 95% CI (determined using the Poisson distribution) ranges from 4.8 to 18.4 raisins per bagel. You can be 95% certain that range includes the overall average number of raisins per bagel.

Radioactive counts

Imagine that you have counted 120 radioactive counts in one minute. Radioactive counts occur randomly and independently, and the average rate doesn't change (within a reasonable time frame much shorter than the isotope's half-life). Thus, the Poisson distribution is a reasonable model. The 95% CI for the average number of counts per minute is 99.5 to 143.5.

Note that you must base this calculation on the actual number of radioactive disintegrations counted. If you counted the tubes for 10 minutes, the calculation must be based on the counts in 10 minutes and not on the calculated number of counts in 1 minute. This point is explained in more detail later in this chapter.

Person-years

Exposure to an environmental toxin caused 1.6 deaths per 1,000 person-years exposure. What is the 95% CI? To calculate the CI, you must know the exact number of deaths that were observed in the study. This study observed 16 deaths in observations of 10,000 person-years (they might have studied 10,000 people for 1 year or 500 people for 20 years). If you set the count (C) to equal 16, the 95% CI for the number of deaths ranges from 9.15 to 25.98. That is the number of deaths per 10,000 person-years. Divide by 10 to return to the original units (per 1,000 person-years). The 95% CI ranges from 0.92 to 2.6 deaths per 1,000 person-years exposure.

HOW TO CALCULATE THE POISSON CI

You need the count from a sample (call this value C) and you can compute the CI of a count. But note:

- If you have multiple samples, just add the counts from each sample to calculate a total count. Then divide both ends of the interval by the number of samples to calculate C.
- It is essential that you base the calculations on the number of events (or objects) *actually observed*. Don't convert to a more convenient time scale or volume until after you have calculated the CI. If you attempt to calculate a CI based on the converted count, the results will be meaningless.

To compute the CI, use one of these methods:

- Use a program or free Web calculator such as www.graphpad.com/quickcalcs/confInterval1/.
- When C is large (greater than 25 or so), the following equation is a useful shortcut approximation for the 95% CI of C. If C is 25, the approximate 95% CI extends from 15.2 to 34.8, whereas the exact CI ranges from 16.2 to 36.9.

$$C - 1.96\sqrt{C} \text{ to } C + 1.96\sqrt{C}$$

THE ADVANTAGE OF COUNTING FOR LONGER TIME INTERVALS (OR IN LARGER VOLUMES)

Figure 6.2 demonstrates the advantage of using a longer time interval. One tube containing a radioactive sample was counted repeatedly. The left side of Figure 6.2 illustrates radioactive decays counted in one-minute intervals. The right side of Figure 6.2 illustrates radioactive decays counted in 10-minute intervals. The graph plots counts per minute, so the number of radioactive counts counted in each 10-minute interval was divided by 10.0 after calculating the CI but before being graphed.

When computing the CIs for Poisson variables, it is essential to do the calculations with the number of objects or events actually counted. If you count 700 radioactive disintegrations in one minute, the 95% CI for the average number of disintegrations per minute will range from about 650 to 754. If you count 7,000 disintegrations in 10 minutes, the 95% CI for the average number of disintegrations per 10 minutes will range from 6,838 to 7,166. Divide those

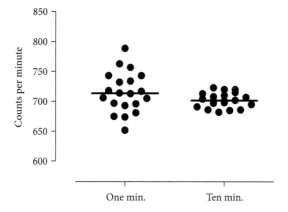

Figure 6.2. The advantages of counting radioactive samples for a longer time interval.
One radioactive sample was repeatedly measured for 1-minute intervals (left) and 10-minute intervals (right). The number of radioactive counts detected in the 10-minute samples was divided by 10, so both parts of the graph show counts per minute. By counting longer, there is less Poisson error. Thanks to Arthur Christopoulos for providing these data.

values by 10 to obtain the 95% CI for the average number of decays per minute, which ranges from 684 to 718. Counting for a longer period of time gives you a more precise assessment of the average number of counts per interval, and thus the CI is narrower.

Let's revisit the example of raisins in bagels. Instead of counting raisins in one bagel, imagine that you dissected seven individual bagels and found counts of 9, 7, 13, 12, 10, 9, and 10 raisins. In total, there were 70 raisins in seven bagels, for an average of 10 raisins per bagel. The CI should be computed using the total counted. Given 70 objects counted in a certain volume, the 95% CI for the average number ranges from 54.57 to 88.44. This is the number of raisins per seven bagels. Divide by seven to express these results in more useful units. The 95% CI ranges from 7.8 to 12.6 raisins per bagel.

Q & A

What is the difference between the binomial and Poisson distributions?
> Both the binomial and Poisson distributions are used for outcomes that are counted, but the two are very different. The binomial distribution describes the distribution of two possible outcomes. The Poisson distribution describes the possible number of objects you'll find in a certain volume or the number of events you'll observe in a particular time span.

Why use 1.96 in the equation?
> You'll learn about this in Chapter 10. In a Gaussian distribution, 95% of the values lie within 1.96 standard deviations of the mean. With large values of C,

the Poisson distribution approximates a Gaussian distribution, with a standard deviation equal to the square root of C.

Is it possible to compute a CI when the observed number of counts is zero?

Yes. When you observe zero objects in a certain volume or zero events in a certain time, the 95% CI for the average number of objects in that volume (or events in that time interval) ranges from 0.0 to 3.69.

CHAPTER SUMMARY

- When events occur randomly, independently of one another, and with an average rate that doesn't change over time, the number of events counted in a certain time interval follows a Poisson distribution.
- When objects are randomly distributed and not clumped, the number of objects counted in a certain volume follows a Poisson distribution.
- From the number of events actually observed (or the number of objects actually counted), a CI can be computed for the average number of events per unit of time (or number of objects per unit volume).
- When using a Poisson distribution to compute a CI, you must base the calculation on the actual number of objects (or events) counted. Any normalization to more convenient units must be done after you have computed the CI.

TERM INTRODUCED IN THIS CHAPTER

- Poisson distribution (p. 55)

PART C

Continuous Variables

CHAPTER 7

Graphing Continuous Data

> When you can measure what you are speaking about and
> express it in numbers, you know something about it; but when
> you cannot measure it, when you cannot express it in numbers,
> your knowledge is of the meager and unsatisfactory kind.
>
> LORD KELVIN

When results are continuous (e.g., blood pressure, enzyme activity, IQ score, blood hemoglobin, oxygen saturation, temperature, etc.), the first and too-often neglected step is to visualize the data. This chapter shows how the actual distribution of values can be graphed without any statistical calculations.

CONTINUOUS DATA

When analyzing data, it is essential to choose methods appropriate for the kind of data with which you are working. To highlight this point, this book began with discussions of three kinds of data. Chapter 4 discussed data expressed as two possible outcomes, summarized as a proportion; Chapter 5 explained survival data and how to account for censored observations; and Chapter 6 discussed data expressed as the actual number of events counted in a certain time or as objects counted in a certain volume.

This chapter begins our discussion of continuous data, such as blood pressures, enzyme activity, weights, and temperature. In many scientific fields, continuous data are more common than other kinds of data.

THE MEAN AND MEDIAN

Mackowiak, Wasserman, and Levine (1992) measured body temperature from many healthy individuals to see what the normal temperature range really is. A subset of 12 of these values is shown in Table 7.1. These values, and the suggestion to use them to explain basic statistical principles, come (with permission) from Schoemaker (1996).

37.0
36.0
37.1
37.1
36.2
37.3
36.8
37.0
36.3
36.9
36.7
36.8

Table 7.1. The body temperature of 12 individuals in degrees centigrade.

These values are used as sample data.

Calculating an arithmetic *mean* or average is easy: add up all the values and divide by the number of observations. The mean of the smaller (n = 12) subset is 36.77°C. If the data contain an outlier (a value far from the others), the mean won't be very representative (learn more about outliers in Chapter 25). For example, if the largest value (37.1) were mistakenly entered into the computer as 371 (i.e., without the decimal point), the mean would increase to 64.6°C, which is larger than all the other values.

The mean is one way to quantify the middle, or central tendency, of the data, but it is not the only way. Here are some others:

- The *median* is the middle value. Rank the values from lowest to highest and identify the middle one. If there is an even number of values, average the two middle ones. For the n = 130 data set, the median is the average of the 65th and 66th ranked values, or 36.85°C. Compared to the mean, the median is not influenced by an outlier, and can be more useful with skewed distributions. For example, the mean cost of a US wedding in 2011 was $27,021, but the median price was $16,886 (Oremus, 2013). A small proportion of really expensive weddings bring up the mean cost but don't affect the median cost.

- The *geometric mean* is calculated by transforming all values into their logarithms, computing the mean of these logarithms, and then taking the antilog of that mean (logarithms and antilogarithms are reviewed in Appendix E). Because a logarithm is only defined for values greater than zero, the geometric mean cannot be calculated if any values are zero or negative. None of the temperature values is zero or negative, so the geometric mean could be computed. However, the geometric mean would not be useful with the sample temperature data, because 0.0°C does not mean "no temperature," and it is certainly possible to have temperatures in °C that are negative (but not body temperatures!). Using terminology you'll encounter in Chapter 8, temperature in degrees centigrade is not a *ratio variable*. Read more about the geometric mean in Chapter 11.

- The *harmonic mean* is calculated by first transforming each value to its reciprocal and then computing the (arithmetic) mean of those reciprocals. The harmonic mean is the reciprocal of that mean. It can't be computed if any values are zero, and it isn't meaningful if any values are negative. The harmonic mean is not commonly used in the biological sciences but is often used in physics, engineering, and finance ("Harmonic Mean," 2017).
- The *trimmed mean* is the mean of most of the values, ignoring the highest and lowest values. Olympic ice skating used to be scored this way, with the largest and smallest score eliminated before averaging the scores from all other judges. Sometimes several of the highest and lowest values are ignored.
- The *mode* is the value that occurs most commonly in the data set. It is not useful with continuous variables that are assessed with at least several digits of accuracy, because each value will be unique. The mode can be useful when working with variables that can only be expressed as integers. Note that the mode doesn't always assess the center of a distribution. Imagine, for example, a medical survey on which one of the questions is "How many times have you had surgery?" In many populations, the most common answer will be zero, so that is the mode. In this case, some values will be higher than the mode, but none will be lower.

LINGO: TERMS USED TO EXPLAIN VARIABILITY

In Table 7.1, the temperatures range from 36.0°C to 37.3°C. There can be several reasons for this kind of variation.

Biological variability

Most of this variation is probably the result of biological variability. People (and animals, and even cells) are different from one another, and these differences are important! Moreover, people (and animals) vary over time because alterations in age, time of day, mood, and diet. In biological and clinical studies, much or most of the scatter is often caused by biological variation.

Precision

Compare repeated measurements to see how precise the measurements are. *Precise* means the same thing as *repeatable* or *reproducible*. A method is precise when repeated measurements give very similar results. The variation among repeated measurements is sometimes called *experimental error*.

Many statistics books (especially those designed for engineers) implicitly assume that most variability is the result of imprecision. In medical studies, biological variation often contributes more variation than does experimental imprecision.

Mistakes and glitches can also contribute to variability. For example, it is possible that a value was written down incorrectly or that the thermometer wasn't positioned properly.

Bias

Biased measurements result from systematic errors. Bias can be caused by any factor that consistently alters the results: the proverbial thumb on the scale, defective thermometers, bugs in computer programs (maybe the temperature was measured in centigrade and the program that converted the values to Fahrenheit was buggy), the placebo effect, and so on. As used in statistics, the word *bias* refers to anything that leads to systematic errors, not only the preconceived notions of the experimenter. Biased data are not accurate.

Accuracy

A result is accurate when it is close to being correct. You can only know if a value is correct, of course, if you can measure that same quantity using another method known to be accurate. A set of measurements can be quite precise without being accurate if the methodology is not working or is not calibrated properly. A measurement, or a method to obtain measurements, can be accurate and precise, accurate (on average) but not precise, precise but not accurate, or neither accurate nor precise.

Error

In ordinary language, the word *error* means something that happened accidentally. But in statistics, *error* is often used to refer to any source of variability, as a synonym for *scatter* or *variability*.

PERCENTILES

You are probably familiar with the concept of a percentile. For example, a quarter (25%) of the values in a set of data are smaller than the 25th percentile, and 75% of the values are smaller the 75th percentile (so 25% are larger).

The 50th percentile is identical to the median. The 50th percentile is the value in the middle. Half the values are larger (or equal to) the median, and half are lower (or equal to). If there is an even number of values, then the median is the average of the middle two values.

Calculating the percentiles is trickier than you'd guess. Eight different equations can be used to compute percentiles (Harter, 1984). All methods compute the same result for the median (50th percentile) but may not compute the same results for other percentiles. With large data sets, however, the results are all similar.

The 25th and 75th percentiles are called *quartiles*. The difference computed as the 75th percentile minus the 25th percentile is called the *interquartile range*. Half the values lie within this interquartile range.

GRAPHING DATA TO SHOW VARIATION

Scatter plots

Figure 7.1 plots the temperature data as a *column scatter plot*. The left side of the graph shows all 130 values. The right side of the graph shows the randomly selected subset of 12 values (Table 7.1), which we will analyze separately. Each value is plotted as a symbol. Within each half of the graph, symbols are moved to the right or left to prevent too much overlap; the horizontal position is arbitrary, but the vertical position, of course, denotes the measured value.

This kind of column scatter graph, also known as a *dot plot*, demonstrates exactly how the data are distributed. You can see the lowest and highest values and the distribution. It is common to include a horizontal line to denote the mean or median, and sometimes error bars are shown too (see Chapter 14).

With huge numbers of values, column scatter graphs get unwieldy with so many overlapping points. The left side of Figure 7.1, with 130 circles, is pushing the limits of this kind of graph. But the right side, with 12 circles, is quite clear and takes the same amount of space as a graph with mean and error bars, while showing more information.

Box-and-whiskers plots

A *box-and-whiskers plot* gives you a good sense of the distribution of data without showing every value (see Figure 7.2). Box-and-whiskers plots work great when you have too many data points to show clearly on a column scatter graph but don't want to take the space to show a full frequency distribution.

A horizontal line marks the median (50th percentile) of each group. The boxes extend from the 25th to the 75th percentiles and therefore contain half the

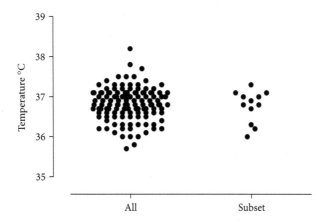

Figure 7.1. Column scatter graph of body temperatures.

(Left) The entire data set, with n = 130. (Right) A randomly selected subset (n = 12). In a column scatter graph, the vertical position of each symbol denotes its value. The horizontal position (within each lane) is adjusted (jittered) to prevent points from overlapping (too much).

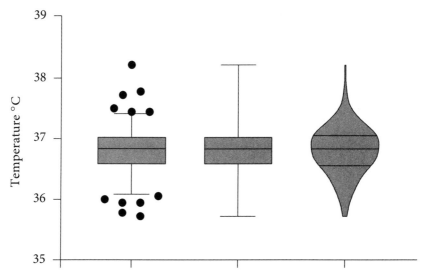

Figure 7.2. Box-and-whiskers and violin plots.

(Left) Box-and-whiskers plots of the entire data set. The whiskers extend down to the 5th percentile and up to the 95th, with individual values showing beyond that. (Middle) The whiskers show the range of all the data. (Right) A violin plot of the same data.

values. A quarter (25%) of the values are higher than the top of the box, and 25% of the values are below the bottom of the box.

The whiskers can be graphed in various ways. The first box-and-whiskers plot in Figure 7.2 plots the whiskers down to the 5th and up to the 95th percentiles and plots individual dots for values lower than the 5th and higher than the 95th percentile. The other box-and-whiskers plot in Figure 7.2 plots the whiskers down to the smallest value and up to the largest, and so it doesn't plot any individual points. Whiskers can be defined in other ways as well.

Violin plots

Figure 7.2 also shows a violin plot of the same data (Hintze & Nelson, 1998). The median and quartiles are shown with black lines. The overall distribution is shown by the violin-shaped gray area. The "violin" is thickest where there are the most values and thinner where there are fewer values, so the shape of the "violin" gives you a sense of the distribution of the values.

GRAPHING DISTRIBUTIONS

Frequency distributions

A frequency distribution lets you see the distribution of many values. Divide the range of values into a set of smaller ranges (bins) and then graph the number of values (or the fraction of values) in each bin. Figure 7.3 displays *frequency distribution graphs* for the temperature data. If you add the height of all the bars, you'll

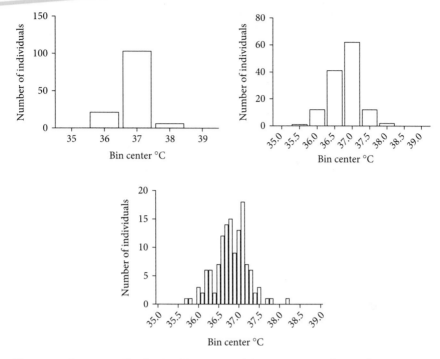

Figure 7.3. Frequency distribution histograms of the temperature data with various bin widths.

With too few bins (top left), you don't get a sense of how the values vary. With many bins (bottom), the graph shows too much detail for most purposes. The top-right graph shows the number of bins that seems about right. Each bar plots the number of individuals whose body temperature is in a defined range (bin). The centers of each range, the bin centers, are labeled.

get the total number of values. If the graph plots fractions or percentages instead of number of values, then the sum of the height of all bars will equal 1.0 or 100%.

The trick in constructing frequency distributions is deciding how wide to make each bin. The three graphs in Figure 7.3 use different bin widths. The graph on the top left has too few bins (each bin covers too wide a range of values), so it doesn't show you enough detail about how the data are distributed. The graph on the bottom has too many bins (each bin covers too narrow a range of values), so it shows you too much detail (in my opinion). The upper-right graph seems the most useful.

Watch out for the term *histogram*. It is usually defined as a frequency distribution plotted as a bar graph, as illustrated in Figure 7.3. Sometimes the term *histogram* is used more generally to refer to any bar graph, even one that is not a frequency distribution.

Cumulative frequency distribution

One way to avoid choosing a bin width is to plot a *cumulative frequency distribution,* in which each Y value is the number of values less than X. The cumulative distribution begins at Y = 0 and ends at Y = n, the number of values in the data set.

This kind of graph can be made without choosing a bin width. Figure 7.4 illustrates a cumulative frequency distribution of the temperature data.

Figure 7.5 shows the same graph, but the Y-axis plots the percentage (rather than the number) of values less than or equal to each X value. The X value when Y is 50% is the median. The right side of Figure 7.5 illustrates the same distribution plotted with the Y-axis transformed in such a way that a cumulative distribution from a Gaussian distribution (see Chapter 10) becomes a straight line.

When graphing a cumulative frequency distribution, there is no need to decide on the width of each bin. Instead, each value can be individually plotted.

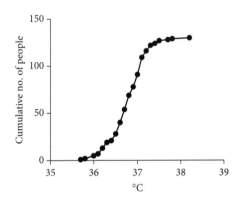

Figure 7.4. Cumulative frequency distribution.
Each circle illustrates the number of people whose temperature was less than (or equal to) the X value. With a cumulative distribution, there is no need to choose a bin width. This graph plots the distribution of 130 values. But since the data were recorded only to within 0.1°, there are only 21 unique values. Accordingly, the graph has 21 points.

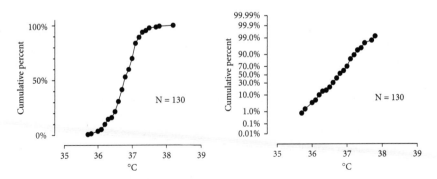

Figure 7.5. Cumulative frequency distribution, shown as percentages.
Left) The graph is the same as Figure 7.4, except that the Y-axis plots the percentage of values less than or equal to each X value (rather than the actual number). (Right) Plot of the data with the Y-axis transformed in such a way that a cumulative Gaussian distribution (to be explained in Chapter 10) becomes a straight line, as seen here.

BEWARE OF DATA MASSAGE

Published graphs sometimes don't plot the data that were actually collected but instead plot the result of computations. Pay attention to the decisions and calculations done between data collection and graphing.

Beware of filtering out impossible values

Data sets are often screened to remove impossible values. Weights can't be negative. The year of death cannot be before the year of birth. The age of a child cannot be greater than the age of the mother. It makes no sense to run statistical analyses on obviously incorrect data. These values must be fixed (if the mistake can be traced) or removed and documented.

But beware! Eliminating "impossible" values can also prevent you from seeing important findings. In 1985, researchers first noticed the decrease in ozone levels over Antarctica. Their software flagged these values as unusual, because they were the lowest ozone levels ever seen. The investigators were wary, so double checked all the instruments and calculations. After convincing themselves that the data were real, they published their findings. Later work discovered that the drop in ozone levels was caused by chlorofluorocarbons, and international efforts to stop using chlorofluorocarbons prevented what could have been a huge environmental problem. The investigators would have made a big mistake and missed an important discovery if they had simply ignored the low values as being impossible or had their software been written to remove these "impossible" values (Pukelshein, 1990; Sparling, 2001).

Beware of adjusting data

The values analyzed by statistical tests are often not direct experimental measurements. It is common to adjust the data, so it is important to think about whether these adjustments are correct and whether they introduce errors. If the numbers going into a statistical test are questionable, then so are the statistical conclusions.

In some cases many adjustments are needed, and these adjustments can have a huge impact. As a case in point, NASA provides historical records of temperature. An examination of how temperature has changed over the past century has convinced most scientists that the world is getting warmer. The temperature record, however, requires many corrections. Corrections are needed to account for the fact that cities are warmer than surrounding areas (because of building heat), that different kinds of thermometers have been used at different historical times, that the time of day that temperatures have been taken is not consistent, and so on (Goddard, 2008). The net result of all these adjustments pushed earlier temperatures down by almost 1°C, which is about the observed temperature increase in the past century. These adjustments require judgment, and different scientists may do the adjustments differently. Anyone interpreting the data must understand the contribution of these adjustments to the overall observed effect, as well as the degree to which these adjustments could possibly be biased by the scientists' desire to make the data come out a certain way.

When interpreting published data, ask about how the data were adjusted before they were graphed or entered into a statistics program.

Beware of smoothing

When plotting data that change over time, it is tempting to remove much of the variability to make the overall trend more visible. This can be accomplished by plotting a *rolling average*, also called a *moving average* or *smoothed data*. For example, each point on the graph can be replaced by the average of it and the three nearest neighbors on each side. The number of points averaged to smooth the data can range from two to many. If more points are included in the rolling average (say 10 on each side, rather than 3), the curve will be smoother. Smoothing methods differ in how neighboring points are weighted.

Smoothed data should never be entered into statistical calculations. If you enter smoothed data into statistics programs, the results reported by most statistical tests will be invalid. Smoothing removes information, so most analyses of smoothed data are not useful.

Look ahead to Figure 33.4 to see an example of how analysis of smoothed data can lead to a misleading conclusion.

Beware of variables that are the ratio of two measurements

Often the value that you care most about is the ratio of two values. For example, divide enzyme activity or the number of binding sites by the cell count or the protein concentration. Calculating the ratio is necessary to express the variable in a way that can be interpreted and compared—for example, as enzyme activity per milligram of protein. The numerator is usually what you are thinking about and what the experimenter spent a lot of time and effort to measure. The denominator can seem like a housekeeping detail. But, of course, the accuracy of the ratio depends on the accuracy of both the numerator and the denominator.

Beware of normalized data

Some scientists transform (normalize) data so that all values are between 0% and 100%. When you see these kinds of data, you should wonder about how the values defining 0% and 100% were chosen. Ideally, the definitions of 0% and 100% are based on theory or on control experiments with plenty of precise replicates. If 0% and 100% are not clearly defined or seem to be defined sloppily, then the normalized values won't be very useful. Ask how many replicates there were of the control measurements that define 0% and 100%.

Q & A

Can the mean or median equal zero? Can it be negative?
> Yes. The mean (or median) can have any value, including negative values and zero.

Can the median and the smallest value be the same?
> Yes. If more than half the values are identical, and the other values are all larger, then the median will equal the smallest value.

Which is larger, the mean or median?

It depends on the data. If the distribution of values is symmetrical, the mean and median will be similar. If the distribution is skewed to the right, with an excess of large values, then the mean will probably be larger than the median (von Hippel, 2005). If the distribution is skewed the other way, with many small values, then the mean will probably be less than the median.

Are *mean* and *average* synonyms?

Yes. As used by statisticians, they are interchangeable. Outside of statistics, some use *average* more generally, and it sometimes refers to the median.

In what units are the mean and median expressed?

The mean and median are expressed in the same units as the data. So are the geometric mean and harmonic mean.

Can the mean and median be computed when n = 1? When n = 2?

Computing a mean or median from two values is no problem. The whole idea of a mean or median makes no sense when there is only one value, but you could say that the mean and the median both equal that value.

Can the mean and median be calculated when some values are negative?

Yes.

Can the geometric and harmonic means be calculated when some values are negative?

No. They can only be calculated when all values are positive. The geometric and harmonic means cannot be computed if any values are negative or zero.

Is the 50th percentile the same as the median?

Yes.

CHAPTER SUMMARY

- Many scientific variables are continuous.
- One way to summarize these values is to calculate the mean, median, mode, geometrical mean, or harmonic mean.
- When graphing this kind of data, consider creating a graph that shows the scatter of the data. Either show every value on a scatter plot or show the distribution of values with a box-and-whiskers plot or a frequency distribution histogram.
- It is often useful to filter, adjust, smooth, or normalize the data before further graphing and analysis. These methods can be abused. Think carefully about whether these methods are being used effectively and honestly.

TERMS INTRODUCED IN THIS CHAPTER

- Arithmetic mean (p. 64)
- Bias (p. 66)
- Box-and-whiskers plot (p. 67)
- Continuous data (p. 63)
- Cumulative frequency distribution (p. 69)

- Dot plot (or scatter plot) (p. 67)
- Error (p. 65)
- Frequency distribution (p. 68)
- Geometric mean (p. 64)
- Harmonic mean (p. 65)
- Histogram (p. 69)
- Interquartile range (p. 66)
- Mean (p. 64)
- Median (p. 64)
- Mode (p. 65)
- Moving average (p. 72)
- Outlier (p. 64)
- Percentile (p. 66)
- Precision (p. 65)
- Quartile (p. 66)
- Rolling average (p. 72)
- Scatter plot (or dot plot) (p. 67)
- Smoothed data (p. 72)
- Trimmed mean (p. 65)
- Violin plot (p. 68)

CHAPTER 8

Types of Variables

Get your facts first, then you can distort them as you please.

MARK TWAIN

The past four chapters have discussed four kinds of data. This chapter reviews the distinctions among different kinds of variables. Much of this chapter is simply terminology, but these definitions commonly appear in exam questions.

CONTINUOUS VARIABLES

Variables that can take on any value (including fractional values) are called *continuous variables*. The next six chapters deal with continuous variables. You need to distinguish two kinds of continuous variables: *interval variables* and *ratio variables*.

Interval variables

Chapter 7 used body temperature in degrees centigrade for its examples. This kind of continuous variable is termed an *interval variable* (but not a ratio variable). It is an interval variable because a difference (interval) of 1°C means the same thing all the way along the scale, no matter where you start.

Computing the difference between two values can make sense when using interval variables. The 10°C difference between the temperatures of 100°C and 90°C has the same meaning as the difference between the temperatures of 90°C and 80°C.

Calculating the ratio of two temperatures measured in this way is not useful. The problem is that the definition of zero is arbitrary. A temperature of 0.0°C is defined as the temperature at which water freezes and certainly does not mean "no temperature." A temperature of 0.0°F is a completely different temperature (−17.8°C). Because the zero point is arbitrary (and doesn't mean no temperature), it would make no sense at all to compute ratios of temperatures. A temperature of 100°C is not twice as hot as 50°C.

Figure 8.1 illustrates average body temperatures of several species (Blumberg, 2004). The platypus has an average temperature of 30.5°C, whereas a canary has an average temperature of 40.5°C. It is incorrect to say that a canary has a temperature 33% higher than that of a platypus. If you did that same calculation using degrees Fahrenheit, you'd get a different answer.

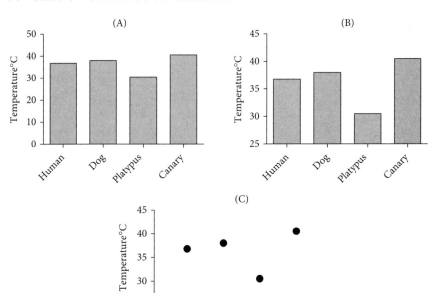

Figure 8.1. Body temperature of four species.
(A) is misleading. It invites you to compare the relative heights of the bars. But because a temperature of 0°C does not mean no temperature, the ratio of bar heights is not a meaningful value. (B) uses a different baseline to emphasize the differences. The bar for the canary is about three times as high as the bar for the platypus, but this ratio (indeed, any ratio) can be misleading. (C) illustrates the most informative way to graph these values.

Figure 8.1A is misleading. The bars start at zero, inviting you to compare their relative heights and think about the ratio of those heights. But that comparison is not useful. This graph is also not helpful because it is difficult to see the differences between values.

Figure 8.1B uses a different baseline to demonstrate the differences. The bar for the canary is about three times as high as the bar for the platypus, but this ratio (indeed, any ratio) is not useful. Figure 8.1C illustrates the most informative way to graph these values. The use of points, rather than bars, doesn't suggest thinking in terms of a ratio. A simple table might be better than any kind of graph for these values.

Ratio variables

With a *ratio variable*, zero is not arbitrary. Zero height is no height. Zero weight is no weight. Zero enzyme activity is no enzyme activity. So height, weight, and enzyme activity are ratio variables.

As the name suggests, it can make sense to compute the ratio of two ratio variables. A weight of 4 grams is twice the weight of 2 grams, because weight is a ratio variable. But a temperature of 100°C is not twice as hot as 50°C, because temperature in degrees centigrade is not a ratio variable. Note, however, that temperature in Kelvin is a ratio variable, because 0.0 in Kelvin really does mean (at least to a physicist) no temperature. Temperatures in Kelvin are far removed from temperatures that we ordinarily encounter, so they are rarely used in biology.

Like interval variables, you can compute the difference between ratio variables. Unlike interval variables, you can calculate the ratio of two ratio variables.

DISCRETE VARIABLES

Variables that can only have a limited set number of possible values are called *discrete variables*.

Ordinal variables

An *ordinal variable* expresses rank. The order matters but not the exact value. For example, pain is expressed on a scale of 1 to 10. A score of 7 means more pain than a score of 5, which is more than a score of 3. But it doesn't make sense to compute the difference between two values, because the difference between 7 and 5 may not be comparable to the difference between 5 and 3. The values simply express an order. Another example would be movie or restaurant ratings from one to five stars.

Nominal and binomial variables

Variables that can only have a set number of discrete possible values are called *nominal variables* or *categorical variables*.

Nominal variables with only two possible values are called *binomial variables*. Examples would include alive or dead, heads or tails, and pass or fail.

WHY IT MATTERS?

Table 8.1 summarizes which kinds of calculations are meaningful with which kinds of variables. It refers to the standard deviation and coefficient of variation, both of which are explained in Chapter 9, as well as the standard error of the mean (explained in Chapter 14).

OK TO COMPUTE	NOMINAL	ORDINAL	INTERVAL	RATIO
Frequency distribution	Yes	Yes	Yes	Yes
Median and percentiles	No	Yes	Yes	Yes
Add or subtract	No	No	Yes	Yes
Ratio	No	No	No	Yes
Mean, standard deviation, standard error of the mean	No	No	Yes	Yes
Coefficient of variation	No	No	No	Yes

Table 8.1. Calculations that are meaningful with various kinds of variables.
The standard deviation and coefficient of variation is explained in Chapter 9, and the standard error of the mean is explained in Chapter 14.

NOT QUITE AS DISTINCT AS THEY SEEM

Note that the categories of variables are nowhere near as distinct as they may sound (Velleman & Wilkinson, 1993). Here are some ambiguous situations:

- Color. In a psychological study of perception, different colors would be regarded as categories, so color would be a nominal variable. But monochromatic colors can be quantified by wavelength and thus can be considered a ratio variable. Alternatively, you could rank the wavelengths and consider color to be an ordinal variable.
- The number of cells actually counted in a certain volume. The number must be an integer, so this is a discrete variable. But it has nearly all the properties of a ratio variable, since it makes sense to compute ratios (with treatment A the cell number is twice what it was with treatment. B). This situation is similar to the examples outlined in Chapter 6.
- Effective concentrations (EC_{50}): An EC_{50} is a measure of a drug's potency. It is the drug concentration that elicits 50% of the maximum response. An EC_{50} cannot equal zero, but it is very useful to calculate the ratios of two EC_{50} values. Therefore, it is sort of a ratio variable but not quite.
- Percentages: Outcomes measured as ratio or interval variables are often transformed so that they are expressed as percentages. For example, a pulse rate (heartbeats per minute, a ratio variable) could be normalized to a percentage of the maximum possible pulse rate. A discrete outcome with a set of mutually exclusive categories can also be expressed as a percentage or proportion—for example, the percentage of transplanted kidneys that are rejected within a year of surgery. But these two situations are very different, and different kinds of statistical analyses are needed.

Q & A

Why is it important to remember these definitions?
> They appear commonly on statistics exams.

Does knowing these definitions help one choose an appropriate test?
> Maybe. It is important that every statistical test be designed to answer the scientific question at hand. Some people say that knowing the definitions of the various kinds of variables will prevent you from choosing the wrong test.

Why does it only make sense to compute a coefficient of variation (CV) of a continuous variable that is a ratio variable?
> The CV (explained in Chapter 9) is the ratio computed by dividing the standard deviation by the mean. If the variable is a ratio variable, then the CV will be the same no matter what units are used. If the variable is not a ratio variable, then changing units (say from °C to °F) will change the value of the CV (which is therefore meaningless).

Why doesn't it make sense to compute the ratio of two continuous variables that are not ratio variables?
> If the variable is not a ratio variable, then changing its units (say from °C to °F) will change the value of the ratio, which is therefore meaningless.

CHAPTER SUMMARY

- Different kinds of variables require different kinds of analyses.
- Prior chapters have already discussed three kinds of data: proportions (binomial variables), survival data, and counts.
- This chapter explains how continuous data can be subdivided into interval and ratio (and ordinal) data.
- Interval variables are variables for which a certain difference (interval) between two values is interpreted identically no matter where you start. For example, the difference between zero and 1 is interpreted the same as the difference between 999 and 1,000.
- Ratio variables are variables for which zero is not an arbitrary value. Weight is a ratio variable because weight = 0 means there is no weight. Temperature in Fahrenheit or centigrade is not a ratio variable, because 0° does not mean there is no temperature.
- It does not make any sense to compute a coefficient of variation, or to compute ratios, of continuous variables that are not ratio variables.
- The distinctions among the different kinds of variables are not quite as crisp as they sound.

TERMS INTRODUCED IN THIS CHAPTER

- Binomial variable (p. 77)
- Continuous variable (p. 75)
- Discrete variable (p. 77)
- Interval variable (p. 75)
- Nominal variable (p. 77)
- Ordinal variable (p. 77)
- Ratio variable (p. 75)

CHAPTER 9

Quantifying Scatter

The average human has one breast and one testicle.

DES MCHALE

Chapter 7 demonstrated various ways to graph continuous data to make it easy to see the degree of variability. This chapter explains how variation can be quantified with the standard deviation, variance, coefficient of variation, interquartile range, and median absolute deviation.

INTERPRETING A STANDARD DEVIATION

The variation among values can be quantified as the *standard deviation* (SD), which is expressed in the same units as the data. Let's first think about how to interpret the SD and then look at how it is calculated.

You can interpret the SD using the following rule of thumb: about two-thirds of the observations in a population usually lie within the range defined by the mean minus 1 SD to the mean plus 1 SD. This definition is unsatisfying because the word *usually is* so vague. Chapter 10 will give a more rigorous interpretation of the SD for data sampled from a Gaussian distribution.

Let's turn to the larger (n = 130) sample in Figure 9.1. The mean temperature is 36.82°C and the SD is 0.41°C. Figure 9.1 plots the mean with error bars extending 1 SD in each direction (error bars can be defined in other ways, as you'll see in Chapter 14).

The range 36.4°C to 37.2°C extends below and above the mean by 1 SD. The left side of Figure 9.1 is a column scatter graph of the body temperature data from Figure 7.1, and you can see that about two-thirds of the values are within that range.

Take a good look at Figure 9.1. Often you'll only see the mean and SD, as shown on the right, and you will need to imagine the actual scatter of the data. If you publish graphs, consider showing the actual data, as on the left side of Figure 9.1, rather than just the mean and the SD.

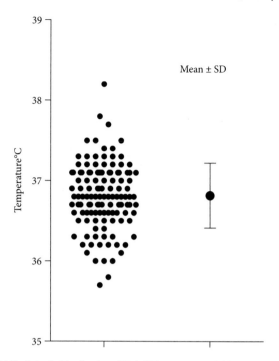

Figure 9.1. (Left) Each individual value. (Right) The mean and SD.

HOW IT WORKS: CALCULATING SD

It would seem that the simplest way to quantify variability would be to ask how far each value is from the mean (or median) of all values and to then report the average or median of those values. That strategy wouldn't work, however, because the positive deviations would balance out the negative deviations so that the average deviation would always equal zero. Another approach might be to take the average, or the median, of the absolute values of the deviations of each value from the mean. Indeed, *median absolute deviation* (MAD) is one way to quantify variability, and it is discussed later in this chapter. However, the most common way to quantify variation is by calculating the SD.

Of course, you'll use computer programs to compute the SD. Even so, the best way to understand the SD is to know how it is calculated.

1. Calculate the mean (average). For the $n = 12$ body temperature sample in Table 7.1, the mean is 36.77°C.
2. Calculate the difference between each value and the mean.
3. Square each of those differences.
4. Add up those squared differences. For the example data, the sum is 1.767.
5. Divide that sum by $(n - 1)$, where n is the number of values. For the example, $n = 12$ and the result is 0.161. This value is called the *variance*.
6. Take the square root of the value you calculated in Step 5. The result is the SD. For this example, the SD = 0.40°C.

These steps can be shown as an equation, where Y_i stands for one of the n values, and $\bar{Y}$ is the mean.

$$SD = \sqrt{\frac{\sum (Y_i - \bar{Y})^2}{n-1}}$$

9.1

WHY n − 1?

When calculating the SD, the sum of squares is divided by (n − 1). This is the definition of the *sample SD*, which is the best possible estimate of the SD of the entire population, as determined from one particular sample. Read on if you are curious to know why the denominator is (n − 1) rather than n.

When it sometimes makes sense to use n in the denominator

A denominator of n is sometimes used when the goal is to quantify the variation in a particular set of data without any extrapolation to make wider conclusions. The resulting SD is correct for those particular values, but it cannot be used to make general conclusions from the data. One example would be when a teacher wants to quantify variation among exam scores. The goal is not to make inferences about a larger population of hypothetical scores but simply to quantify the variation among those particular scores.

Why it usually makes sense to use n − 1 in the denominator

Most often, a scientist's goal is to compute SD from a sample of data and make inferences about the population from which the data were sampled. Depending on which values were randomly selected to be in the sample, the SD you calculate may be higher or lower than the true population SD, which you don't know. This is analogous to computing a mean from a sample of data. The result might be lower or higher than the true population mean, which you don't know. All you can do is use a method that will be correct on average in the long run.

When you go to calculate the SD using Equation 9.1, the numerator is the sum of the squares of the difference between each value and the mean of those values. Except for the rare case in which the sample mean happens to equal the population mean, the data will be closer to the sample mean than they will be to the true population mean (which you can't know). Therefore, the sum of squares will probably be a bit smaller (and can't be larger) than what it would be if you computed the difference between each value and the true population mean. Because the numerator is a bit too small, the denominator must be made smaller too.

But why n − 1? If you knew the sample mean and all but one of the values, you could calculate what that last value must be. Only n − 1 of the values are free to assume any value. Therefore, we calculate the average of the squared deviations by dividing by (n − 1) and say that there are n − 1 *degrees of freedom* (df). Many people find the concept of df to be quite confusing. Fortunately, being confused about df is not a big handicap! You can choose statistical tests and interpret the results with only a vague understanding of df.

The fine print

This SD computed with n – 1 is said to be the best possible estimate of the SD of the entire population, as determined from one particular sample. But best for what purpose? It is "best" because it will lead to accurate inferences when used to compute confidence intervals (CIs) or P values. However, the SD computed with n – 1 as the denominator is *not* the most accurate estimate of the population SD. On the contrary, it is a biased estimate of the population SD. *Biased* means that, on average, the computed SD will not equal (in this example, will be smaller than) the population SD, especially with small samples.

So why is the SD computed with (n – 1) used routinely? Because the sample variance (the square of the SD; discussed later in this chapter) computed using (n – 1) as the denominator is an unbiased estimate of the population variance and the theory of CIs and much of statistics depends on variances. This book, like most others, ignores the fact that the sample SD is a biased estimate of the population SD, because that discrepancy doesn't affect any statistical conclusions.

SITUATIONS IN WHICH n CAN SEEM AMBIGUOUS

When calculating SD (and various other statistical tests), the definition of the sample size n can seem ambiguous in some situations.

Replicate measurements within repeat experiments

Table 9.1 presents a common situation. Data for five replicate measurements (rows) were collected from three animals (represented by the three columns). In total, 15 values were collected. But these are not 15 independent samples from one population. There are two sources of variability here: variation between animals (called *biological replicates*) and variation between replicate measurements in one animal (called *technical replicates*). Lumping these samples together would lead to invalid analyses. It is not correct to compute an SD (or CI; see Chapter 12) using n = 15. This mistake is called *pseudoreplication* (Lazic, 2010).

	ANIMAL A	ANIMAL B	ANIMAL C
	47.7	64.7	39.3
	43.1	65.4	40.0
	52.3	88.3	23.9
	55.2	64.0	36.6
	42.5	71.9	48.9
Mean	48.2	70.9	37.7

Table 9.1. What is n?

Data for five replicate measurements (rows) taken from three animals (columns). Because there are not 15 independent values here, it is not correct to compute an SD and CI using n = 15. Instead, average the values from each animal and then compute the SD and CI from those three means using n = 3.

The simplest way to analyze such data is to average the values from each animal. For this example, there are three means (one for each animal). You would then compute the SD (and CI) from those three means using n = 3. The results can then be extrapolated to the population of animals you could have studied.

Eyes, ears, and elbows

Let's say you study 10 people and make a measurement in both eyes (or ears, elbows, etc.), so have 20 measurements from 10 people. Computing the SD (or doing any statistical calculation) using n = 20 isn't quite right. Biological variation between people is much greater than variation between the two eyes (or two ears, two elbows, etc.) of one person. If the study involves any drug therapy, the distinction is even more important as people vary in drug compliance, absorption, elimination and action, and these differences will affect both eyes (or ears, elbows, etc.) equally. Using n = 20 in statistical analyses is an example of *pseudoreplication* and will lead to misleading results.

The simplest solution is to average the results for the two eyes (or ears, elbows, etc.) to get an average response for that person. Then compute the mean and SD of those 10 averages.

The larger problem in this example is that the 20 measurements are not *independent*. This concept is discussed in Chapters 4 and 5. The interpretation of the SD (and nearly any statistical calculation) depends on the assumption that each value contributes independent information. If you follow the recommendation of the last paragraph, you have 10 independent measurements (unless some of the people were siblings). But the 20 measurements (counting each eye separately) are not independent, and any interpretation that depends on the assumption of independence will be incorrect.

Representative experiments

Some investigators prefer to just show data from one "representative" experiment. The error bars in the table or graph are calculated only from the replicates in that one experiment. If the data were collected in triplicate, the graph might be labeled "n = 3," but, in fact, all the data come from one experiment, not three. It can be useful to report the SD of replicate data within a single experiment as a way to demonstrate the precision of the method and to spot experimental problems. But substantive results should be reported with data from multiple experiments (Vaux, Fidler, & Cumming, 2012).

Trials with one subject

Chapter 3 presented an example of an experiment conducted on one person with repeated samples (see Roberts, 2004); such experiments are sometimes called "n = 1" or "n of 1" studies. But despite the name, the n value used in statistical analysis is not 1 but rather is the number of values collected. The results can then be extrapolated to the population of results you could have collected from that one person.

SD AND SAMPLE SIZE

For the full body temperature data set (n = 130), as shown in Figure 7.1, SD = 0.41°C. For the smaller sample (n = 12, randomly sampled from the larger sample), SD = 0.40°C. Many find it surprising that the SD is so similar in samples of such different sizes. But, in fact, this is to be expected. The SD quantifies the variation within a population. As you collect larger samples, you'll be able to quantify the variability more precisely, but collecting more data doesn't change the variability among the values.

OTHER WAYS TO QUANTIFY AND DISPLAY VARIABILITY

The SD is not the only way to quantify variability.

Coefficient of variation

For ratio variables, variability can be quantified as the *coefficient of variation* (CV), which equals the SD divided by the mean. If the CV equals 0.25, you know that the SD is 25% of the mean.

Because the SD and the mean are both expressed in the same units, the CV is a fraction with no units. Often the CV is expressed as a percentage.

For the preceding temperature example, the CV would be completely meaningless. Temperature is an interval variable, not a ratio variable, because zero is defined arbitrarily (see Chapter 8). A CV computed from temperatures measured in degrees centigrade would not be the same as a CV computed from temperatures measured in degrees Fahrenheit. Neither CV would be meaningful, because the idea of dividing a measure of scatter by the mean only makes sense with ratio variables, for which zero really means zero.

The CV is useful for comparing scatter of variables measured in different units. You could ask, for example, whether the variation in pulse rate is greater than or less than the variation in the concentration of serum sodium. The pulse rate and sodium are measured in completely different units, so comparing their SDs would make no sense. Comparing their coefficients of variation might be useful to someone studying homeostasis.

Variance

The variance equals the SD squared and so is expressed in the same units as the data but squared. In the body temperature example, the variance is 0.16°C squared.

Statistical theory is based on variances, rather than SD, so mathematical statisticians routinely think about variances. Scientists analyzing data do not often have to encounter variances, not even when using analysis of variance (ANOVA), explained in Chapter 39.

Interquartile range

You probably are already familiar with the concept of percentiles. The 25th percentile is a value below which 25% of the values in your data set lie. The *inter-quartile range* is defined by subtracting the 25th percentile from the 75th percentile. Because both percentile values are expressed in the same units as the data, the interquartile range is also expressed in the same units.

For the body temperature data (n = 12 subset), the 25th percentile is 36.4°C and the 75th percentile is 37.1°C, so the interquartile range is 0.7°C. For the full (n = 130) data set, the 25th percentile is 36.6°C and the 75th percentile is 37.1°C, so the interquartile range is 0.5°C.

Five-number summary

The distribution of a set of numbers can be summarized with five values, known as the *five-number summary*: the minimum, the 25th percentile (first quartile), the median, the 75th percentile (the third quartile), and the maximum.

Median absolute deviation

The median absolute deviation (MAD) is a simple way to quantify variation. It is the value such that half of the values are closer to the median than it and half are farther away.

The best way to understand the MAD is to understand the "recipe" for how it is calculated. First, compute the median of all values. The median is the 50th percentile. Then calculate how far each value is from the median of all values. Regardless of whether the value is greater than or less than the median, express the distance between it and the median as a positive value (i.e., take the absolute value of the difference between the value and the median). Now find the median of that set of differences. The result is the MAD.

For the body temperature data (n = 12 subset), the median is 36.85°C and the MAD is 0.2°C. For the full (n = 130) data set, the median is 36.8°C and the MAD is 0.3°C.

Note two points of confusion:

- There are two distinct computations of the median. First, you compute the median (the 50th percentile) of the actual data. Then you compute the median of the absolute values of the distances of the data from the 50th percentile.
- Some investigators take the mean (rather than the median) of the set of differences and call the result the *mean absolute deviation*, also abbreviated MAD.)

Half of the values are within 1 MAD of the median. Therefore, a symmetrical range that extends one MAD in each direction from the median will contain about half of the values. The interquartile range also contains half of the values. The distinction is that the interquartile range can be asymmetrical around the median. Like the interquartile range, but unlike the SD, computation of the MAD is resilient to the presence of outliers (see Chapter 25).

Q & A

Can the SD ever equal zero? Can it be negative?

> The SD will equal zero if all the values are identical. The SD can never be a negative number.

In what units do you express the SD?

> The SD is expressed in the same units as the data.

Can the SD be computed when n = 1?

> The SD quantifies variability, so it cannot be computed from a single value.

Can the SD be computed when n = 2?

> Yes, the SD can be computed from two values (n = 2).

Is the SD the same as the standard error of the mean?

> No. They are very different. See Chapter 14.

Can the SD be computed if the data clearly do not come from a Gaussian distribution?

> Yes. The SD can be computed from any set of values. The Gaussian distribution is explained in Chapter 10. If the data are sampled from a Gaussian population, you can use the rule of thumb that roughly two-thirds of the values are expected to be within 1 SD of the mean, and about 95% are expected to be within 2 SD of the mean. If you can't assume sampling from a Gaussian distribution, you can use this alternative rule of thumb: At least 75% of the values lie within 2 SD of the mean.

Is the SD larger or smaller than the CV?

> The SD is expressed in the same units as the data. The CV is a unitless ratio, often expressed as a percentage. Because the two are not expressed in the same units, it makes no sense to ask which is larger.

Is the SD larger or smaller than the variance?

> The variance is the SD squared, so it is expressed in different units. It makes no sense to ask which is larger.

Will all programs compute SD the same way?

> The only ambiguity is whether to use n or (n − 1) in the denominator.

Will all programs compute the same value for the interquartile range?

> No. Eight different equations can be used to compute percentiles (Harter, 1984), and the difference between these methods can be noticeable with small samples.

What is the best way to report a SD in a publication?

> Many people report a mean and a SD something like this: "115±10 mmHg," with a footnote or statement in the Methods section defining the second value as a SD. Some journals prefer this style: "115 mmHg (SD 10)."

CHAPTER SUMMARY

- The most common way to quantify scatter is with a SD.
- A useful rule of thumb is that about two-thirds of the observations in a population usually lie within the range defined by the mean minus 1 SD to the mean plus 1 SD.
- Other variables used to quantify scatter are the variance (SD squared), the CV (which equals the SD divided by the mean), the interquartile range, and the median absolute deviation.
- While it is useful to quantify variation, it is often easiest to understand the variability in a data set by seeing a graph of every data point or a graph of the frequency distribution.

TERMS INTRODUCED IN THIS CHAPTER

- Biological replicates (p. 83)
- Coefficient of variation (CV) (p. 85)
- Degrees of freedom (df) (p. 82)
- Five-number summary (p. 86)
- Interquartile range (p. 85)
- Median absolute deviation (MAD) (p. 81)
- Pseudoreplication (p. 83)
- Sample standard deviation (p. 82)
- Standard deviation (SD) (p. 80)
- Technical replicates (p. 83)
- Variance (p. 81)

CHAPTER 10

The Gaussian Distribution

Everybody believes in the normal approximation, the experimenters because they think it is a mathematical theorem, the mathematicians because they think it is an experimental fact.

G. LIPPMAN

Many statistical methods assume that data follow a Gaussian distribution. This chapter briefly explains the origin and use of the Gaussian distribution. Chapter 24 will explain how to test for deviations from a Gaussian distribution.

THE NATURE OF THE GAUSSIAN DISTRIBUTION

The Gaussian bell-shaped distribution, also called the *normal distribution*, is the basis for much of statistics. It arises when many random factors create variability. This happens because random factors tend to offset each other. Some will push the value higher, and others will pull it lower. The effects usually partly cancel one another, so many values end up near the center (the mean). Sometimes many random factors will tend to work in the same direction, pushing a value away from the mean. Only rarely do almost all random factors work in the same direction, pushing that value far from the mean. Thus, many values are near the mean, some values are farther from the mean, and very few are quite far from the mean. When you plot the data on a frequency distribution, the result is a symmetrical, bell-shaped distribution, idealized as the *Gaussian distribution.*

Variation among values will approximate a Gaussian distribution when there are many sources of variation, so long as the various contributors to that variation are added up to get the final result and the sample size is large. As you have more and more sources of scatter, the predicted result approaches a Gaussian distribution.

Interpreting many commonly used statistical tests requires an assumption that the data were sampled from a population that follows a Gaussian distribution. This is often a reasonable assumption. For example, in a laboratory experiment, variation between experiments might be caused by several factors: imprecise weighing of reagents, imprecise pipetting, the random nature of radioactive decay, nonhomogenous suspensions of cells or membranes, and so on. Variation in a clinical value might be caused by many genetic and environmental factors. When scatter is the result of many independent additive causes, the distribution will tend to follow a bell-shaped Gaussian distribution.

SD AND THE GAUSSIAN DISTRIBUTION

Figure 10.1 illustrates an ideal Gaussian distribution. The horizontal axis shows various values that can be observed, and the vertical axis quantifies their relative frequency. The mean, of course, is the center of the Gaussian distribution. The Gaussian distribution is high near the mean, because that is where most of the values are. As you move away from the mean, the distribution follows its characteristic bell shape. The distribution is symmetrical, so the median and the mean are identical.

The SD is a measure of the spread or width of the distribution. The area under the entire curve represents the entire population. The left side of Figure 10.1 shades the area under the curve within 1 SD of the mean. You can see that the shaded portion is about two-thirds (68.3%) of the entire area, demonstrating that about two-thirds of the values in a Gaussian population are within 1 SD of the mean. The right side of the figure demonstrates that about 95% of the values in a Gaussian population are within 2 SD of the mean (the actual multiplier is 1.96).

Scientific papers and presentations often show the mean and SD but not the actual data. If you assume that the distribution is approximately Gaussian, you can recreate the distribution in your head. Going back to the sample body temperature data in Figure 9.1, what could you infer if you knew only that the mean is 36.82°C and its SD is 0.41°C (n = 130)? If you assume a Gaussian distribution, you could infer that about two-thirds of the values lie between 36.4 and 37.2°C and that 95%

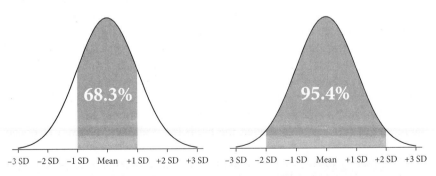

Figure 10.1. Ideal Gaussian distributions (also called normal distributions).
The horizontal axis plots various values, and the vertical axis plots their relative abundance. The area under the curve represents all values in the population. The fraction of that area within a range of values tells you how common those values are. (Left) About two-thirds of the values are within 1 SD of the mean. (Right) Slightly more than 95% of the values are within 2 SD of the mean.

of the values lie between 36.0 and 37.6°C. If you look back at Figure 9.1, you can see that these estimates are fairly accurate.

THE STANDARD NORMAL DISTRIBUTION

When the mean equals zero and the SD equals 1.0, the Gaussian (or normal) distribution is called a *standard normal distribution*. Figure 10.1 would be a standard normal distribution if the labels went from −3 to +3 without using the labels SD and mean.

All Gaussian distributions can be converted to a standard normal distribution. To do this, subtract the mean from each value and divide the difference by the SD.

$$z = \frac{\text{Value} - \text{Mean}}{\text{SD}}$$

The variable z is the number of SD away from the mean. When $z = 1$, a value is 1 SD above the mean. When $z = -2$, a value is 2 SD below the mean. Table 10.1 tabulates the fraction of a normal distribution between $-z$ and $+z$ for various values of z.

If you work in pharmacology, don't confuse this use of the variable z with the specialized Z-factor used to assess the quality of an assay used to screen drugs (Zhang, Chung, & Oldenburg, 1999). The two are not related.

THE NORMAL DISTRIBUTION DOES NOT DEFINE NORMAL LIMITS

A Gaussian distribution is also called a *normal distribution*. This is another case in which statistics has endowed an ordinary word with a special meaning. Don't mistake this special use of the word *normal* with its more ordinary meaning to

z	PERCENTAGE OF STANDARD NORMAL DISTRIBUTION BETWEEN −z AND z
0.67	50.00
0.97	66.66
1.00	68.27
1.65	90.00
1.96	95.00
2.00	95.45
2.58	99.00
3.00	99.73

Table 10.1. The standard normal distribution.

Table 10.1 is best understood by example. The range between z = −1 and z = +1 contains 68.27% of a standard normal distribution. Calculate additional values using this Excel formula: = 1 − 2*(1-NORM.S.DIST(z, TRUE)).

describe acommonly observed value, or a value of a lab test or clinical measurement that does not indicate disease.

Here is a simple but conceptually flawed approach that is often used to determine the normal range of a measured value (in this case, body temperature): assume that the population's body temperatures (from the examples of the previous chapters) follow a Gaussian distribution and define 5% of the population to be abnormal. Using the sample mean and SD from our study, the normal range can be (incorrectly) defined as the mean plus or minus 1.96 SD; that is, $36.816 \pm (1.96 \times 0.405)$, which ranges from 36.0 to 37.6°C.

There are a number of problems with this approach:

- It really doesn't make sense to define normal and abnormal just in terms of the distribution of values in the general population. We know that high body temperature can be an indication of infection or inflammatory disease. So, the question we really want to answer is this: When is a temperature high enough that it needs investigation? The answer to this question requires scientific or clinical context. It cannot be answered by statistical calculations.
- The definitions of *normal* and *abnormal* really should depend on other factors such as age and sex. What is abnormal for a 25-year-old may be normal for an 80-year-old.
- In many cases, we don't really want a crisp boundary between normal and abnormal. It often makes sense to label some values as "borderline."
- Even if the population is approximately Gaussian, it is unlikely to follow a Gaussian distribution exactly. The deviations from a Gaussian distribution are likely to be most apparent in the tails (extreme values) of the distribution, where the abnormal values lie.
- This approach defines just as many people as having abnormally high values as those having abnormally low values. But, in fact, having a temperature in the lowest 2.5% is not an indication that something is wrong requiring medical attention. There is no reason to define the normal limits symmetrically around the mean or median.

Defining the normal limits of a clinical measurement is not straightforward and requires clinical thinking. Simple statistics rules based on the mean, SD, and Gaussian distribution are rarely useful except as a starting point before defining smarter limits.

WHY THE GAUSSIAN DISTRIBUTION
IS SO CENTRAL TO STATISTICAL THEORY

The Gaussian distribution plays a central role in statistics because of a mathematical relationship known as the *central limit theorem*. You'll need to read a more theoretical book to really understand this theorem, but the following explanation gives you the basics.

To understand this theorem, follow this imaginary experiment:

1. Create a population with a known distribution that is not Gaussian.
2. Randomly pick many samples of equal size from that population.
3. Tabulate the means of these samples.
4. Graph the frequency distribution of those means.

The central limit theorem says that if your sample size is large enough, the distribution of means will approximate a Gaussian distribution, even though the population is not Gaussian. Because most statistical tests (such as the t test and analysis of variance) are concerned only with differences between means, the central limit theorem explains why these tests work well even when the populations are not Gaussian.

Q & A

Who was Gauss?

Karl Gauss was a mathematician (one of the greatest of all time) who used this distribution in 1809 to analyze astronomical data. Although his name is now attached to the distribution, others (Laplace and de Moivre) actually used it earlier.

Is the Gaussian distribution the same as a normal distribution?

Yes, the two terms are used interchangeably. It is sometimes written as *normal distribution* and sometimes as *Normal distribution.*

Are all bell-shaped distributions Gaussian?

As you can see in Figure 10.1, the Gaussian distribution is bell shaped. But not all bell-shaped curves are Gaussian.

Will numerous sources of scatter always create a Gaussian distribution?

No. A Gaussian distribution is formed only when each source of variability is independent and additive with the others and no one source dominates. Chapter 11 discusses what happens when the sources of variation multiply.

CHAPTER SUMMARY

- The Gaussian bell-shaped distribution is the basis for much of statistics. It arises when many random factors create variability.
- With a Gaussian distribution, about two-thirds of the values are within 1 SD of the mean, and about 95% of the values are within 2 SD of the mean.
- The Gaussian distribution is also called a normal distribution. But this use of *normal* is very different than the usual use of that word to mean ordinary or abundant.
- The central limit theorem explains why Gaussian distributions are central to much of statistics. Basically, this theorem says that the distribution of many sample means will tend to be Gaussian, even if the data are not sampled from a Gaussian distribution.

TERMS INTRODUCED IN THIS CHAPTER

- Central limit theorem (p. 92)
- Gaussian distribution (p. 89)
- Normal distribution (p. 89)
- Standard normal distribution (p. 91)
- z (p. 91)

CHAPTER 11

The Lognormal Distribution and Geometric Mean

42.7 percent of all statistics are made up on the spot.

STEVEN WRIGHT

Lognormal distributions are not an obscure mathematical quirk. On the contrary, they are very common in many fields of science. This chapter explains how lognormal distributions arise and how to analyze lognormal data. Read this chapter to avoid making the common mistake of choosing analyses that assume sampling from a Gaussian distribution when working with data that are actually sampled from a lognormal distribution.

THE ORIGIN OF A LOGNORMAL DISTRIBUTION

Chapter 10 explained that a Gaussian distribution arises when variation is caused by many factors that are additive. Some factors push a value higher and some pull it lower, and the cumulative result is a symmetrical bell-shaped distribution that approximates a Gaussian distribution.

But what if the factors act in a multiplicative, rather than additive, manner? If a factor works multiplicatively, it is equally likely to double a value as to cut it in half. If that value starts at 100 and is multiplied by 2, it ends up at 200. If it is divided by 2, it ends up at 50. Consequently, that factor is equally likely to increase a value by 100 or decrease it by 50. The effect is not symmetrical so the variation of data is not symmetrical. This distribution is called a *lognormal distribution*.

Figure 11.1 shows an example. Frazier, Schneider, and Michel (2006) measured the ability of isoprenaline (a drug that acts much like the neurotransmitter norepinephrine) to relax the bladder muscle. And you thought statistics was never relaxing!

The results are expressed as the EC_{50}, which is the concentration required to relax the bladder halfway between its minimum and maximum possible relaxation. The graph on the left side of Figure 11.1 illustrates the data plotted on a linear scale. The distribution is far from symmetrical and is quite skewed. One value is far from the rest and almost looks like a mistake.

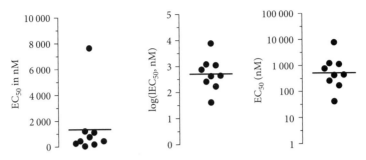

Figure 11.1. Lognormal data.

These data demonstrate the EC_{50} of isoprenaline for relaxing bladder muscles (Frazier, Schneider, & Michel, 2006). Each dot indicates data from the bladder of a different animal. The EC_{50} is the concentration required to relax the bladder halfway between its minimum and maximum possible relaxation, measured in nanomoles per liter (nM). The graph on the left plots the original concentration scale. The data are far from symmetrical, and the highest value appears to be an outlier. The middle graph plots the logarithm (base 10) of the EC50 and is symmetrical. The graph on the right plots the raw data on a logarithmic axis. This kind of graph is a bit easier to read.

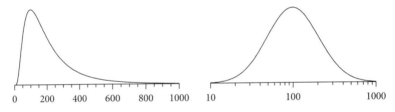

Figure 11.2. Lognormal distribution.

(Left) Lognormal distribution. The distribution appears Gaussian when plotted on a logarithmic axis (right) or when all values are transformed to their logarithm.

Figure 11.2 left shows an ideal lognormal frequency distribution. If many random factors act in a multiplicative manner, this is what the distribution of all the values looks like.

LOGARITHMS?

Logarithms? How did logarithms get involved? Briefly, it is because the logarithm of the product of two values equals the sum of the logarithm of the first value plus the logarithm of the second. So logarithms convert multiplicative scatter (lognormal distribution) to additive scatter (Gaussian). Logarithms (and antilogarithms) are reviewed in Appendix E.

If you transform each value sampled from a lognormal distribution to its logarithm, the distribution becomes Gaussian. Thus, the logarithms of the values follow a Gaussian distribution when the raw data are sampled from a lognormal distribution.

The middle graph in Figure 11.1 plots the logarithm of the EC_{50} values. Note that the distribution is symmetrical. The graph on the right illustrates an alternative way to plot the data. The axis has a logarithmic scale. Note that every major tick on the axis represents a value 10 times higher than the previous value. The distribution of data points is identical to that in the middle graph, but the graph on the right is easier to comprehend because the Y values are labeled in natural units of the data, rather than logarithms.

If data are sampled from a lognormal distribution, convert to logarithms before using a statistical test that assumes sampling from a Gaussian distribution, such as a t test or ANOVA.

GEOMETRIC MEAN

The mean of the data from Figure 11.1 is 1,333 nM. This is illustrated as a horizontal line in the left panel of Figure 11.1. The mean is larger than all but one of the values, so it is not a good measure of the central tendency of the data.

The middle panel plots the logarithms of the values on a linear scale. The horizontal line is at the mean of the logarithms, 2.71. About half of the values are higher and half are smaller.

The right panel in Figure 11.1 uses a logarithmic axis. The values are the same as those of the graph on the left, but the spacing of the values on the axis is logarithmic. The horizontal line is at the antilog of the mean of the logarithms. This graph uses logarithms base 10, so the antilog is computed by calculating $10^{2.71}$, which equals 513. The value 513 nM is called the *geometric mean* (GM).

To compute a GM, first transform all the values to their logarithms and then calculate the mean of those logarithms. Finally, transform that mean of the logarithms back to the original units of the data.

GEOMETRIC SD

To calculate the GM, we first computed the mean of the logarithms and then computed the antilogarithm (power of 10) of that mean. Similar steps compute the *geometric standard deviation*. First compute the standard deviation of the logarithms, which for the example in Figure 11.1 equals 0.632. Then take the antilogarithm (10 to the power) of that value, which is 4.29. That is the geometric standard deviation.

When interpreting the SD of values from a Gaussian distribution, you expect about two-thirds of the values to in the range that goes from the mean minus SD to the mean plus SD. Remember that logarithms essentially convert multiplication to addition, so the geometric SD must be multiplied times, or divided into, the GM.

The GM equals 513 nM and the geometric SD is 4.29. You expect two-thirds of the values in this distribution to lie in the range 513/4.29 to 513*4.29, which is 120 nM to 2201 nM. This range will appear symmetrical when plotted on a logarithmic axis but asymmetrical when plotted on a linear axis. You'll see this in Figure 14.3 (in the chapter on error bars).

The geometric SD has no units. Because the geometric SD is multiplied by or divided into the GM, it is sometimes called the *geometric SD factor*. The geometric SD was defined by Kirkwood (1979) and is not commonly used (but should be). Limpert and Stahel (2011) propose reporting the GM and geometric standard deviation factor (GSD) as GM $^\times/$ GSD, read as "times or divided by". This contrasts with mean ± SD, read as "plus or minus".

COMMON MISTAKES: LOGNORMAL DISTRIBUTIONS

Mistake: Being inconsistent with the use of common and natural logarithms

To compute the GM, you need to first convert all the values to their logarithms, then compute the mean of those logarithms, and finally compute the antilogarithm of that mean. It is critical that the same base be used for computing the logarithm and antilogarithm. If you use log10 common logarithms, compute the antilogarithm by taking 10 to the power of the mean. If you use natural logarithms, compute the antilogarithm using the exp() function.

Mistake: Converting data to logarithms when some values are zero or negative

The logarithm is simply not defined for zero or negative values. If you transform a set of values to their logarithms, any values are zero or negative will be lost to the analysis.

Mistake: Not recognizing a lognormal distribution and eliminating some high values as outliers

A common mistake is to view the left graph of Figure 11.1 and conclude that these values are sampled from a Gaussian distribution but are contaminated by the presence of an outlier. Running an outlier test would confirm that conclusion. But that would be misleading. Outlier tests assume that all the values (except for any outliers) are sampled from a Gaussian distribution. When presented with data from a lognormal distribution, outlier tests are likely to incorrectly flag very high values as outliers, when, in fact, those high values are expected in a lognormal distribution. Chapter 25 shows some examples.

Q & A

Where can I review logarithms and antilogarithms?
> In Appendix E.

Lognormal or log-normal?
> Both forms are commonly used. This book uses "lognormal."

Are values in a lognormal distribution always positive?

Yes. The logarithm of zero and negative numbers is simply not defined. Samples that contain zero or negative values cannot be sampled from lognormal distributions, so those data should not be analyzed as if they are sampled from lognormal distributions.

Can the geometric mean be computed if any values are zero or negative?

No.

When computing the geometric mean, should I use natural logarithms or logarithms base 10?

It doesn't matter as long as you are consistent. More often, scientists use logarithms base 10, so the reverse transform is the power of 10. The alternative is to use natural logarithms, so the reverse transform is taking e to that power. Whichever log base is used, the geometric mean will have exactly the same value. See Appendix E.

Are lognormal distributions common?

Yes, they occur commonly (Limpert, Stahel, & Abbt, 2001; Limpert & Stahel, 2011). For example, the potency of a drug (assessed as EC_{50}, IC_{50}, Km, Ki, etc.) is almost always lognormal. For this reason, it makes sense to compare treatment groups using ratios rather than differences and to summarize data with a geometric mean. Other examples of lognormal distributions are the blood serum concentrations of many natural or toxic compounds.

When variables follow a lognormal distribution, does it make sense to report the data on a log scale?

Yes, this is commonly done. Some examples are sound intensity (dB), earthquake strength (Richter scale), and acidity (pH). These variables are converted to a log scale before being reported. The dB quantifies the logarithm of sound intensity; the Richter scale quantifies the logarithm of the intensity of an earthquake; and pH is -1 times the logarithm of the concentration of hydrogen ions. There is no need to transform from the logarithmic scale back to the original scale, because these logarithmic scales are used commonly and are understood much better than the antilogarithm would be.

Are lognormal distributions always skewed to the right as shown in Figure 11.2?

Yes, the longer stretched-out tail always extends to the right, representing larger values.

Should I use a logarithmic axis when plotting data from a lognormal distribution?

Yes, this can make it easier to understand. See Figure 11.1 (right).

What happens if you analyze data from a lognormal distribution as if they were sampled from a Gaussian distribution?

It depends on the details of the data and the sample size. It is likely that the results of any statistical analysis that assumes a Gaussian distribution will be misleading. The confidence intervals of differences will be much wider than they should be, so P values are higher.

What units are used to express the geometric mean?

The same units that are used with the values being analyzed. Thus, the mean and geometric mean are expressed in the same units.

Is the geometric mean larger or smaller than the regular mean?

The geometric mean is always smaller. (The trivial exception: if all the values are identical, then the mean and geometric mean will be equal.)

What units are used to express the geometric SD?

> The geometric SD has no units. It is a factor you multiply or divide the geometric mean by

Is the geometric SD larger or smaller than the regular SD?

> The regular SD is in the same units as the data. The geometric mean is a unitless factor. The two cannot be meaningfully compared.

Is there a distinction between the geometric standard deviation and the geometric standard deviation factor?

> No. The geometric standard deviation always is a factor that the geometric mean is multiplied or divided by. But the word "factor" is often omitted.

CHAPTER SUMMARY

- Lognormal distributions are very common in many fields of science.
- Lognormal distributions arise when multiple random factors are multiplied together to determine the value. This is common in biology. In contrast, Gaussian distributions arise when multiple random factors are added together.
- Lognormal distributions have a long right tail (are said to be skewed to the right).
- The center of a lognormal distribution is quantified with a geometric mean, measured in the same units as the data.
- The variation of a lognormal distribution is quantified with a standard deviation factor, a unitless value.
- With data sampled from a Gaussian distribution, you think about the mean plus or minus the SD. With data sampled from a lognormal distribution, you think about the geometric mean multiplied or divided by the geometric SD factor.
- You may get misleading results if you make the common mistake of choosing analyses that assume sampling from a Gaussian distribution when in fact your data are actually sampled from a lognormal distribution.
- In most cases, the best way to analyze lognormal data is to take the logarithm of each value and then analyze those logarithms.

TERMS INTRODUCED IN THIS CHAPTER

- Geometric mean (p. 97)
- Geometric SD factor (p. 98)
- Lognormal distribution (p. 95)

CHAPTER 12

Confidence Interval of a Mean

It is easy to lie with statistics. It is hard to tell the truth
without it.

ANDREJS DUNKELS

Chapters 4 through 6 explained confidence intervals (CIs) of proportions, survival fractions, and counts. This chapter extends
those concepts to the CI of a mean, a calculation that depends on the
size of the sample and the variability of the values (expressed as the
standard deviation).

INTERPRETING A CI OF A MEAN

For our ongoing n = 130 body temperature example (see Figure 7.1), any statistics program will calculate that the 95% CI of the mean ranges from 36.75°C
to 36.89°C. For the smaller n = 12 subset, the 95% CI of the mean ranges from
36.51°C to 37.02°C.

Note that there is no uncertainty about the sample mean. We are 100% sure
that we have calculated the sample mean correctly. Any errors in recording the
data or computing the mean will not be accounted for in computing the CI of the
mean. By definition, the CI is always centered on the sample mean. The population mean is not known and can't be known. However, given some assumptions
outlined in the following discussion, we can be 95% sure that the calculated interval contains it.

What exactly does it mean to be "95% sure"? When you have measured only
one sample, you don't know the value of the population mean. The population mean
either lies within the 95% CI or it doesn't. You don't know, and there is no way to
find out. If you calculate a 95% CI from many independent samples, the population
mean will be included in the CI in 95% of the samples but will be outside of the CI
in the other 5% of the samples. Using data from one sample, therefore, you can say
that you are 95% confident that the 95% CI includes the population mean.

The correct syntax is to express the CI as "36.75 to 36.89" or as
"[36.75, 36.89]." It is considered bad form to express the CI as "36.75–36.89,"
because the hyphen would be confusing when the values are negative. Although
it seems sensible to express the CI as "36.82 ± 0.07," that format is rarely used.

WHAT VALUES DETERMINE THE CI OF A MEAN?

The CI of a mean is computed from four values:

- The sample mean. Our best estimate of the population mean is the sample mean. Accordingly, the CI is centered on the sample mean.
- The SD. If the data are widely scattered (large SD), then the sample mean is likely to be farther from the population mean than if the data are tight (small SD). The width of the CI, therefore, is proportional to the sample SD.
- Sample size. Our sample has 130 values, so the sample mean is likely to be quite close to the population mean, and the CI will be very narrow. With tiny samples, the sample mean is likely to be farther from the population mean, so the CI will be wider. The width of the CI is inversely proportional to the square root of the sample size. If the sample were four times larger, the CI would be half as wide (assuming the same SD). Note that the CI from the n = 12 sample is wider than the CI for the n = 130 sample (see Figure 12.1).
- Degree of confidence. Although CIs are typically calculated for 95% confidence, any value can be used. If you wish to have more confidence (e.g., 99% confidence), you must generate a wider interval. If you are willing to accept less confidence (e.g., 90% confidence), you can generate a narrower interval.

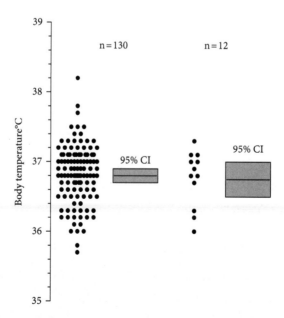

Figure 12.1. The 95% CI does not contain 95% of the values, especially when the sample size is large.

ASSUMPTIONS: CI OF A MEAN

To interpret a CI of a mean, you must accept the following assumptions.

Assumption: Random (or representative) sample

The 95% CI is based on the assumption that your sample was randomly selected from the population. In many cases, this assumption is not strictly true. You can still interpret the CI as long as you assume that your sample is representative of the population.

In clinical studies, it is not feasible to randomly select patients from the entire population of similar patients. Instead, patients are selected for the study because they happened to be at the right clinic at the right time. This is called a *convenience sample* rather than a *random sample*. For statistical calculations to be meaningful, we must assume that the convenience sample adequately represents the population and that the results are similar to what would have been observed had we used a true random sample.

This assumption would be violated in the body temperature example if the people who participated chose to join a study of body temperature because they knew (or suspected) that their own temperature was consistently higher or lower than that of most other people.

Assumption: Independent observations

The 95% CI is only valid when all subjects are sampled from the same population and each has been selected independently of the others. Selecting one member of the population should not change the chance of selecting any other person. This assumption would be violated if some individuals' temperatures were measured twice and both values were included in the sample. The assumption would also be violated if several of the subjects were siblings, because it is likely that genetic factors affect body temperature.

Assumption: Accurate data

The 95% CI is only valid when each value is measured correctly. This assumption would be violated if subjects didn't place the thermometer in their mouths correctly or if the thermometer was misread.

Assumption: Assessing an event you really care about

The 95% CI allows you to extrapolate from the sample to the population for the event that you tabulated. But sometimes you might really care about a different event. In this example, what you really want to know is the core body temperature. Instead, temperature under the tongue was measured. The difference in this example is trivial, but it is always worth thinking about the distinction between what is measured and what one really wants to know. Chapter 45 gives two examples in which drugs changed a surrogate marker as desired—fewer arrhythmias in one example, higher "good cholesterol" in the other—but increased the number of deaths.

Assumption: The population is distributed in a Gaussian manner, at least approximately

The most common method of computing the CI of a mean is based on the assumption that the data are sampled from a population that follows a Gaussian distribution. This assumption is important when the sample is small but doesn't matter much when samples are large. How big must a sample be for us to disregard this assumption? It depends on the shape of the non-Gaussian distribution. See the discussion of the central limit theorem at the end of Chapter 10.

An alternative resampling method of computing the 95% CI that does not assume sampling from a Gaussian distribution is explained in Chapter 13.

What if the assumptions are violated?

In many situations, these assumptions are not strictly true. The patients in your study may be more homogeneous than the entire population of patients. Measurements made in one lab will have a smaller SD than measurements made in other labs at other times. More generally, the population you really care about may be more diverse than the population from which the data were sampled. Furthermore, the population may not be Gaussian. If any assumption is violated, the CI will probably be too optimistic (too narrow). The true CI (taking into account any violation of the assumptions) is likely to be wider than the calculated CI.

HOW TO CALCULATE THE CI OF A MEAN

Although computers will do the calculations, it is easier to understand a CI of a mean once you know how it is computed. The CI is centered on the sample mean (m). To calculate the width requires taking account of the SD (s), the number of values in the sample (n), and the degree of confidence you desire (usually 95%).

Use Table 12.1 (reprinted in Appendix D) to determine the value of a constant based on the sample size and the degree of confidence you desire. This value is called a *constant from the t distribution* (explained in Chapter 13), denoted in this book by the nonstandard abbreviation t*. For the body temperature example of the right side of Figure 12.1, n = 12, so df = 11 and t* (for 95% confidence) = 2.201.

Calculate the margin of error of the CI, W, which is half the width of the CI:

$$W = \frac{t^* \cdot s}{\sqrt{n}}$$

For the example data, s = 0.40°C, so W = 0.254 (for 95% confidence). The CI covers this range:

$$m - W \text{ to } m + W$$

For the example data, m = 36.77°C, so the 95% CI extends from 36.52°C to 37.02°C. If you want 90% confidence, t* = 1.7959 (n = 12, so df = 11), W = 0.207, and the 90% CI extends from 36.56 to 36.98°C.

df	\multicolumn{4}{c}{DESIRED CONFIDENCE LEVEL}	df	\multicolumn{4}{c}{DESIRED CONFIDENCE LEVEL}
	80%	90%	95%	99%		80%	90%	95%	99%
1	3.0777	6.3138	12.7062	63.6567	27	1.3137	1.7033	2.0518	2.7707
2	1.8856	2.9200	4.3027	9.9248	28	1.3125	1.7011	2.0484	2.7633
3	1.6377	2.3534	3.1824	5.8409	29	1.3114	1.6991	2.0452	2.7564
4	1.5332	2.1318	2.7764	4.6041	30	1.3104	1.6973	2.0423	2.7500
5	1.4759	2.0150	2.5706	4.0321	35	1.3062	1.6896	2.0301	2.7238
6	1.4398	1.9432	2.4469	3.7074	40	1.3031	1.6839	2.0211	2.7045
7	1.4149	1.8946	2.3646	3.4995	45	1.3006	1.6794	2.0141	2.6896
8	1.3968	1.8595	2.3060	3.3554	50	1.2987	1.6759	2.0086	2.6778
9	1.3830	1.8331	2.2622	3.2498	55	1.2971	1.6730	2.0040	2.6682
10	1.3722	1.8125	2.2281	3.1693	60	1.2958	1.6706	2.0003	2.6603
11	1.3634	1.7959	2.2010	3.1058	65	1.2947	1.6686	1.9971	2.6536
12	1.3562	1.7823	2.1788	3.0545	70	1.2938	1.6669	1.9944	2.6479
13	1.3502	1.7709	2.1604	3.0123	75	1.2929	1.6654	1.9921	2.6430
14	1.3450	1.7613	2.1448	2.9768	80	1.2922	1.6641	1.9901	2.6387
15	1.3406	1.7531	2.1314	2.9467	85	1.2916	1.6630	1.9883	2.6349
16	1.3368	1.7459	2.1199	2.9208	90	1.2910	1.6620	1.9867	2.6316
17	1.3334	1.7396	2.1098	2.8982	95	1.2905	1.6611	1.9853	2.6286
18	1.3304	1.7341	2.1009	2.8784	100	1.2901	1.6602	1.9840	2.6259
19	1.3277	1.7291	2.0930	2.8609	150	1.2872	1.6551	1.9759	2.6090
20	1.3253	1.7247	2.0860	2.8453	200	1.2858	1.6525	1.9719	2.6006
21	1.3232	1.7207	2.0796	2.8314	250	1.2849	1.6510	1.9695	2.5956
22	1.3212	1.7171	2.0739	2.8188	300	1.2844	1.6499	1.9679	2.5923
23	1.3195	1.7139	2.0687	2.8073	350	1.2840	1.6492	1.9668	2.5899
24	1.3178	1.7109	2.0639	2.7969	400	1.2837	1.6487	1.9659	2.5882
25	1.3163	1.7081	2.0595	2.7874	450	1.2834	1.6482	1.9652	2.5868
26	1.3150	1.7056	2.0555	2.7787	500	1.2832	1.6479	1.9647	2.5857

Table 12.1. Critical values of the t distribution for computing CIs.
When computing a CI of a mean, df equals (n − 1). More generally, df equals n minus the number of parameters being estimated.

MORE ABOUT CONFIDENCE INTERVALS

One-sided CIs

CIs are almost always calculated and presented as described in the previous section: as an interval defined by two limits. But it is also possible to create *one-sided CIs*.

For the example, the 90% CI extends from 36.56°C to 36.98°C. Because this is a 90% two-sided CI, there is room for 5% error on each end. There is a 5% chance that the upper limit (36.98) is less than the population mean. There is also a 5% chance that the lower limit (36.56) is greater than the population mean. That leaves a 90% (100% minus 5% minus 5%) chance that the interval 36.56°C to 36.98°C includes the true population mean.

But what if we are clinically interested only in fevers and want to know only the upper confidence limit? Because there is a 5% chance that the population mean is greater than 36.98°C (see previous paragraph), that leaves a 95% chance that the true population mean is less than 36.98°C. This is a one-sided 95% CI. More precisely, you could say that the range from minus infinity to 36.98°C is 95% likely to contain the population mean.

CI of the SD

A CI can be determined for nearly any value you calculate from a sample of data. With the n = 12 body temperature data, the SD was 0.40°C. It isn't done often, but it is possible to compute a 95% CI of the SD itself. In this example, the 95% CI of the SD extends from SD = 0.28°C to SD = 0.68°C (calculated using a Web-based calculator at www.graphpad.com/quickcalcs/CISD1/).

The interpretation is straightforward. Given the same assumptions as those used when interpreting the CI of the mean, we are 95% sure that the calculated interval contains the true population SD. With the n = 130 body temperature data, the SD is 0.41°C. Because the sample size is so much larger, the sample SD is a more precise estimate of the population SD, and the CI is narrower, ranging from SD = 0.37°C to SD = 0.47°C.

Table 12.2 presents the CI for a standard deviation for various sample sizes.

CI of a geometric mean

Chapter 11 explained how to compute the geometric mean. This process is easily extended to computing the 95% CI of the geometric mean. The first step in

n	95% CI OF SD
2	0.45 to 31.9 SD
3	0.52 to 6.29 SD
5	0.60 to 2.87 SD
10	0.69 to 1.83 SD
25	0.78 to 1.39 SD
50	0.84 to 1.25 SD
100	0.88 to 1.16 SD
500	0.94 to 1.07 SD
1,000	0.96 to 1.05 SD

Table 12.2. The CI of an SD.
Assuming random sampling from a Gaussian population, this table gives the 95% CI for the population SD, given the sample SD and sample size (n). These values were computed using equations from Sheskin (2007, p. 197).

computing a geometric mean is to transform all the values to their logarithms. Then the mean of the logarithms is computed and reverse transformed to its antilog. The result is the geometric mean.

To create a CI of the geometric mean, compute the CI of the mean of the logarithms and then reverse transform each confidence limit.

For the EC_{50} example in Chapter 11, the mean of the logarithms is 2.71, and the 95% CI of this mean extends from 2.22 to 3.20. Reverse transform each of these values, and the 95% CI of the geometric mean (which equals 512 nM) ranges from 166 to 1585 nM. Note that the CI of the geometric mean is not symmetrical around the geometric mean.

The CI and the SEM

Looking ahead, Chapter 14 will define the standard error of the mean (SEM) as the SD divided by the square root of sample size

$$SEM = \frac{s}{\sqrt{n}}$$

Recall that W, the margin of error of the confidence interval, is defined as

$$W = \frac{t^* \cdot s}{\sqrt{n}}$$

Combine those two equations to simplify calculation of W

$$W = t^* \cdot SEM$$

Q & A

Why 95% confidence?

CIs can be computed for any degree of confidence. By convention, 95% CIs are presented most commonly, although 90% and 99% CIs are sometimes published.

Is a 99% CI wider or narrower than a 95% CI?

To be more confident that your interval contains the population value, the interval must be made wider. Thus, 99% CIs are wider than 95% CIs, and 90% CIs are narrower than 95% CIs (see Figure 12.2).

Does a CI quantify variability?

No. The CI of the mean tells you how precisely you have determined the population mean. It does not tell you about scatter among values.

When can I use the rule of thumb that a 95% CI equals the mean plus or minus 2 SD?

Never! In a Gaussian distribution, you expect to find about 95% of the individual values within 2 SD of the mean. But the idea of a CI is to define how precisely you know the population mean. For that, you need to take into account sample size.

When can I use the rule of thumb that a 95% CI equals the mean plus or minus 2 SEM?

When n is large. Chapter 14 explains the SEM.

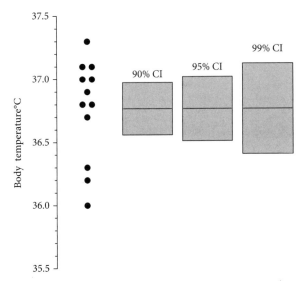

Figure 12.2. If you want to be more confident that a CI contains the population mean, you must make the interval wider.

The 99% interval is wider than the 95% interval, which is wider than the 90% CI.

Do 95% of values lie within the CI?

No! The CI quantifies how precisely you know the population mean. With large samples, you can estimate the population mean quite accurately, so the CI is quite narrow and includes only a small fraction of the values (see Figure 12.1).

If I collect more data, should I expect 95% of the new values to lie within this CI?

No! The CI quantifies how precisely you know the population mean. With large samples, you can estimate the population mean quite accurately, so the CI is quite narrow and will include only a small fraction of the values you sample in the future.

You compute a 95% CI from one experiment. Now you repeat that experiment a bunch of times. What fraction of the means of the new experiments will be expected to lie within the 95% CI from the first experiment?

It seems as though the answer ought to be 95%, but it isn't. If the first experiment had a sample mean close to the population mean, then a high fraction of repeat experiments will have sample means within the 95% CI. But if that first experiment happened to have a sample mean far from the population mean, then a much smaller fraction of repeat experiments will have a sample mean within the first 95% CI. Cumming and Maillardet (2016) have shown that on average, the first 95% CI will include the sample means of 83% of repeat experiments.

Is the 95% CI of the mean always symmetrical around the mean?

It depends on how it is computed. But with the standard method, explained in this chapter, the CI always extends equally above and below the mean.

How is a CI distinct from confidence limits?

The CI is a range of values. The two values that define that range are called *confidence limits*.

CHAPTER SUMMARY

- A CI of the mean shows you how precisely you have determined the population mean.
- If you compute 95% CIs from many samples, you expect that 95% will include the true population mean and 5% will not. You'll never know whether a particular CI includes the population mean.
- The CI of a mean is computed from the sample mean, the sample SD, and the sample size.
- The CI does not display the scatter of the data. In most cases, the majority of the data values will lie outside the CI.
- If you desire more confidence (99% rather than 95%), the CI will be wider.
- Larger samples have narrower CIs than smaller samples with the same SD.
- Interpreting the CI of a mean requires accepting a list of assumptions.

TERMS INTRODUCED IN THIS CHAPTER

- Confidence limit (p. 106)
- Constant from the t distribution (p. 104)
- Convenience sample (p. 103)
- One-sided confidence interval (p. 105)
- Random sample (p. 103)

CHAPTER 13

The Theory of Confidence Intervals

> Confidence is what you have before you understand the
> problem.
>
> WOODY ALLEN

The statistical theory underlying confidence intervals (CIs) is hard to understand. Probability theory starts with a known population and then computes the probabilities of obtaining various possible samples. Statistical analysis flips this theory: it starts with data and then computes the likelihood that the data were sampled from various populations. This chapter tries to give you a general sense of how the theory is flipped. It includes more equations than the rest of the book. Later chapters do not assume you have mastered this one.

CI OF A MEAN VIA THE t DISTRIBUTION

What is the t distribution?

Chapter 12 demonstrated how to calculate a CI of a mean using a method that required us to find a value from the *t distribution*. A full understanding of the t distribution requires a mathematical approach beyond the scope of this book, but the following provides an overview.

Let's assume we know that a population follows a Gaussian distribution. The mean of this population is designated by the Greek letter μ (mu). Let's assume we also know the SD of this population, designated by the Greek letter σ (sigma).

Using a computer program, one can randomly choose values from this defined population. Pick n values and compute the mean of this sample, which we'll abbreviate m. Also compute the SD of this sample, which we'll abbreviate s. This sample was chosen randomly, so m and s won't equal the population values μ and σ.

Repeat these computations thousands of times. For each random sample, compute the ratio t:

$$t = \frac{m - \mu}{s / \sqrt{n}} \qquad\qquad \textbf{13.1}$$

Because these are simulated data from a hypothetical population, the value of μ is known and constant, as is n. For each sample, compute m and s and then use Equation 13.1 to compute t.

For each random sample, the sample mean is equally likely to be larger or smaller than the population mean, so the t ratio is equally likely to be positive or negative. It will usually be fairly close to zero, but it can also be far away. How far? It depends on sample size (n), the SD (s), and chance. Figure 13.1 illustrates the distribution of t computed for a sample size of 12. Of course, there is no need to actually perform these simulations. Mathematical statisticians have derived the distribution of t using calculus.

Sample size, n, is included in the equation that defines t, so you might expect the distribution of t to be the same regardless of sample size. In fact, the t distribution depends on sample size. With small samples, the curve in Figure 13.1 will be wider; with large samples, the curve will be narrower. With huge samples, the curve in Figure 13.1 will become indistinguishable from the Gaussian distribution shown in Figure 10.1.

The critical value of t

The area under the curve in Figure 13.1 represents all possible samples. Chop off the 2.5% tails on each side (shaded) to determine a range of t ratios that includes 95% of the samples. For this example, with n = 12 and df = 11, the t ratio has a value between −2.201 and 2.201 in 95% of the samples (see Table 12.1). Let's define the variable t* to equal 2.201 (this abbreviation is not standard).

The value of t* depends on sample size and the degree of confidence you want (here we chose the standard, 95%). Note that its value does not depend on the actual data we are analyzing. You can look up this value in Appendix D.

The flip!

Here comes the most important step, in which the math is flipped around to compute a CI.

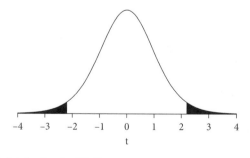

Figure 13.1. The t distribution for 11 df.
The shaded areas in the two tails each represent 2.5% of the area under the curve, which leaves 95% of the area unshaded. Because 95% of the time a value sampled from this t distribution is between −2.201 and 2.201, a 95% CI is computed by setting t* equal to 2.201.

In any one sample, the sample mean (m) and SD (s) are known, as is the sample size (n). What is not known is the population mean. That is why we want to generate a CI.

Rearrange Equation 13.1 to solve for μ (i.e., put μ on the left side of the equation):

$$\mu = m \pm t * \frac{s}{\sqrt{n}}$$

For the example, n = 12, m = 36.77°C, and s = 0.40°C. The t ratio has a 95% probability of being between –2.201 and +2.201, so t* = 2.201. Plug in those values and compute the population mean twice. The first time use the minus sign, and the second time do the calculations again using the plus sign. The results are the limits of the 95% CI, which ranges from 36.52°C to 37.02°C.

How it works

The t distribution is defined by assuming the population is Gaussian and investigating the variation among the means of many samples. The math is then flipped to make inferences about the population mean from the mean and SD of one sample.

This method uses probability calculations that are logically simple and flips them to answer questions about data analysis. Depending on how you look at it, this flip is either really obvious or deeply profound. Thinking about this kind of inverse probability is tricky and has kept statisticians and philosophers busy for centuries.

CI OF A MEAN VIA RESAMPLING

The resampling approach

One of the assumptions of the usual (t distribution) method for computing a CI of a mean is that the values were sampled from a Gaussian distribution. What if you can't support that Gaussian assumption?

Resampling is an alternative approach to statistics that does not assume a Gaussian (or any other) distribution.

Create many pseudosamples via resampling

The first step is to generate many more pseudosamples via resampling.

The data set we are analyzing has 12 values, so each of our "resampled samples" will also have 12 values. To find the first value, pick a random integer between 1 and 12. If it is 3, pick the third value from the original sample. To find the second value, again pick a random integer between 1 and 12 and then pick the corresponding value. It might be the third value again, but it is more likely to be one of the other 11 values. Repeat this process again until you have a new *pseudosample* with 12 values.

Repeat this process many times to create, say, 500 pseudosamples. These new samples are created by sampling from the same set of values (12 in this example). Those same 12 values appear over and over, and no other values ever appear. But the samples aren't all identical. In each sample, some values are repeated and others are left out.

For each of the 500 new samples, compute the mean (because we want to know the CI of the population mean). Next, determine the 2.5th and 97.5th percentiles of that list of means. For this example, the 2.5th percentile is 36.55°C and the 97.5th percentile is 36.97°C. Because the difference between 97.5 and 2.5 is 95, we can say that 95% of the means of the pseudosamples are between 36.55°C and 36.97°C.

The flip!

Statistical conclusions require flipping the logic from the distribution of multiple samples to the CI of the population mean. With the resampling approach, the flip is simple. The range of values that contains 95% of the resampled means (36.55°C to 36.97°C) is also the 95% CI of the population mean.

For this example, the resampled CI is nearly identical to the CI computed using the conventional method (which assumes a Gaussian distribution). But the resampling method does not require the assumption of a Gaussian distribution and is more accurate when that assumption is not justified. The only assumption used in the resampling approach is that the values in the sample vary independently and are representative of the population.

That's it?

This resampling method seems too simple to be useful! It is surprising that randomly resampling from the same set of values gives useful information about the population from which that sample was drawn. But it does create useful results. Plenty of theoretical and practical (simulation) work has validated this approach, which some statisticians think should be widely used.

Learn more

To learn more about the resampling approach—also called the *bootstrapping* or *computer-intensive method*—start by reading books by Wilcox (2010) and Manly (2006). The method explained here is called a *percentile-based resampling confidence interval.* Fancier methods create slightly more accurate CIs.

A huge advantage of the resampling approach is that it is so versatile. It can be used to obtain the CI of the median, the interquartile range, or almost any other parameter. Resampling approaches are extensively used in the analysis of genomics data.

CI OF A PROPORTION VIA RESAMPLING

Let's revisit the voting example from Chapter 4. When polled, 33 of 100 people said they would vote a certain way. Our goal is to find the 95% CI for the true population proportion.

The resampling approach is easier to use for proportions than for means. The sample proportion is 33% with n = 100. To resample, simply assume the population proportion is 0.33, draw numerous random samples, and find the range of proportions that includes 95% of those samples.

Observed proportion (33/100)

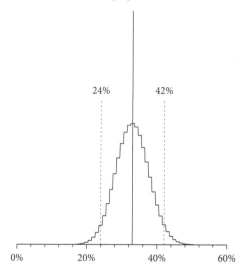

Figure 13.2. A resampling approach to computing the CI of a proportion.

If the true proportion of "success" (in this example, voting for your candidate) were 33%, the graph illustrates the probability distribution of the success rate in many samples of 100 people. After removing the 2.5% tails on each side, you can see that 95% of samples will have success rates between 24% and 42%. Surprisingly, you can flip the logic and use this graph as a way to derive CIs.

Figure 13.2 illustrates the answer. If 33% of the entire population of voters is in favor of your candidate, then if you collect many samples of 100 voters, in 95% of those samples, between 24% and 42% of people will be in favor of your candidate.

That range tells you about multiple samples from one population. We want to know what can be inferred about the population from one sample. It turns out that that same range is the CI. Given our one sample, we are 95% sure the true population value is somewhere between 24% and 42%.

With continuous data, the resampling approach is more versatile than the t distribution approach, because it doesn't assume a Gaussian (or any other) distribution. With binomial data, there is no real advantage to the resampling approach, except that it is a bit easier to understand than the approach based on the binomial distribution (see the next section).

CI OF A PROPORTION VIA BINOMIAL DISTRIBUTION

When polled, 33 of 100 people said they would vote a certain way. The goal is to find two confidence limits, L (lower) and U (upper), so that the 95% CI of the proportion ranges from L to U. This section demonstrates one approach, summarized in Figure 13.3. As mentioned in Chapter 4, there are several approaches to computing CIs of proportions, and they don't give the same exact result.

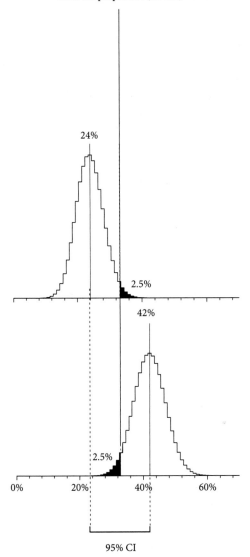

Figure 13.3. How the CI of a proportion is derived.

The solid vertical line is at 33%, which is the sample proportion (33/100). The top probability distribution demonstrates the probability of various outcomes if the true population proportion were 24%. This distribution is centered on 24%, but the right tail of the distribution indicates that there is a 2.5% chance that the observed proportion in one particular sample will be 33% or higher. The bottom distribution indicates the probability of various outcomes if the population proportion was 42%. The left tail demonstrates that there is a 2.5% chance that the observed proportion in one particular sample will be 33% or lower. Subtracting 2.5% from 100% twice, we are left with a 95% interval ranging from 24% to 42%.

The lower 95% CI (L) can be determined by an indirect approach. We can use probability equations to answer the question, If the population proportion equals L, what is the probability that a sample proportion (n = 100) will equal 0.33 (as we observed in our sample) or more? To compute a 95% CI, each tail of the distribution must be 2.5%, so we want to find the value of L that makes the answer to that question 2.5%.

Determining the upper 95% confidence limit of a proportion (U) uses a similar approach. If the population proportion equals U, what is the probability that a sample proportion (n = 100) will equal 0.33 or less? We want the answer to be 2.5% and solve for U.

Using the binomial distribution to solve for L and U is not a straightforward task. A brute force approach is to try lots of possible values of L or U to find values for which the cumulative binomial distribution gives a result of 2.5%. Excel's Solver can automate this process (Winston, 2004).

For this example, L equals 0.24, or 24%. If fewer than 24% of voters were truly in favor of your candidate, there would be less than a 2.5% chance of randomly choosing 100 subjects of which 33% (or more) are in favor of your candidate.

For the example, U equals 0.42, or 42%. If more than 42% of voters really were in favor of your candidate, there would be less than a 2.5% chance of randomly choosing 100 subjects of which 33% (or less) were in favor of your candidate.

So far, we have been asking about which sample proportions are likely given known population proportions (L and U). Now we want to make inferences about the population from one sample. That requires a flip in reasoning. Here it comes.

Calculate 100% – 2.5% – 2.5%, which leaves 95%. Therefore, there is a 95% chance that the true percentage of voters in favor of your candidate is between 24% and 42%.

Q & A

How can resampling from one sample tell you anything useful about the population from which that sample was drawn?

It is amazing that resampling works. It is also called bootstrapping, which comes from the phrase "pulling oneself up by the bootstrap," which is impossible. It is not obvious that bootstrap/resample methods produce valid inferences about the population the sample was drawn from, but this has been proven by both mathematical proofs and simulations.

Is it correct to say that there is a 95% chance that the population mean (or proportion) lies within the computed CI?

No. The population mean (or proportion) has a fixed value (which we never know). So it is not appropriate to ask about the chance that the population mean (or proportion) has any particular value. It is what it is with no chance involved. In contrast, the computed CIs depend on which sample of data you happened to choose, so vary from sample to sample based on random sampling. It is correct to say that there is a 95% chance that a 95% CI includes the population mean (or proportion).

Why the focus on parameters like the mean? I don't know want to know the precision of the mean. I want to know about the distribution of values in the population.

> The 95% *prediction interval* answers your question. It is the range of values that is 95% sure to contain 95% of the values in the entire population. I'm not sure why, but prediction intervals are not often used in most fields of biology. Prediction intervals are much wider than CIs.

CHAPTER SUMMARY

- You can understand CIs without understanding how they are computed. This chapter gives you a peek at how the math works. You don't need to understand this chapter to understand how to interpret CIs.
- The math works by flipping around (solving) equations that predict samples from a known population to let you make inferences about the population from a single sample.
- Depending on your frame of mind, this flip can seem really obvious or deeply profound.
- An alternative approach is to use resampling (also called bootstrapping) methods. The idea is that analyzing a set of bootstrapped samples created from the original data will generate a CI for the population mean or proportion.

TERMS INTRODUCED IN THIS CHAPTER

- Bootstrapping (p. 113)
- Computer-intensive method (p. 113)
- Percentile-based resampling confidence interval (p. 113)
- Prediction intervals (p. 117)
- Pseudosample (p. 112)
- Resampling (p. 112)
- t distribution (p. 110)

CHAPTER 14

Error Bars

The only role of the standard error . . . is to distort and conceal the data. The reader wants to know the actual span of the data; but the investigator displays an estimated zone for the mean.

A. R. FEINSTEIN

Scientific papers often tabulate or graph results as mean ± SD or mean ± SEM. This chapter explains the difference, as well as how to compute the SD from the SEM and sample size.

SD VERSUS SEM

What is the SD?
The standard deviation, abbreviated SD or s, was explained in Chapters 9 and 10. It is expressed in the same units as the data and quantifies variation among the values. If the data are sampled from a Gaussian distribution, you expect about two-thirds of the values to lie within 1 SD of the mean.

What is the SEM?
The ratio of the SD divided by the square root of the sample size (n) is called the *standard error of the mean*, often abbreviated SEM:

$$\text{SEM} = \frac{\text{SD}}{\sqrt{n}}$$

Often the SEM is referred to as the *standard error*, with the word *mean* missing but implied, and some journals abbreviate it SE. Like many statistical terms, this jargon can be confusing. The SEM has nothing to do with standards or errors, and standard errors can be computed for values other than the mean (e.g., the standard error of the best-fit value of a slope in linear regression; see Chapter 33).

The SEM does not quantify the variability among values
Note that the SEM does not directly quantify scatter or variability among values in the population. Many scientists misunderstand this point. The SEM will be small when the SD is large, provided that the sample size is also large. With huge samples, the SEM is always tiny. For the n = 130 sample of the body temperature data, the SEM equals 0.0359°C.

The SEM quantifies how precisely you know the population mean

Imagine taking many samples of sample size n from your population. These sample means will not be identical, so you could quantify their variation by computing the SD of the set of means. The SEM computed from one sample is your best estimate of what the SD among sample means would be if you collected an infinite number of samples of a defined size.

Chapter 12 defined the margin of error (W) of a CI of a mean as follows:

$$W = \frac{t^* \cdot SD}{\sqrt{n}}$$

Substituting the definition of the SEM,

$$W = t^* \cdot SEM$$

So the width of a CI of the mean is proportional to the SEM.

How to compute the SD from the SEM

Many scientists present the SEM in papers and presentations. Remember this equation so you can easily calculate the SD:

$$SD = SEM \cdot \sqrt{n} \qquad\qquad 14.1$$

For the n = 12 body temperature sample, the SEM equals 0.1157°C. If that was all you knew, you could compute the SD. Multiply 0.1157 times the square root of 12, and the SD equals 0.40°C. This is an exact calculation, not an approximation.

WHICH KIND OF ERROR BAR SHOULD I PLOT?

An *error bar* displays variation or uncertainty. Your choice of error bar depends on your goals.

Goal: To show the variation among the values

If each value represents a different individual, you probably want to show the variation among values. Even if each value represents a different lab experiment, it often makes sense to show the variation.

With fewer than 100 or so values, you can create a scatter plot that shows every value. What better way to show the variation among values than to show every value? If your data set has more than 100 or so values, a scatter plot becomes messy. Alternatives include a box-and-whiskers plot, a violin plot, a frequency distribution (histogram), or a cumulative frequency distribution (see Chapter 7).

What about plotting the mean and SD? The SD does quantify scatter, so this is indeed one way to graph variability. But since the SD is only one value, plotting it is a pretty limited way to show variation. A graph showing mean and SD *error bars* is less informative than any of the other alternatives. The advantages of plotting a mean with SD error bars (as opposed to a column scatter graph, a box-and-whiskers plot, or a frequency distribution) are that the graph is simpler and the approach is conventional in many fields.

Of course, if you do decide to show SD error bars, be sure to say so in the figure legend, so they won't be confused with SEM error bars.

If you are creating a table, rather than a graph, there are several ways to show the data, depending on how much detail you want to show. One choice is to show the mean as well as the smallest and largest values. A more compact choice is to tabulate the mean ± SD.

Goal: To show how precisely you have determined the population mean

If your goal is to compare means with a t test (Chapter 30), ANOVA (Chapter 39), or to show how closely your data come to the predictions of a model, you are likely trying to show how precisely the data define the population mean, rather than to show variability among values. In this case, the best approach is to plot the 95% CI of the mean (or perhaps a 90% or 99% CI).

What about the SEM? Graphing the mean with SEM error bars is a method commonly used to show how well you know the mean. The only advantages of SEM error bars versus CI error bars is that SEM error bars are shorter and are more conventional in some fields. But it is harder to interpret SEM error bars than it is to interpret a CI.

Whatever error bars you choose to show, be sure to state your choice.

Goal: To create persuasive propaganda

Simon (2005) "advocated" the following approach in his excellent blog. Of course, he meant it as a joke!

- When you are trying to emphasize small and unimportant differences in your data, show your error bars as SEM and hope that your readers think they are SD. (The readers will believe that the standard deviation is much smaller than it really is so will incorrectly conclude that the difference among means is large compared to variation among replicates.)
- When you are trying to cover up large differences, show the error bars as SD and hope that your readers think they are SEM. (The readers will believe that the standard deviation is much larger than it really is so will incorrectly conclude that the difference among means is small compared to the variation among replicates.)

THE APPEARANCE OF ERROR BARS

Frazier, Schneider, and Michel (2006) measured how well the neurotransmitter norepinephrine relaxes bladder muscles. Chapter 11 examined the concentrations required to get half-maximal bladder relaxation (EC_{50}) as an example of lognormal data. Now let's look at the maximal relaxation (% E_{max}) that can be achieved with large doses of norepinephrine in old rats.

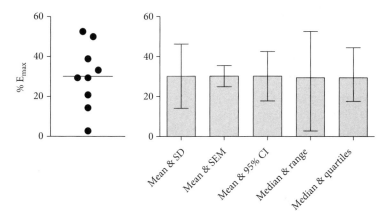

Figure 14.1. (Left) The actual data. (Right) Five kinds of error bars, each representing a different way to portray variation or precision.

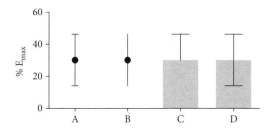

Figure 14.2. Four different styles for plotting the mean and SD.

Figure 14.1 (left) illustrates the raw data plotted as a column scatter plot. Figure 14.1 (right) shows the same data with several kinds of error bars representing the SD, SEM, 95% CI, range, and interquartile range. When you create graphs with error bars, be sure to state clearly how they were computed. These methods all plot the same values, and there is no real reason to prefer one style over another.

Figure 14.2 shows that error bars (in this case, representing the SD) can be plotted with or without horizontal caps, and (with bar graphs) in one direction or both directions. The choice is a matter of preference and style. When error bars with caps are placed on bars and only extend above the bar (bar C in Figure 14.2), the resulting plot is sometimes called a *dynamite plunger plot*, sometimes shortened to *dynamite plot*.

HOW ARE SD AND SEM RELATED TO SAMPLE SIZE?

If you increase the sample size, is the SEM expected to get larger, get smaller, or stay about the same?

It is expected to get smaller. The SEM quantifies how precisely the population mean has been determined. Larger samples have smaller SEMs, because the mean of a large sample is likely to be closer to the true population mean than is the mean of a small sample.

Of course, in any particular experiment, you can't be sure that increasing n will decrease SEM. It is a general rule, but it is also a matter of chance. It is possible, but unlikely, that the SEM might increase if you collect a larger sample size.

If you increase the sample size, is the SD expected to get larger, get smaller, or stay about the same?

The SD quantifies the scatter of the data. Whether increasing the size of the sample increases or decreases the scatter in that sample depends on the random selection of values. The SD, therefore, is equally likely to get larger or to get smaller as the sample size increases.

That statement, however, has some fine print: for all sample sizes, the sample variance (i.e., the square of the sample SD) is the best possible estimate of the population variance. If you were to compute the sample variance from many samples from a population, on average, it would be correct. Therefore, the sample variance is said to be *unbiased*, regardless of n. In contrast, when n is small, the sample SD tends to slightly underestimate the population SD. Increasing n, therefore, is expected to increase the SD a bit. Because statistical theory is based on the variance rather than the SD, it is not worth worrying about (or correcting for) the bias in the sample SD. Few statistics books even mention it. The expected increase in the sample SD as you increase sample size is tiny compared to the expected decrease in the SEM.

GEOMETRIC SD ERROR BARS

Chapter 11 explained lognormal distributions and the geometric SD. Figure 14.3 is the same as Figure 11.1 but includes geometric SD error bars. They are asymmetrical when the Y axis is linear but symmetrical when the Y axis is logarithmic.

COMMON MISTAKES: ERROR BARS

Mistake: Plotting mean and error bar instead of plotting a frequency distribution

Figure 14.4 shows the number of citations (references in other papers within five years of publication) received by 500 papers randomly selected from the journal *Nature*

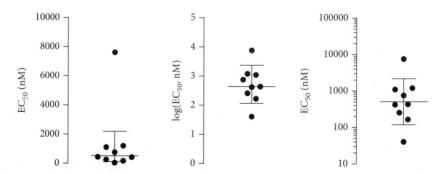

Figure 14.3. Geometric SD error bars.

This is the same as Figure 11.1 but with geometric SD error bars included.

(Colquhoun, 2003). The first (leftmost) bar shows that there are 71 (Y value) papers that had between zero and 20 citations (since the X value is centered at 10 and each bar represents a bin width of 20 citations). The last (rightmost) bar shows that a single paper (of the 500 studied) had 2,364 citations. This frequency distribution is easy to understand even though it is very asymmetrical, even more asymmetrical than a log-normal distribution (see Chapter 11). With a glance at the graph, you can see that many papers are only cited a few times (or not at all), while only a few are heavily cited.

The mean number of citations is approximately 115, with a SD of approximately 157 ("approximately" because I only have access to the frequency distribution, not the raw data). Viewing a graph showing that the mean is 115 and the SD is 157 would not help you to understand the data. Seeing that the mean is 115 and the SEM is 7 would be less helpful and might be quite misleading.

The median number of citations is approximately 65, and the interquartile range runs from about 20 to 130. Reporting the median with that range would be more informative than reporting the mean and SD or SEM. However, it seems to me that the best way to summarize these data is to show a frequency distribution. No numerical summary really does a good job.

Mistake: Assuming that all distributions are Gaussian

What would you think about the distribution of a data set if you were told only that the mean is 101 and the SD is 43 (or if you saw the bar graph of Data Set A in Figure 14.5)? You'd probably imagine that the data look something like Data Set B, and that would often be correct. But Data Sets C and D of Figure 14.5 also share that same mean and SD (and SEM). The distributions of values in Data Sets

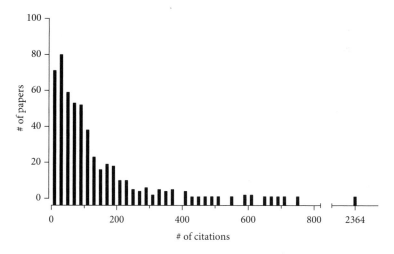

Figure 14.4. A very asymmetrical distribution.

This figure, remade from Colquhoun (2003), shows the number of citations (X-axis) received (within five years of publication) by 500 papers randomly chosen from the journal Nature. The bin width is 20 citations. The first bar shows that 74 papers received between zero and 20 citations, the second bar shows that 80 papers received between 21 and 40 citations, and so on.

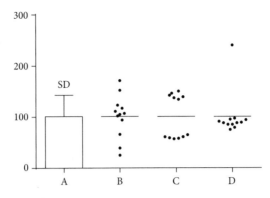

Figure 14.5. The mean and SD can be misleading.
All four data sets in the graph have approximately the same mean (101) and SD (43). If you were only told those two values or only saw the bar graph of Data Set A, you'd probably imagine that the data look like Data Set B. Data Sets C and D have very different distributions, but the same mean, SD, and SEM as Data Set B.

C and D are extremely different from the distribution of values in Data Set B. It can be useful to summarize data as mean and SD (or SEM), but it can also be misleading. Look at the raw data when possible.

When graphing your own data, keep in mind that it is easy to show the actual data on dot plots (column scatter graphs) with up to 100 or even a few hundred values. With more points, it is easy to plot a box-and-whiskers graph, a violin plot, or a frequency distribution histogram. Don't plot mean and an error bar without thinking first about alternative ways to plot the data.

Mistake: Plotting a mean and error bar without defining how the error bars were computed

If you make a graph showing means and error bars (or a table with means plus or minus error values), it is essential that you state clearly what the error bars are— SD, SEM, CI, range, or something else? Without a clear legend or a statement in the methods section, the graph or table is ambiguous.

Q & A

What does the SD quantify?
The SD quantifies scatter—how much the values vary from one another.

What does the SEM quantify?
The SEM quantifies how precisely you know the true mean of the population. The value of the SEM depends upon both the SD and the sample size.

Are the SD and SEM expressed in the same units?
Yes. Both are expressed in the same units as the data.

Which is smaller, the SD or the SEM?

> The SEM is always smaller than the SD.

SEM or SE?

> SEM means standard error of the mean and is the abbreviation I prefer. SE means standard error, and some journals use that abbreviation. You can compute the standard error of many computed values, but it is usually obvious in context when SE refers to the standard error of the mean.

When can I use the rule of thumb that the 95% CI of a mean equals the mean plus or minus 2 SEMs?

> With large samples. Calculating a CI requires using a multiplier from the t distribution, abbreviated (in this book) as t*. With large samples, t* has a value close to 2.0. Hence the rule of thumb that a 95% CI of a mean extends about 2 SEM in each direction from the mean. With small sample sizes, this approximate CI will be too narrow because the correct multiplier is greater than 2.0. For example, if n = 3, the distance a 95% CI extends in each direction from the mean equals 4.30 times the SEM.

Graphs often show an error bar extending 1 SEM in each direction from the mean. What does that denote?

> The range defined by the mean plus or minus 1 SEM has no simple interpretation. With large samples, that range is a 68% CI of the mean. But the degree of confidence depends on sample size. When n = 3, that range is only a 58% CI. You can think of the mean plus or minus 1 SEM as covering approximately a 60% CI.

Why are SEM error bars used so often?

> SEM error bars are popular because they are always smaller than the SD and because they are a compact way to show how precisely you know the mean.

What can I learn by observing whether two error bars overlap?

> Not much. See Chapter 30.

What error bars can be created when there are only two values?

> With n = 2, there isn't much point in showing error bars. Just show both values. But if you do create error bars from two values, know that all these will be identical: the 50% CI, the range, and the mean plus/minus the SEM. The mean plus/minus SD will extend beyond the range.

CHAPTER SUMMARY

- The SD quantifies variation.
- The SEM quantifies how precisely you know the population mean. It does not quantify variability. The SEM is computed from the SD and the sample size.
- The SEM is always smaller than the SD. Both are measured in the same units as the data.
- Graphs are often plotted as means and error bars, which are usually either the SD or the SEM.
- Consider plotting each value, or a frequency distribution, before deciding to graph a mean and error bar.
- You can calculate the SD from the SEM and n using Equation 14.1.

- Show the SD when your goal is to show variability and show the SEM (or, better, the confidence interval) when your goal is to show how precisely you have determined the population mean.
- The choice of SD or SEM is often based on traditions in a particular scientific field. Often, SEMs are chosen simply because they are smaller.
- Graphs with error bars (or tables with error values) should always contain labels, whether the error bars are SD, SEM, or something else.

TERMS INTRODUCED IN THIS CHAPTER

- Dynamite plunger plot (p. 121)
- Error bar (p. 119)
- Standard error of the mean (SEM) (p. 118)
- Unbiased (p. 122)

P Values and Statistical Significance

CHAPTER 15

Introducing P Values

> Imperfectly understood CIs are more useful and less dangerous
> than incorrectly understood P values.
>
> HOENIG AND HEISEY (2001)

Many statistical analyses are reported with P values, which are explained in this essential chapter. P values are often misunderstood because they answer a question you probably never thought to ask. This chapter explains what P values are and the many incorrect ways in which they are interpreted.

INTRODUCING P VALUES

You've probably already encountered P values. They appear in many statistical results, either as a number ("P = 0.0432") or as an inequality ("P < 0.01"). They are also used to make conclusions about statistical significance, as is explained in Chapter 16.

P values are difficult to understand and are often misinterpreted. This chapter starts with several examples to explain what P values are and then identifies the common incorrect beliefs many have about P values.

EXAMPLE 1: COIN FLIPPING

You flip a coin 20 times and observe 16 heads and 4 tails. Because the probability of heads is 50%, you'd expect about 10 heads in 20 flips. How unlikely is it to find 16 heads in 20 flips? Should you suspect that the coin is unfair?

Calculating the P value

To evaluate our suspicion that the coin might be unfair, we need to define what a fair coin is. A coin flip is fair when each coin flip has a 50% chance of landing on heads and a 50% chance of landing on tails, and each result is accurately recorded.

What probability do we want to compute? We could compute the chance of observing 16 heads in 20 flips, but that isn't sufficient. We would have been even more suspicious that the coin flipping wasn't fair had the coins landed on heads 17 times, or 18 times, and so on. Similarly, we would have been suspicious if the coin had landed on tails 16 or more times (and thus on heads 4 or fewer times). Putting all this together, we want to answer the following question: If the coin tosses are random and the answers are recorded correctly, what is the chance that when you flip the coin 20 times, you'll observe either 16 or more heads or 4 or fewer heads (meaning 16 or more tails)?

The values in the third column can also be computed using the Excel function Binom.dist. For example, the value for 10 heads is computed as =BINOM. DIST(10, 20, 0.5, FALSE). The first value is the number of "successes" (heads in this case); the second value is the number of tries (20 in this case); the third number is the fraction success for each try (0.5 for a coin flip); and the last value is false because you don't want a cumulative probability. Older versions of Excel offer the function BinomDist, which works the same way but is said to be less accurate.

Table 15.1 shows all the possible probabilities when you flip a fair coin 20 times. This book doesn't explain how to compute these answers, but the table legend points you to a Web page and an Excel function that does these calculations.

HEADS	TAILS	PROBABILITY (%)
0	20	0.000
1	19	0.002
2	18	0.018
3	17	0.109
4	16	0.462
5	15	1.479
6	14	3.696
7	13	7.393
8	12	12.013
9	11	16.018
10	10	17.620
11	9	16.018
12	8	12.013
13	7	7.393
14	6	3.696
15	5	1.479
16	4	0.462
17	3	0.109
18	2	0.018
19	1	0.002
20	0	0.000

Table 15.1. Probabilities when flipping a fair coin 20 times.

When you flip a fair coin, there is a 50% chance of heads and a 50% chance of tails. This table shows all the probabilities of flipping a fair coin 20 times. This table was created using the free binomial distribution web calculator at www.graphpad.com/quickcalcs/probability1/.

To answer the current question, we need to sum the probability of seeing zero, 1, 2, 3, 4, 16, 17, 18, 19, or 20 heads in 20 coin flips. The answer is 0.0118, or 1.18%. This result is called a *P value*. It is always reported as a fraction rather than a percentage.

Interpreting the P value

The P value answers this question: If the coin tosses were random and the answers are recorded correctly, what is the chance that when you flip a coin 20 times, you'll observe results as extreme (or more extreme) than those you observed in the previous example—that is, that you would observe 16 or more, or 4 or fewer, heads? The answer, the P value, is 0.0118. You'd only see results this far from an even split of heads to tails in 1.18% of 20-coin runs.

Should you conclude that the coin is unfair? There are two possible explanations for your observations. One possibility is that it is a coincidence. The other possibility is that the coin flipping wasn't fair. Which is it? All statistics can do is tell you how rare it is that such a coincidence would happen (1.18% of the time). What you conclude depends on the context. Consider these three scenarios:

- You've examined the coin and are 100% certain it is an ordinary coin, and you did the coin flips and recorded the results yourself, so you are sure there is no trickery. In this case, you can be virtually 100% certain that this streak of many heads is just a chance event. The P value won't change that conclusion.
- If the coin flips were part of a magic show, you can be pretty sure that trickery invalidates the assumptions that the coin is fair, the tosses are random, and the results are accurately recorded. You'll conclude that the coin flipping was not fair.
- If everyone in a class of 200 students had flipped the coin 20 times and the run of 16 heads was the most heads obtained by any student, you wouldn't conclude that the coin was unfair. It is not surprising to see a coincidence that happens 1.18% of the time when you run the test (coin flipping) 200 times.

EXAMPLE 2: ANTIBIOTICS ON SURGICAL WOUNDS

Heal and colleagues (2009) tested whether applying an antibiotic (chloramphenicol) to surgical wounds would reduce the incidence of wound infection. They randomly assigned 972 surgical patients to receive either an antibiotic ointment or an ointment with no active medication.

Infections occurred in 6.6% of the patients who received the antibiotic, compared with 11.0% of the patients who received inactive ointment. In other words, infections occurred in about 40% fewer patients who received the antibiotic. Does this mean that the antibiotic worked in this situation? Or could the results be due to a coincidence of random sampling?

To assess this question, we first must ask what would have happened if the risk of infections in patients who receive an antibiotic ointment is the same as the risk in those who receive an inactive ointment, so any discrepancy in the incidence of infections observed in this study is the result of chance. Then we can ask the following question: If the risk of infection overall is identical in the two groups and the experiment was performed properly, what is the chance that random sampling would lead to a difference in incidence rates equal to, or greater than, the difference actually observed in this study?

The authors did the calculation and published the answer, which is 0.010. Since the study was well designed, the P value was low, and the hypothesis that an antibiotic would prevent infection is quite sensible, the investigators concluded that the antibiotic worked to prevent wound infection.

EXAMPLE 3: ANGIOPLASTY AND MYOCARDIAL INFARCTION

Cantor and colleagues (2009) studied the optimal way to treat patients who had experienced myocardial infarctions (heart attacks) but were admitted to a hospital that could not do percutaneous coronary intervention (PCI, also known as angioplasty). They randomized 1,059 such patients into two groups. One group was immediately transferred to a different hospital and received PCI. The other half was given standard therapy at the admitting hospital. The researchers assessed the fraction of patients who died or experienced worsening heart disease (another myocardial infarction or worsened congestive heart failure) within 30 days of being admitted to the hospital.

By the end of 30 days, one (or more) of these outcomes had occurred in 11.0% of the patients who were transferred to a different hospital and 17.2% of the patients who received standard therapy. The risk ratio was 11.0/17.2 = 0.64. That means that the risk of death or worsening heart disease in patients who were transferred was 64% of the risk for patients who stayed at the original hospital.

Is this a real difference? Or is it just a coincidence of random sampling, such that sicker or unluckier patients in this study tended to be assigned to stay in the same hospital? To answer this question, we first must ask what would have happened if the risk of death or worsening heart disease was identical in the two populations. In this situation, what is the chance that the risk ratio would have been 0.64 or less in a study this size? The investigators state that the answer to this question, the P value, is 0.004. You don't need to know how they calculated the P value to interpret it.

There are two possibilities. One is that the sicker or unluckier patients in this study just happened to be assigned to stay in the same hospital. That coincidence would happen in 0.4%, which is 1 in 250, of studies of this size. The other possibility is that patients who are transferred to a hospital that can do angioplasty on average have better outcomes than those who stay at a hospital that cannot do angioplasty. Since the P value is so low and since that hypothesis makes sense, the investigators concluded that transferring patients is the safer protocol.

LINGO: NULL HYPOTHESIS

In all three examples, the P value definition requires calculating what results would be in a simplified situation.

- In the coin-flipping example, the situation is that the coin is fair so that the observed deviation from an outcome of half heads and half tails occurred randomly.
- In the antibiotic example, the situation is that the antibiotic used does not actually prevent infection in the experimental situation, so that the observed difference in infection rates was due to random sampling of subjects.
- In the angioplasty example, the situation is that patients transferred to another hospital have the same risk of death or worsening heart disease as do those who remain in the existing hospital, so that the observed difference in risk of death or heart disease was a coincidence.

In each case, the simplified situation is called a *null hypothesis*. P values are calculated from the hypothetical results you would observe if the null hypothesis is true. If you can't state what the null hypothesis is, you can't understand the P value.

In most cases, the null hypothesis is that there is no difference between means in the populations, no correlation in the population, or a slope of zero in the population. In these cases, the null hypothesis can also be called a *nil hypothesis*. Most, but not all, null hypotheses are also nil hypotheses. However, you can set a null hypothesis to any size difference or association you want (not just zero).

WHY P VALUES ARE CONFUSING

P values are confusing for several reasons:

- The null hypothesis is usually the opposite of the hypothesis the experimenter expects or hopes to be true.
- In many situations, you know before collecting any data that the null hypothesis is almost certainly false. The null hypothesis is usually that in the populations being studied there is zero difference between the means, zero correlation, or a relative risk or odds ratio of 1.0000. People rarely conduct experiments or studies in which it is even conceivable that the null hypothesis is true. The difference or correlation or association in the overall population might be trivial, but it almost certainly isn't zero. So the P value asks a question that it rarely makes sense to ask.
- Clinicians and scientists find it strange to calculate the probability of obtaining results that weren't actually obtained. The math of theoretical probability distributions is beyond the easy comprehension of most scientists.
- The logic of P values goes in a direction that seems backward. In conducting a study, you observe a sample of data and ask about the populations from which the data are sampled. The definition of the P value starts with a

provisional assumption about the population (the null hypothesis) and asks about possible data samples. It's backwards! Thinking about P values seems quite counterintuitive, except maybe to lawyers or Talmudic scholars used to this sort of argument by contradiction.

- The P value answers a question that few scientists would ever think to ask!

ONE- OR TWO-TAILED P VALUE?

Example 1: Coin flipping

Let's return to the coin-flipping example from the beginning of the chapter. The P value we've already defined is known as a *two-tailed P value*. It answers the following quesion. If the null hypothesis (fair coin) were true, what is the chance of observing 16 or more, or 4 or fewer, heads in 20 flips? It is two-tailed because it asks about both extremes: lots of heads or very few heads. This two-tailed P value is 0.0118.

To define a *one-tailed P value*, you must predict which way the data will go before collecting them. Perhaps someone told you they suspected that this coin comes up heads too often and so the question you want to ask is whether that claim is true. In that case, the P value answers the question: If the null hypothesis were true, what is the chance of observing 16 or more heads in 20 flips? Find out by adding up the bottom five rows of Table 15.1. The answer, the one-tailed P value, equals 0.0059. Note that this is half the two-tailed P value.

What if someone had suggested that this coin was biased the other way, with tails coming up too often? In that case, the one-tailed P value would answer this question: If the null hypothesis were true, what is the chance of observing 16 or fewer heads in 20 flips? Find out the answer by adding up the top 17 rows in Table 15.1 (which starts with 0 heads). The result, the one-tailed P value, is 0.9940. You expected to have fewer heads than tails, but in fact, the results went the other way, so the one-tailed P value is high.

Figure 15.1 shows the definitions of the two-tailed P value and the two distinct one-tailed P values. Note that the two one-tailed P values add up to 1.00.

Example 2: Antibiotics on Surgical Wounds

The null hypothesis for the antibiotics example is that the antibiotic ointment does not change the risk of infection. In the patients studied, there were 40% fewer infections among those treated with the antibiotic compared to those treated with the inactive ointment.

The two-tailed P value answers this question: If the null hypothesis were true, what is the chance that random sampling would lead to either a 40% reduction (or more) in infection rate in those treated with antibiotics or a 40% reduction (or more) in those treated with inactive ointment? The answer, as calculated by the investigators and included in the published paper, is 0.010.

To compute a one-tailed P value, you must have made a prediction about the direction of the findings. Let's say you had predicted that there would be fewer infections in the antibiotic-treated patients. In that case, the one-tailed P value answers this question: If the null hypothesis were true, what is the chance that random

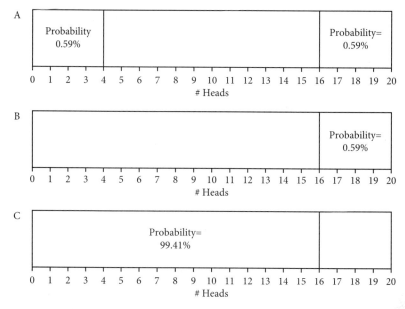

Figure 15.1. Defining one- and two-tailed P values.
You flip a coin 20 times, see 16 heads, and want to compute a P value testing the null hypothesis that the coin flips are fair. Part A defines the two-tailed P value, the probability (under the null hypothesis) of observing 16 or more heads, plus the probability (the other tail) of observing 4 or fewer heads. Part B shows the one-tailed P value if you had predicted that the coin would land on heads more than half the time. Since the actual result (16 heads) corresponds to the hypothesis, the P value is less than 0.50. Part C shows the one-tailed P value if you had predicted that the coin would land on heads less than half the time. Since the actual data (16 heads) was opposite to the prediction, the P value is greater than 0.50. The P values in the figure are shown as percentages to make them more compact. However, P values are almost always presented as proportions, not percentages.

sampling would lead to a 40% reduction (or more) in infection rate in those treated with antibiotics? The answer is 0.005, which is half the two-tailed P value.

What if you had predicted that there would be more infections in the patients treated with the antibiotic than in those treated with the inactive ointment? Sure, that is a pretty unlikely prediction for this example, but go with it. Now the one-tailed P value answers this question: If the null hypothesis were true, what is the chance that random sampling would lead to a 40% reduction (or less) in infection rate in those treated with the inactive ointment? Since the prediction runs opposite to the data, this one-tailed P value, 0.995, is high.

What is the advantage of a one-tailed P value?
When planning a study, choosing a one-tailed P value gives you a more focused hypothesis and so reduces the necessary sample size. A study designed to use a one-tailed P value requires about 20% fewer subjects than the same study designed to use a two-tailed P value (Moyé & Tita, 2002). Even though that is not a huge difference, it does decrease cost and risk and so is a reason to prefer one-tailed P values (Knottnerus & Bouter, 2001).

Once the data are collected, the appeal of a one-tailed P value is clear. Assuming you predicted the direction of the effect correctly, a one-tailed P value equals one-half of the two-tailed P value (this is not always exactly true, but it is true for most statistical tests).

When is a one-tailed P value appropriate?

Use of a one-tailed P value requires planning. If you haven't given any thought as to whether you will use one- or two-tailed P values until after you have begun to analyze the data, it is too late. Use two-tailed P values. You should only report a one-tailed P value when you have predicted which treatment group will have the larger mean (or proportion, survival time, etc.) before you collect any data and have recorded this prediction (so you can't be tempted to change your "prediction" after viewing the data).

Here is an example in which one could argue that it is appropriate to choose a one-tailed P value: Imagine you are testing whether a new antibiotic impairs kidney function, as measured by serum creatinine. Many antibiotics poison kidney cells, resulting in reduced glomerular filtration and increased serum creatinine. As far as I know, no antibiotic is known to decrease serum creatinine. Before collecting any data, it would not be unreasonable to state that there are two possibilities: Either the drug will not change the mean serum creatinine of the population, or it will increase the mean serum creatinine in the population. Accordingly, it makes sense to calculate a one-tailed P value that tests the null hypothesis that the drug does not increase the creatinine level. If the creatinine level goes down, no matter how much, you'd attribute that decrease to chance.

The problem with one-tailed P values

Before committing to a decision to use a one-tailed P value, think about what you would do if the data surprised you and the treatment effect went in the direction opposite to what you predicted. In this case, the one-tailed P value would be larger than 0.50.

Would you be tempted to switch to a two-tailed P value? Would you be tempted to change your "prediction" about the direction of the effect? If you answer yes to either of these questions, then you should plan to use a two-tailed P value.

One famous example of this problem is the Cardiac Arrhythmia Suppression Trial (CAST) study of the use of anti-arrhythmic drugs following myocardial infarction (heart attack). These drugs suppress premature ventricular contractions (extra heart beats), which can occur after a heart attack and were assumed to also prevent fatal arrhythmias (including sudden death) and thus prolong life after a heart attack. For many years, these drugs were the standard therapy for anyone who survived a heart attack. The CAST study tested the hypothesis that anti-arrhythmic drugs prolonged life after a heart attack (CAST Investigators, 1989). The people planning the study expected that the drugs would either prolong the life of the patients or have no effect, and so they chose to calculate and report a one-tailed P value.

The results were a surprise to the cardiologists who designed the study. Patients treated with anti-arrhythmic drugs turned out to be four times more likely to die than patients given a placebo. The results went in the direction opposite to what was predicted.

When in doubt, use two-tailed P values

When in doubt, I suggest that you use a two-tailed P value for the following reasons:

- The relationship between P values and CIs is more straightforward with two-tailed P values.
- Some tests compare three or more groups, in which case the concept of one or two tails makes no sense.
- The practical effect of choosing a one-tailed P value is to make it appear that the evidence is stronger (because the P value is lower). But this is not due to any data collected in the experiment, but rather to a choice (or perhaps a belief) of the experimenter (Goodman, 2008).
- Some reviewers or editors may criticize any use of a one-tailed P value, no matter how well you justify it.

P VALUES ARE NOT VERY REPRODUCIBLE

I simulated multiple data sets randomly sampled from the same populations. The top portion of Figure 15.2 shows four simulated experiments each analyzed with an unpaired t test. The four P values vary considerably. The bottom part of the figure shows the distribution of P values from 2,500 such simulated experiments. Leaving out the 2.5% highest and lowest P values, the middle 95% of the P values range from 0.0001517 to 0.6869—a span covering more than three orders of magnitude!

This example uses an experiment in which the median P value is about 0.05. Simulation with a smaller SD resulted in smaller P values, but the range still covered more than three orders of magnitude. Boos and Stefanski (2011) have shown that P values from repeated simulated experiments vary over more than three orders of magnitude in a number of situations.

THERE IS MUCH MORE TO STATISTICS THAN P VALUES

Many statistical analyses generate both P values and CIs. Many scientists report the P value and ignore the CI. I think that this is a mistake.

Interpreting P values is tricky, requiring one to define a null hypothesis and think about the distribution of the results of hypothetical experiments. Interpreting CIs, in contrast, is quite simple. You collect some data and do some calculations to quantify a treatment effect. Then you report that value (perhaps a difference or a relative risk) along with a CI that quantifies the precision of that value. This approach is very logical and intuitive. You report what you calculated along with an interval that shows how sure you are.

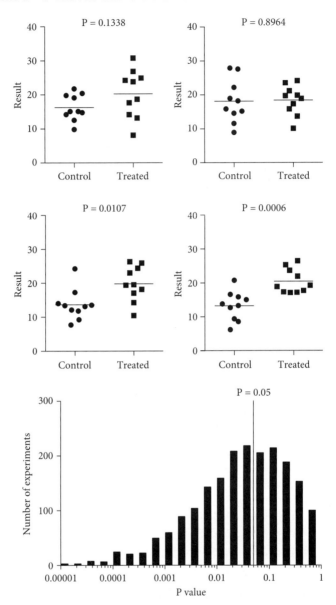

Figure 15.2. P values are not very reproducible.

The top panels show four pairs of data sets simulated (using GraphPad Prism) from Gaussian distributions with an SD equal to 5.0. The population means of the two populations differ by 5.0. Ten values were simulated in each group, and an unpaired t test was used to compare the sample means. Note how far apart the four P values are, even though the data in each of the four experiments were sampled from the same populations. The histogram at the bottom shows the distribution of P values from 2,500 simulated data sets sampled from those same populations. Note that the X-axis is logarithmic. The P values span a very wide range.

Rothman (2016) pointed out a huge distinction between P values and CIs. P values are computed by combining two separate values: the size of the effect and the precision of the effect (partly determined by sample size). In contrast, a CI shows the two separately. The location of the center of the CI tells you about the observed effect size. It answers this question: How big is the observed effect in this study? The width of the CI tells you about precision. It answers this question: How precisely has this study determined the size of the effect? Both questions are important, and the CI answers both in an understandable way. In contrast, the P value is computed by combining the two concepts and answers a single question that is less interesting: If there were no effect in the overall population (if the null hypothesis were actually true), what is the chance of observing an effect this large or larger?

When reading papers, I suggest that you don't let yourself get distracted by P values. Instead, look for a summary of how large the effect was and its CI.

When writing a paper, first make sure that the results are shown clearly. Then compute CIs to quantify the precision of the effect sizes. Consider stopping there. Only add P values to your manuscript when you have a good reason to think that they will make the paper easier to understand.

COMMON MISTAKES: P VALUES

Many of the following points come from Kline (2004) and Goodman (2008).

The most common misconception about P values

Many people misunderstand what a P value means. Consider a situation where investigators compared a mean in groups given two alternative treatments and reported that the P value equals 0.03. Here are two correct definitions of this P value:

- If the two population means are identical (the null hypothesis is true), there is a 3% chance of observing a difference as large as you observed (or larger).
- Random sampling from identical populations would lead to a difference smaller than you observed in 97% of experiments and larger than you observed in 3% of experiments.

Many people want to believe the following definition, which appears crossed out to emphasize that it is wrong:

~~There is a 97% chance that there is a real difference between populations and a 3% chance that the difference is a random coincidence.~~

The P value is not the probability that the result was due to sampling error

The P value is computed from figuring out what results you would see if the null hypothesis is true. In other words, the P value is computed from the results you would see if observed differences were only due to randomness in selecting subjects—that is, to sampling error. Therefore, the P value cannot tell you the probability that the result is due to sampling error.

The P value is not the probability that the null hypothesis is true

The P value is computed from the set of results you would see if the null hypothesis is true, so it cannot calculate the probability that the null hypothesis is true.

The probability that the alternative hypothesis is true is not 1.0 minus the P value

If the P value is 0.03, it is very tempting to think that if there is only a 3% probability that my difference would have been caused by random chance, then there must be a 97% probability that it was caused by a real difference. But this is wrong!

What you *can* say is that if the null hypothesis were true, then 97% of experiments would lead to a difference smaller than the one you observed and 3% of experiments would lead to a difference as large or larger than the one you observed.

Calculation of a P value is predicated on the assumption that the null hypothesis is true. P values cannot tell you whether this assumption is correct. Instead, the P value tells you how rarely you would observe a difference as large or larger than the one you observed if the null hypothesis were true.

The probability that the results will hold up when the experiment is repeated is not 1.0 minus the P value

If the P value is 0.03, it is tempting to think that this means there is a 97% chance of getting similar results in a repeated experiment. Not so. The P value does not itself quantify reproducibility.

A high P value does not prove that the null hypothesis is true

A high P value means that if the null hypothesis were true, it would not be surprising to observe the treatment effect seen in a particular experiment. But that does not prove that the null hypothesis is true. It just says the data are consistent with the null hypothesis.

"P = 0.05" is not the same as "P < 0.05"

"P = 0.05" means what it says: that the P value equals 0.05. In contrast, "P < 0.05" means that P is less than 0.05.

A P value does not always have to be written as an inequality

A P value is a number, and it should be expressed as such. When very small, it can make sense to simply say, for example, that P < 0.0001. But it is more informative to state the P value as a number (e.g., 0.0345) than to just state some value it is less than (e.g., P < 0.05).

The P value tests more than the null hypothesis

The idea of a P value is to test the ability of the null hypothesis to explain the data. But the P value tests more than the null hypothesis. To calculate a P value, you first have to create a hypothetical model. The P value answers the question: If that model were correct, how unlikely are the data we observed? That model specifies the null hypothesis of course, but it also specifies (often implicitly) many other assumptions about experimental design and the distribution of values. If any of these assumptions are substantially wrong, then the P value will be misleading. The P value tests all the assumptions about how the data were generated, not just the null hypothesis (Greenland et al., 2016).

Q & A

Can P values be negative?
> No. P values are fractions, so they are always between 0.0 and 1.0.

Can a P value equal 0.0?
> A P value can be very small but can never equal 0.0. If you see a report that a P value equals 0.0000, that really means that it is less than 0.0001.

Can a P value equal 1.0?
> A P value would equal 1.0 only in the rare case in which the treatment effect in your sample precisely equals the one defined by the null hypothesis. When a computer program reports that the P value is 1.0000, it often means that the P value is greater than 0.9999.

Should P values be reported as fractions or percentages?
> By tradition, P values are always presented as fractions and never as percentages.

Is a one-tailed P value always equal to half the two-tailed P value?
> Not always. Some distributions are asymmetrical. For example, a one-tailed P value from a Fisher's exact test (see Chapter 27) is usually not exactly equal to half the two-tailed P value. With some data, in fact, the one- and two-tailed P values can be identical. This event is rare.
>
> Even if the distribution is symmetrical (as most are), the one-tailed P value is only equal to half the two-tailed value if you correctly predicted the direction of the difference (correlation, association, etc.) in advance. If the effect actually went in the opposite direction to your prediction, the one-tailed P will not be equal to half the two-tailed P value. If you were to calculate this value, it would be greater than 0.5 and greater than the two-tailed P value.

Can a one-tailed P value have a value greater than 0.50?
> Yes, this happens when the direction of the treatment effect is opposite to the prediction.

Is a P value always associated with a null hypothesis?
> Yes. If you can't state the null hypothesis, then you can't interpret the P value.

Shouldn't P values always be presented with a conclusion about whether the results are statistically significant?

No. A P value can be interpreted on its own. In some situations, it can make sense to go one step further and report whether the results are statistically significant (as will be explained in Chapter 16). But this step is optional.

Should it be P or p? Italicized or not? Hyphenated or not?

Different books and journals use different styles. This book uses "P value," but "*p*-value" is probably more common.

I chose to use a one-tailed P value, but the results came out in the direction opposite to my prediction. Can I report the one-tailed P value calculated by a statistical program?

Probably not. Most statistical programs don't ask you to specify the direction of your hypothesis and report the one-tailed P value assuming you correctly predicted the direction of the effect. If your prediction was wrong, then the correct one-tailed P value equals 1.0 minus the P value reported by the program.

Why do some investigators present the negative logarithm of P values?

When presenting many tiny P values, some investigators (especially those doing genome-wide association studies; see Chapter 22) present the negative logarithm of the P value to avoid the confusion of dealing with tiny fractions. For example, if the P value is 0.01, its logarithm (base 10) is −2, and the negative logarithm is 2. If the P value is 0.00000001, its logarithm is −8, and the negative logarithm is 8. When these are plotted on a bar graph, it is called a *Manhattan plot* (Figure 15.3). Why that name? Because the arrangement

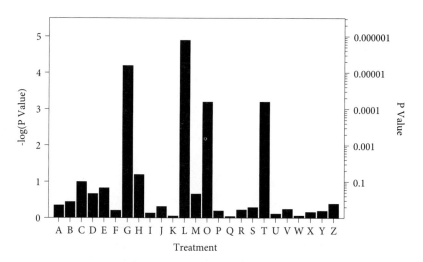

Figure 15.3. A Manhattan plot.

Each bar represents a P value computed from a different treatment, different gene, different study, and so on. The left Y axis is the negative logarithm of the P value; the right Y axis shows the P value itself (but this is often not shown on Manhattan plots). For example, if the P values is 0.001, its logarithm is −3, so the negative logarithm is 3. If the P values is 0.00000001, its logarithm is −8, so the negative logarithm is 8. Note a point of potential confusion. P values are always fractions, so their logarithms are always negative numbers, so their negative logarithms are always positive numbers.

of high and low bars make the overall look vaguely resembles the skyline of Manhattan, New York (maybe you need to have a few drinks before seeing that resemblance).

Is the null hypothesis ever true?

Rarely.

Is it possible to compute a CI of a P value?

No. CIs are computed for parameters like the mean or a slope. The idea is to express the likely range of values that includes the true population value for the mean or slope etc. The P value is not an estimate of a value from the population. It is computed for that one sample. Since it doesn't make sense to ask what the overall P value is in the population, it doesn't make sense to compute a CI of a P value.

If the null hypothesis is really true and you run lots of experiments, will you expect mostly large P values?

No. If the null hypothesis is true, the P value is equally likely to have any value. You'll find just as many less than 0.10 as greater than 0.90.

This chapter takes the point of view that P values are often misunderstood and over-emphasized. Is this a mainstream opinion of statisticians?

Yes. The American Statistical Association released a report in 2016 expressing many of the same concerns as this chapter (Wasserstein & Lazar, 2016). But there is plenty of disagreement on the nuances, as you can see from the fact that the report came with 20 accompanying commentaries!

CHAPTER SUMMARY

- P values are frequently reported in scientific papers, so it is essential that every scientist understand exactly what a P value is and is not.
- All P values are based on a null hypothesis. If you cannot state the null hypothesis, you can't understand the P value.
- A P value answers the following general question: If the null hypothesis is true, what is the chance that random sampling would lead to a difference (or association, etc.) as large or larger than that observed in this study?
- P values are calculated values between 0.0 and 1.0. They can be reported and interpreted without ever using the word "significant."
- When interpreting published P values, note whether they are calculated for one or two tails. If the author doesn't say, the result is somewhat ambiguous.
- If you repeat an experiment, expect the P value to be very different. P values are much less reproducible than you would guess.
- One-tailed P values have their advantages, but you should always use a two-tailed P value unless you have a really good reason to use a one-tailed P value.
- There is much more to statistics than P values. When reading scientific papers, don't get mesmerized by P values. Instead, focus on what the investigators found and how large the effect size is.

TERMS INTRODUCED IN THIS CHAPTER

- Alternative hypothesis (p. 140)
- Manhattan plot (p. 142)
- Null hypothesis (p. 133)
- One-tailed P value (p. 134)
- P value (p. 131)
- Two-tailed P value (p. 134)

CHAPTER 16

Statistical Significance and Hypothesis Testing

For the past eighty years, it appears that some of the sciences have
made a mistake by basing decisions on statistical significance.

ZILIAK AND MCCLOSKEY (2008)

The phrase "statistically significant" is commonly misunderstood, because this use of the word "significant" has almost no relationship to
the conventional meaning of the word as used to describe something that
is important or consequential. This chapter explains that the concept is
quite straightforward once you get past the confusing terminology.

STATISTICAL HYPOTHESIS TESTING

Making a decision is the primary goal when analyzing some kinds of data. In a
pilot experiment of a new drug, the goal may be to decide whether the results are
promising enough to merit a second experiment. In a Phase III drug study, the goal
may be to decide whether the drug should be recommended. In quality control, the
goal is to decide whether a batch of product can be released.

Statistical hypothesis testing automates decision-making. First, define a threshold P value for declaring whether a result is statistically significant. This threshold is
called the *significance level* of the test; it is denoted by α (*alpha*) and is commonly
set to 0.05. If the P value is less than α, conclude that the difference is *statistically
significant* and decide to reject the null hypothesis. Otherwise, conclude that the
difference is not statistically significant and decide to not reject the null hypothesis.

Note that statistics gives the term *hypothesis testing* a unique meaning that is
very different than the meaning most scientists think of when they go about testing a scientific hypothesis.

ANALOGY: INNOCENT UNTIL PROVEN GUILTY

Scientist as juror

The steps that a jury must follow to determine criminal guilt are very similar to the
steps that a scientist follows when using hypothesis testing to determine statistical
significance.

A juror starts with the presumption that the defendant is innocent. A scientist starts with the presumption that the null hypothesis of "no difference" is true.

A juror bases his or her decision only on factual evidence presented in the trial and should not consider any other information, such as newspaper stories. A scientist bases his or her decision about statistical significance only on data from one experiment, without considering what other experiments have concluded.

A juror reaches the verdict of guilty when the evidence is inconsistent with the assumption of innocence. Otherwise, the juror reaches a verdict of not guilty. When performing a statistical test, a scientist reaches a conclusion that the results are statistically significant when the P value is small enough to make the null hypothesis unlikely. Otherwise, a scientist concludes that the results are not statistically significant.

A juror does not have to be convinced that the defendant is innocent to reach a verdict of not guilty. A juror reaches a verdict of not guilty when the evidence is consistent with the presumption of innocence, when guilt has not been proven. A scientist reaches the conclusion that results are not statistically significant whenever the data are consistent with the null hypothesis. The scientist does not have to be convinced that the null hypothesis is true.

A juror can never reach a verdict that a defendant is innocent. The only choices are guilty or not guilty. A statistical test never concludes the null hypothesis is true, only that there is insufficient evidence to reject it (but see Chapter 21).

A juror must try to reach a conclusion of guilty or not guilty and can't conclude "I am not sure." Similarly, each statistical test leads to a crisp conclusion of statistically significant or not statistically significant. A scientist who strictly follows the logic of statistical hypothesis testing cannot conclude "Let's wait for more data before deciding."

Scientist as journalist

Jurors aren't the only people who evaluate evidence presented at a criminal trial. Journalists also evaluate evidence presented at a trial, but they have very different goals than jurors. A journalist's job is not to reach a verdict of guilty or not guilty but rather to summarize the proceedings.

Scientists in many fields are often more similar to journalists than jurors. If you don't need to make a clear decision based on one P value, you don't need to use the term *statistically significant* or use the rubric of statistical hypothesis testing.

EXTREMELY SIGNIFICANT? BORDERLINE SIGNIFICANT?

Tiny P values

Is a result with $P = 0.004$ more statistically significant than a result with $P = 0.04$? By the strict rules of statistical hypothesis testing, the answer is no. Once you have established a significance level, every result is either statistically significant or not statistically significant. Because the goal of statistical hypothesis testing is to make a crisp decision, only two possible conclusions are possible.

SYMBOLS	PHRASE	P VALUE
ns	Not significant	$P > 0.05$
*	Significant	$P < 0.05$
**	Significant Highly	$P < 0.01$
***	Extremely significant	$P < 0.001$

Table 16.1. Using asterisks to denote statistical significance.
The scheme shown here is commonly used but is not universal. If you see graphs decorated with asterisks to denote statistical significance, be sure to check the legend to see how the authors defined the symbols.

Most scientists are less rigid and refer to *very significant* or *extremely significant* results when the P value is tiny. When showing P values on graphs, investigators commonly use asterisks to create a scale such as that used in Michelin Guides or movie reviews (see Table 16.1). When you read this kind of graph, make sure that you look at the key that defines the symbols, because different investigators can use different threshold values.

Borderline statistical significance

If you follow the strict paradigm of statistical hypothesis testing and set α to its conventional value of 0.05, then a P value of 0.049 denotes a statistically significant difference and a P value of 0.051 does not. This arbitrary distinction is unavoidable, because the whole point of statistical hypothesis testing is to reach a crisp conclusion from every P value, without exception.

When a P value is just slightly greater than α, some scientists avoid the phrase "not significantly different" and instead refer to the result as "marginally significant" or "almost statistically significant." Some get more creative and use phrases like "a clear trend," "barely missed statistical significance," "not exactly significant," "not currently significant," "on the cusp of significance," "provisionally significant," "trending toward significance," "verging on significance," "weakly significant," and so on. Hankin (2013) lists 509 such phrases he found in published papers! Rather than deal with linguistic tricks, it is better in many cases to just report the actual P value and not worry about whether it is above or below some arbitrary threshold value.

When a two-tailed P value is between 0.05 and 0.10, you may be tempted to switch to a one-tailed P value. Because the one-tail P value (with rare exceptions) equals half the two-tailed P value (as explained in Chapter 15), it will be smaller than 0.05. Your results will have become statistically significant, as if by magic. Obviously, this is not an appropriate reason to choose a one-tailed P value! The decision to use one-tail P values should have be made before the data were collected.

LINGO: TYPE I AND TYPE II ERRORS

Statistical hypothesis testing makes a decision based on the results of one comparison. When you make this decision, there are two kinds of mistakes you can make (see Table 16.2).

	DECISION: REJECT NULL HYPOTHESIS	DECISION: DO NOT REJECT NULL HYPOTHESIS
Null hypothesis is true	Type I error	(No error)
Null hypothesis is false	(No error)	Type II error

Table 16.2. Definition of Type I and Type II errors.

Type I error

When there really is no difference (or association or correlation) between the populations, random sampling can lead to a difference (or association or correlation) large enough to be a statistically significant. This is a *Type I error*. It occurs when you decide to reject the null hypothesis when in fact the null hypothesis is true. It is a false positive.

Type II error

When there really is a difference (or association or correlation) between the populations, random sampling (and small sample size) can lead to a difference (or association or correlation) small enough to be not statistically significant. This is a *Type II error*. It occurs when you decide not to reject the null hypothesis when in fact the null hypothesis is false. It is a false negative.

A worse kind of error

What if you conclude that a difference (or association or correlation) is statistically significant but are wrong about the direction of the effect? You are correct that the null hypothesis truly is wrong, so you haven't made a Type I error. But you've got the direction of the effect backwards. In your sample, the new drug works better than the standard drug, but in fact (if you had infinite sample size) it works (on average) worse. Or the data in your sample show that people taking a certain drug lose weight, when in fact (if you had an infinite sample) that drug causes people (on average) to gain weight. Gelman and Tuerlinckx (2000) call this a *Type S error*, because the sign of the difference is backwards. Hsu (1996) calls this a *Type III error*. Neither term is widely used.

You won't know

Type I and II (and III) errors are theoretical concepts. When you analyze data, you don't know whether the populations are identical, so can't know whether you made one of these errors.

TRADEOFFS WHEN CHOOSING A SIGNIFICANCE LEVEL

The distinction between the P value and α

The P value and α are not the same:

- The significance level α is chosen by the experimenter as part of the experimental design before collecting any data. If the null hypothesis is true, α is the probability of rejecting the null hypothesis.

- A P value is computed from the data. You reject the null hypothesis (and conclude that the results are statistically significant) when the P value from a particular experiment is less than the significance level α set in advance.

The trade-off

A result is deemed statistically significant when the P value is less than a significance level (α) set in advance. So you need to choose that level. By tradition, α is usually set to equal 0.05, but you can choose whatever value you want. When choosing a significance level, you confront a trade-off.

If you set α to a very low value, you will make few Type I errors. That means that if the null hypothesis is true, there will be only a small chance that you will mistakenly call a result statistically significant. However, there is also a larger chance that you will not find a significant difference, even if the null hypothesis is false. In other words, reducing the value of α will decrease your chance of making a Type I error but increase the chance of a Type II error.

If you set α to a very large value, you will make more Type I errors. If the null hypothesis is true, there is a large chance that you will mistakenly conclude that the effect is statistically significant. But there is a small chance of missing a real difference. In other words, increasing the value of α will increase your chance of making a Type I error but decrease the chance of a Type II error. The only way to reduce the chances of both a Type I error and a Type II error is to collect bigger samples (see Chapter 26).

Example: Detecting spam

Table 16.3 recasts Table 16.2 in the context of detecting spam (junk) email. Spam filters use various criteria to evaluate the probability that a particular email is spam and then use this probability to decide whether to deliver that message to the inbox or to the spam folder.

The null hypothesis is that an email is good (not spam). A Type I error occurs when a good email is mistakenly sent to the spam folder, where it probably will never be seen. A Type II error occurs when spam is delivered to the inbox.

When deciding how aggressively to define spam, people designing the spam filters have to decide on the relative consequences of the two types of errors. Some spam-filtering software lets the user adjust how aggressively it defines spam, so that each user can choose whether he or she would rather have more spam in the inbox or more good mail in the spam folder.

Example: Trial by jury

Table 16.4 continues the analogy between statistical significance and the legal system. The relative consequences of Type I and Type II errors depend on the type of trial.

	DECISION: SEND TO SPAM FOLDER	DECISION: PLACE IN INBOX
Good email	Type I error	(No error)
Spam	(No error)	Type II error

Table 16.3. Type I and Type II errors in the context of spam filters.

	VERDICT: GUILTY	VERDICT: NOT GUILTY
Did not commit the crime	Type I error	(No error)
Did commit the crime	(No error)	Type II error

Table 16.4. Type I and Type II errors in the context of trial by jury in a criminal case.

In the United States (and many other countries), a defendant in a criminal trial is considered innocent until proven guilty "beyond a reasonable doubt." This system is based on the principle that it is better to let many guilty people go free than to falsely convict one innocent person. The system is designed to avoid Type I errors in criminal trials, even at the expense of many Type II errors. You could say that α is set to a very low value.

In civil trials in the United States (and many other countries), the court or jury rules in favor of the plaintiff if the evidence shows that the plaintiff is "more likely than not" to be right. The thinking is that it is no worse to falsely rule in favor of the plaintiff than to falsely rule in favor of the defendant. The legal system in these countries attempts to equalize the chances of Type I and Type II errors in civil trials.

Example: Detecting athletes that dope

When an athletic association tests an athlete for using a banned drug (doping), there are two kinds of mistakes it can make (Fung, 2011). A false positive (Type I error) is concluding that the athlete had used a banned drug when in fact he or she hadn't done so. This can ruin an athlete's reputation and career. A false negative (Type II error) is concluding that an athlete had not been using a banned drug when she or he actually had done so. This is unfair to other athletes who follow the rules. The officials for each sport need to decide where to draw the line.

WHAT SIGNIFICANCE LEVEL SHOULD YOU CHOOSE?

P < 0.05?

Many scientific fields strictly and consistently define statistical significance as $P < 0.05$. This threshold came from Ronald Fisher (1935), an early pioneer of statistics, but he did not intend this definition to be strict:

> . . . it is convenient to draw the line at about the level at which we can say: "Either there is something in the treatment, or a coincidence has occurred such as does not occur more than once in twenty trials". If one in twenty does not seem high enough odds, we may, if we prefer it, draw the line at one in fifty (the 2 per cent point), or one in a hundred (the 1 per cent point).

P < 0.005?

Benjamin (2017) and 71 colleagues urge scientists to use a stricter threshold, "We propose to change the default P value threshold for statistical significance for claims of new discoveries from 0.05 to 0.005." Their goal is to reduce the False Positive Reporting Rate (FPRP) as will be explained in the next chapter. The tradeoff, of course, is that statistical power will be reduced unless sample size is increased. They note that increasing sample size by about 70% can maintain statistical power while changing the significance level from 0.05 to 0.005. Johnson (2013) made a similar suggestion.

P < 0.0000003?

A large international group of physicists announced in 2012 the discovery of the Higgs boson. The results were not announced in terms of a P value or statistical significance. Instead, the group announced that two separate experiments met the *five-sigma threshold* (Lamb, 2012).

Meeting the five-sigma threshold means that the results would occur by chance alone as rarely as a value sampled from a Gaussian distribution would be 5 SD (sigma) away from the mean. How rare is that? The probability is 0.0000003, or 0.00003%, or about 1 in 3.5 million, as computed with the Excel function "=1-normsdist(5)." In other words, the one-tailed P value is less than 0.0000003. If the Higgs boson doesn't exist, that probability is the chance that random sampling of data would give results as striking as the physicists observed. Two separate groups obtained data with this level of evidence.

The standard of statistical significance in most fields is that the two-tailed P is less than 0.05. About 5% of a Gaussian distribution is more than 2 SD away from the mean (in either direction). So the conventional definition of statistical significance can be called a *two-sigma threshold.*

The tradition in particle physics is that the threshold to report "evidence of a particle" is P < 0.003 (three sigma), and the standard to report a "discovery of a particle" is P < 0.0000003 (five sigma). These are one-tailed P values.

Choose and justify the significance level

The key point is this: There is no need for one significance threshold for all purposes. A large international group of physicists announced in 2012 the discovery of the Higgs boson. The results were not announced in terms of a P value or statistical significance. Instead, the group announced that two separate experiments met the *five-sigma threshold* (Lamb, 2012).

Meeting the five-sigma threshold means that the results would occur by chance alone as rarely as a value sampled from a Gaussian distribution would be 5 SD (sigma) away from the mean. How rare is that? The probability is 0.0000003, or 0.00003%, or about 1 in 3.5 million, as computed with the Excel function "=1-normsdist(5)." In other words, the one-tailed P value is less than 0.0000003. If the Higgs boson doesn't exist, that probability is the chance that random sampling of data would give results as striking as the physicists observed. Two separate groups obtained data with this level of evidence.

The standard of statistical significance in most fields is that the two-tailed P is less than 0.05. About 5% of a Gaussian distribution is more than 2 SD away from the mean (in either direction). So the conventional definition of statistical significance can be called a *two-sigma threshold.*

The tradition in particle physics is that the threshold to report "evidence of a particle" is P < 0.003 (three sigma), and the standard to report a "discovery of a particle" is P < 0.0000003 (five sigma). These are one-tailed P values.

INTERPRETING A CI, A P VALUE, AND A HYPOTHESIS TEST

Table 16.5 points out the very different questions that are answered by a CI, by a P value, and by a statistical hypothesis test (adapted from Goodman 2016).

STATISTICAL SIGNIFICANCE VS. SCIENTIFIC SIGNIFICANCE

A result is said to be statistically significant when the calculated P value is less than an arbitrary preset value of α. A conclusion that a result is statistically significant does not mean the difference is large enough to be interesting or intriguing enough to be worthy of further investigation so does not mean the finding is scientifically or clinically significant.

In an article that defines many new "diseases" that scientists are plagued by, Casadevall and Fang (2014) defined the disease *significosis*, which is a condition you should avoid! Significosis is manifested by a failure to discern between biological and statistical significance. Individuals with significosis fail to realize that just because something is unlikely to have occurred by chance doesn't mean it's important.

To avoid the ambiguity in the word *significant*, never use that word by itself. Always use the phrase "statistically significant" when referring to statistical hypothesis testing and phrases such as "scientifically significant" or "clinically significant" when referring to the size of a difference (or association).

Even better, avoid the word "significant" altogether. It is not needed. When writing about statistical hypothesis testing, state the P value and whether the null hypothesis can be rejected. When discussing the importance or impact of data, banish the word *significant* and instead use words such as *consequential, eventful, meaningful, momentous, large, big, important, substantial, remarkable, valuable, worthwhile, impressive, major*, or *prominent*.

COMMON MISTAKES: STATISTICAL HYPOTHESIS TESTING

Mistake: Believing that statistical hypothesis testing is an essential part of all statistical analyses

The whole point of statistical hypothesis testing is to make a crisp decision from one result. Make one decision when the results are statistically significant and another decision when the results are not statistically significant. This situation is common in quality control but rare in exploratory basic research.

Approach	Question answered
Confidence interval	What range of true effects is consistent with the data?
P value	How much evidence is there that the true effect is not zero?
Hypothesis test	Is there enough evidence to make a decision based on a conclusion that the effect is not zero.

Table 16.5. Questions answered by a CI, a P value and a hypothesis tests

In many scientific situations, it is not necessary—and can even be counterproductive—to reach a crisp conclusion that a result is either statistically significant or not statistically significant. P values and CIs can help you assess and present scientific evidence without ever using the phrase *statistically significant*.

When reading the scientific literature, don't let a conclusion about statistical significance stop you from trying to understand what the data really show. Don't let the rubric of statistical hypothesis testing get in the way of clear scientific thinking. Because statistical significance is often shown on graphs and tables as asterisks or stars, the process of overemphasizing statistical significance is sometimes referred to as *stargazing*. To avoid stargazing, ignore statements of statistical significance when you first read a paper and focus instead on the size of the effect (difference, association, etc.) and its CI.

Mistake: Believing that if a result is statistically significant, the effect must be large

The conclusion that something is statistically significant applies to the strength of the evidence, not the size of the effect. Statistical hypothesis testing answers this question: Is there significant evidence of a difference? It does not answer this question: Is there evidence of a scientifically significant difference? (Marsupial, 2013). If the sample size is large, even tiny, inconsequential differences will be statistically significant.

Once some people hear the word *significant*, they often stop thinking about what the data actually show. Belief in the usefulness of statistical significance has even been called a cult (Ziliak & McCloskey, 2008). Be sure you know what a conclusion that a result is statistically significant does not mean. This conclusion:

- Does *not* mean the difference is large enough to be interesting
- Does *not* mean the results are intriguing enough to be worthy of further investigations
- Does *not* mean that the finding is scientifically or clinically significant

Mistake: Not realizing how fragile statistical significance can be

Walsh and colleagues (2014) defined the concept of fragility of a conclusion that a result is statistically significant. They reviewed many clinical trials where the outcome was binary and the treatment was found to have a statistically significant effect on that outcome. In about 25% of these trials, if the result in three or fewer of the subjects had gone the other way, the results would have been not statistically significant. Because the conclusion that the results are statistically significant could so easily have gone the other way, they called those conclusions "fragile."

Mistake: P-hacking to obtain statistical significance

One problem with statistical hypothesis testing is that investigators may be tempted to work hard to push their P value to be low enough to be declared significant. Gotzsche (2006) used a clever approach to quantify this. He figured that if results were presented honestly, the number of P values between 0.04 and 0.05 would be similar to the number of P values between 0.05 and 0.06. However, in 130 abstracts of papers published in 2003, he found five times more P values between 0.04 and 0.05 than between 0.05 and 0.06.

Masicampo and Lalande (2012) found similar results. They reviewed papers in three highly regarded, peer-reviewed psychology journals. The distribution of 3,627 P values shows a clear discontinuity of 0.05. There are too many P values just smaller than 0.05 and too many greater than 0.05. They concluded, "The number of P values in the psychology literature that barely meet the criterion for statistical significance (i.e., that fall just below .05) is unusually large, given the number of P values occurring in other ranges."

Some of the investigators probably did the analyses properly and simply didn't write up results when the P value was just a bit higher than 0.05. And some may have had the borderline results rejected for publication. You can read more about publication bias in Chapter 43. It also is likely, I think, that many investigators cheat a bit by:

- Tweaking. The investigators may have played with the analyses. If one analysis didn't give a P value less than 0.05, then they tried a different one. Perhaps they switched between parametric and nonparametric analysis or tried including different variables in multiple regression models. Or perhaps they reported only the analyses that gave P values less than 0.05 and ignored the rest.
- Dynamic sample size. The investigators may have analyzed their data several times. Each time, they may have stopped if the P value was less than 0.05 but collected more data when the P value was above 0.05. This approach would yield misleading results, as it is biased toward stopping when P values are small.
- Slice and dice. The investigators may have analyzed various subsets of the data and only reported the subsets that gave low P values.
- Selectively report results of multiple outcomes. If several outcomes were measured, the investigators may have chosen to only report the ones for which the P value was less than 0.05.
- Play with outliers. The investigators may have tried using various definitions of outliers, reanalyzed the data several times, and only reported the analyses with low P values.

Simmons, Nelson, and Simonsohn (2012) coined the term *P-hacking* to refer to attempts by investigators to lower the P value by trying various analyses or by analyzing subsets of data.

Q & A

Is it possible to report scientific data without using the word *significant*?
Yes. Report the data, along with CIs and perhaps P values. Decisions about statistical significance are often not helpful.

Is the concept of statistical hypothesis testing about making decisions or about making conclusions?
Decision-making. The system of statistical hypothesis testing makes perfect sense when it is necessary to make a crisp decision based on one statistical analysis. If you have no need to make a decision from one analysis, then it may not be helpful to use the term *statistically significant*.

But isn't the whole point of statistics to decide when an effect is statistically significant?

No. The goal of statistics is to quantify scientific evidence and uncertainty.

Why is statistical hypothesis testing so popular?

There is a natural aversion to ambiguity. The crisp conclusion "The results are statistically significant" is more satisfying to many than the wordy conclusion "Random sampling would create a difference this big or bigger in 3% of experiments if the null hypothesis were true."

Who invented the threshold of P < 0.05 as meaning statistically significant?

That threshold, like much of statistics, came from the work of Ronald Fisher.

Are the P value and α the same?

No. A P value is computed from the data. The significance level α is chosen by the experimenter as part of the experimental design before collecting any data. A difference is termed statistically significant if the P value computed from the data is less than the value of α set in advance.

Is α the probability of rejecting the null hypothesis?

Only if the null hypothesis is true. In some experimental protocols, the null hypothesis is often true (or close to it). In other experimental protocols, the null hypothesis is almost certainly false. If the null hypothesis is actually true, α is the probability that random sampling will result in data that will lead you to reject the null hypothesis, thus making a Type I error.

If I perform many statistical tests, is it true that the conclusion "statistically significant" will be incorrect 5% of the time?

No! That would only be true if the null hypothesis is, in fact, true in every single experiment. It depends on the scientific context.

My two-tailed P value is not low enough to be statistically significant, but the one-tailed P value is. What do I conclude?

Stop playing games with your analysis. It is only OK to compute a one-tailed P value when you decided to do so as part of the experimental protocol (see Chapter 15).

Isn't it possible to look at statistical hypothesis testing as a way to choose between alternative models?

Yes. See Chapter 35.

What is the difference between Type I and Type II errors?

Type I errors reject a true null hypothesis. Type II errors accept a false null hypothesis.

My P value to four digits is 0.0501. Can I round to 0.05 and call the result statistically significant?

No. The whole idea of statistical hypothesis test is to make a strict criterion (usually at P=0.05) between rejecting and not rejecting the null hypothesis. Your P value is greater than α, so you cannot reject the null hypothesis and call the results statistically significant.

My P value to nine digits is 0.050000000. Can I call the result statistically significant?

Having a P value equal 0.05 (to many digits) is rare, so this won't come up often. It is just a matter of definition. But I think most would reject the null hypothesis when a P value is exactly equal to α.

CHAPTER SUMMARY

- Statistical hypothesis testing reduces all findings to two conclusions, "statistically significant" or "not statistically significant."
- The distinction between the two conclusions is made purely based on whether the computed value of a P value is less than or not less than an arbitrary threshold, called the significance level.
- A conclusion of statistical significance does not mean the difference is large enough to be interesting, does not mean the results are intriguing enough to be worthy of further investigations, and does not mean that the finding is scientifically or clinically significant.
- The concept of statistical significance is often overemphasized. The whole idea of statistical hypothesis testing only is useful when a crisp decision needs to be made based on one analysis.
- Avoid the use of the term *significant* when you can. Instead, talk about whether the P value is low enough to reject the null hypothesis and whether the effect size (or difference or association) is large enough to be important.
- Some scientists use the phrase *very significant* or *extremely significant* when a P value is tiny and *borderline significant* when a P value is just a little bit greater than 0.05.
- While it is conventional to use 0.05 as the threshold P value (called α) that separates statistically significant from not statistically significant, this value is totally arbitrary and some scientists use a much smaller threshold.
- Ideally, one should choose α based on the consequences of Type I (false positive) and Type II (false negative) errors.
- It is important to distinguish the P value from the significance level α. A P value is computed from data, while the value of α is (or should be) decided as part of the experimental design.
- There is more—a lot more—to statistics than deciding if an effect is statistically significant or not statistically significant.

TERMS DEFINED IN THIS CHAPTER

- Alpha (α) (p. 145)
- Five-sigma threshold (p. 151)
- P-hacking (p. 154)
- Significance level (p. 145)
- Stargazing (p. 153)
- Statistical hypothesis testing (p. 145)
- Statistically significant (p. 145)
- Type I error (p. 148)
- Type II error (p. 148)
- Type III error (p. 148)
- Type S error (p. 148)

CHAPTER 17

Comparing Groups with Confidence Intervals and P Values

Reality must take precedence over public relations, for Nature cannot be fooled.

RICHARD FEYNMAN

This book has presented the concepts of CIs and statistical hypothesis testing in separate chapters. The two approaches are based on the same assumptions and the same mathematical logic. This chapter explains how they are related.

CIS AND STATISTICAL HYPOTHESIS TESTING ARE CLOSELY RELATED

CIs and statistical hypothesis testing are based on the same statistical theory and assumptions, so they are closely linked.

The CI computes a range that 95% of the time will contain the population value (given some assumptions).

The hypothesis testing approach computes a range that you can be 95% sure would contain experimental results if the null hypothesis were true. Any result within this range is considered not statistically significant, and any result outside this range is considered statistically significant.

When a CI includes the null hypothesis

Figure 17.1 shows the results of comparing observed body temperatures (n = 12) with the hypothetical mean value of 37°C (see Chapter 7).

The bottom bar of Figure 17.1 shows the 95% CI of the mean. It is centered on the sample mean. It extends in each direction a distance computed by multiplying the SEM by the critical value of the t distribution from Appendix D (for 95% confidence).

The top bar in Figure 17.1 shows the range of results that would not be statistically significant (P > 0.05). It is centered on the null hypothesis, in this case, that the mean body temperature is 37°C. It extends in each direction exactly the same distance as the other bar, equal to the product of the SEM times the critical value of the t distribution.

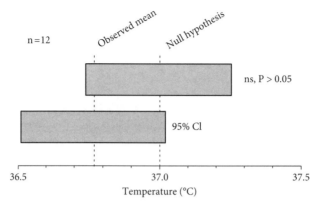

Figure 17.1. Comparing observed body temperatures (n = 12) with the hypothetical mean value of 37°C.

The top bar shows the range of results that are not statistically significant (P > 0.05). The bottom bar shows the 95% CI of the mean and is centered on the sample mean. The lengths of the two bars are identical. Because the 95% CI contains the null hypothesis, the zone of statistically not significant results must include the sample mean.

The two bars have identical widths but different centers. In this example, the 95% CI contains the null hypothesis (37°C). Therefore, the zone of statistically not significant results must include the sample result (in this case, the mean).

When a CI does not include the null hypothesis

Figure 17.2 shows the n = 130 data but is otherwise similar to Figure 17.1.

The bottom bar of Figure 17.2 shows the 95% CI of the mean. It is centered on the sample mean. It extends in each direction a distance computed by

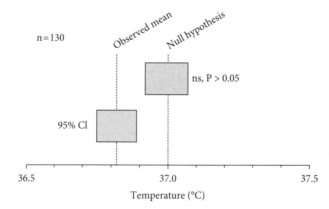

Figure 17.2. Comparing observed body temperatures (n = 130) with the hypothetical mean value of 37°C.

Because the sample size is larger, the CI is narrower than it is in Figure 17.1, as is the zone of possible results that would not be statistically significant. Because the 95% CI does not contain the null hypothesis, the zone of not statistically significant results cannot contain the sample mean. Whether the two boxes overlap is not relevant.

multiplying the SEM by the critical value of the t distribution from Appendix D (for 95% confidence). This bar is much shorter than the corresponding bar in Figure 17.1. The main reason for this is that the SEM is much smaller because of the much larger sample size. Another reason is that the critical value from the t distribution is a bit smaller (1.98 vs. 2.20) because the sample size (and number of dfs) is larger.

The top bar in Figure 17.2 shows the range of results that would not be statistically significant (P > 0.05). It is centered on the null hypothesis, in this case, that the mean body temperature is 37°C. It extends in each direction exactly the same distance as the other bar, equal to the product of the SEM times the critical value of the t distribution.

In this example, the 95% CI does not contain the null hypothesis. Therefore, the zone of statistically not significant results cannot include the sample mean.

A rule that links CIs and statistical significance

Here is a general rule that you should remember:

- If a 95% CI does not contain the value of the null hypothesis, then the result must be statistically significant, with P < 0.05.
- If a 95% CI does contain the value of the null hypothesis, then the result must not be statistically significant, with P > 0.05.

There is nothing special about 95% CIs and a significance level of 5%. The rule also works like this: if the 99% CI does not contain the null hypothesis value, then the P value must be less than 0.01.

In the examples, the outcome was the difference between the sample mean and a hypothetical population mean. This rule works for many other kinds of outcomes. For example:

- If the CI for the difference between two means does not include zero (the null hypothesis), then the result must be statistically significant (P < 0.05).
- If the CI for the ratio of two proportions does not include 1.0 (the null hypothesis), then the result must be statistically significant (P < 0.05).
- If you are comparing a set of percentages against a hypothetical value of 100 and the 95% CI of the mean of the percentages does not include 100, then the discrepancy must be statistically significant (P < 0.05).

FOUR EXAMPLES WITH CIS, P VALUES, AND CONCLUSION ABOUT STATISTICAL SIGNIFICANCE

Previous chapters explained the concept of using a CI to quantify how precisely you have determined a proportion or mean from a sample of data. When comparing two treatments, you can quantify the precision of the effect of the treatment, often a difference or ratio.

Example 1: Does apixaban prevent thromboembolism?

Background

Agnelli and colleagues (2013) tested whether apixaban (a new anticoagulant or "blood thinner") would be beneficial in extended treatment of patients with a venous thromboembolism (a clot in the leg vein), which can be dangerous if it breaks loose and goes to the lung. After all the patients completed the standard anticoagulation therapy, the investigators randomly divided them into two groups that were assigned to receive either placebo or apixaban. The investigators followed these patients for a year and recorded whether they experienced another thromboembolism.

CI

The thromboembolism recurred in 8.8% of the patients receiving the placebo (73/829) but in only 1.7% of the people receiving apixaban (14/840). So patients who received apixaban were a lot less likely to have a recurrent thromboembolism. The drug worked!

As explained in Chapter 4, we could compute the CI for each of those proportions (8.8% and 1.7%), but the results are better summarized by computing the ratio of the two proportions and its CI. This ratio is termed the *risk ratio*, or the *relative risk* (to be discussed in Chapter 27). The ratio is 1.7/8.8, which is 0.19. Patients receiving the drug had 19% the risk of a recurrent thromboembolism compared to patients receiving the placebo.

That ratio is an exact calculation for the patients in the study. A 95% CI generalizes the results to the larger population of patients with thromboembolism similar to those in the study. This book won't explain how the calculation is done, but many programs will report the result. The 95% CI of this ratio extends from 0.11 to 0.33. If we assume the patients in the study are representative of the larger population of adults with thromboembolism, then we are 95% confident that treatment with apixaban will reduce the incidence of disease progression to somewhere between 11% and 33% of the risk in untreated patients. In other words, patients in the study taking apixaban had about one-fifth the risk of another clot than those taking placebo did, and if we studied a much larger group of patients (and accepted some assumptions), we could be 95% confident that the risk would be somewhere between about one-ninth and one-third the risk of those taking placebo.

Is reducing the risk of thromboembolism to one-fifth its control value a large, important change? That is a clinical question, not a statistical one, but I think anyone would agree that is a substantial effect. The 95% CI goes as high as 0.33. If that were the true effect, a reduction in risk down to only one-third of its control value, it would still be considered substantial. When interpreting any results, you also have to ask whether the experimental methods were sound (I don't see any problem with this study) and whether the results are plausible (yes, it makes sense that extending treatment with a blood thinner might prevent thromboembolism). Therefore, we can conclude with 95% confidence that apixaban substantially reduces the recurrence of thromboembolism in this patient population.

P value

To interpret the P value, we first must tentatively assume that the risk of thromboembolism is the same in patients who receive an anticoagulant as in those who receive placebo and that the discrepancy in the incidence of thromboembolism observed in this study was the result of chance. This is the null hypothesis. The P value answers the following question:

> If the risk of thromboembolism overall is identical in the two groups, what is the chance that random sampling (in a study as large as ours) would lead to a ratio of incidence rates as far or farther from 1.0 as the ratio computed in this study? (Why 1.0? Because a ratio of 1.0 implies no treatment effect.)

The authors did the calculation using a test called the *Fisher exact test* and published the answer as P < 0.0001. When P values are tiny, it is traditional to simply state that they are less than some small value—here, 0.0001. Since this P value is low and since the hypothesis that an anticoagulant would prevent thromboembolism is quite sensible, the investigators concluded that the drug worked to prevent recurrent thromboembolism. This example is discussed in more detail in Chapter 27.

Statistical significance

The P value is less than 0.05 and the CI of the relative risk does not include 1.0 (no increased risk). So the results can be deemed to be statistically significant.

Example 2: Is the maximal bladder relaxation different in old and young rats?

Background

Frazier, Schneider, and Michel (2006) compared old and young rats to see how well the neurotransmitter norepinephrine relaxes bladder muscles. Figure 17.1 shows the maximal relaxation that can be achieved by large concentrations of norepinephrine in old and young rats. These values are the percentage of relaxation, where the scale from 0% to 100% is defined using suitable experimental controls.

CI

The values for the two groups of rats overlap considerably, but the two means are distinct. The mean maximum response in old rats is 23.5% lower than the mean maximum response in young rats. That value is exactly correct for our sample of data, but we know the true difference in the populations is unlikely to equal 23.5%. To make an inference about the population of all similar rats, look at the 95% CI of the difference between means. This is done as part of the calculations of a statistical test called the *unpaired t test* (not explained in detail in this book). This CI (for the mean value measured in the young rats minus the mean value observed in the old rats) ranges from 9.3 to 37.8, and it is centered on the difference we observed in our sample. Its width depends on the sizes (number of rats) and the variability (standard deviation) of the two samples and on the degree of confidence you want (95% is standard).

Because the 95% CI does not include zero, we can be at least 95% confident that the mean response in old rats is less than the mean response in young ones. Beyond that, the interpretation needs to be made in a scientific context. Is a difference of 23.5% physiologically trivial or large? How about 9.3%, the lower limit of the CI? These are not statistical questions. They are scientific ones. The investigators concluded this effect is large enough (and is defined with sufficient precision) to perhaps explain some physiological changes with aging.

P value

The null hypothesis is that both sets of data are randomly sampled from populations with identical means. The P value answers the following question:

> If the null hypothesis is true, what is the chance of randomly observing a difference as large as or larger than the one observed in this experiment of this size?

The P value depends on the difference between the means, on the SD of each group, and on the sample sizes. A test called the *unpaired t test* (also called the *two-sample t test*) reports that the P value equals 0.003. If old and young rats have the same maximal relaxation of bladder muscle by norepinephrine overall, there is only a 0.03% chance of observing such a large (or a larger) difference in an experiment of this size. This example will be discussed in more detail in Chapter 31.

Statistical significance

The P value is less than 0.05 and the CI of the difference between means does not include 0.0. So the results can be deemed to be statistically significant.

Example 3: Do selective serotonin reuptake inhibitors increase the risk of autism?

CI

Hviid and colleagues (2013) asked whether there is an association between use of selective serotonin reuptake inhibitors (SSRIs, a treatment for depression) by pregnant women and autism in their children. They followed 626,875 babies from birth and identified 3,892 kids diagnosed with autism and 6,068 mothers who had used SSRIs during pregnancy. They used a fairly sophisticated method to analyze the data, but it is easy to understand the results without understanding the details of the methodology. The risk of autism was 1.2 times higher among children of women who took SSRIs compared to the risk in children of those who didn't.

To generalize to the larger population, the investigators computed a 95% CI for that risk ratio, which ranged from 0.9 to 1.6. A ratio of 1.0 would mean that the risk of having a child with autism is the same in women who took SSRIs as in those who didn't. The CI extends from a risk of less than 1.0 (which would mean SSRIs are associated with a lower risk of autism) to a value of greater than 1.0 (which would mean SSRIs are associated with a higher risk of autism).

Clearly, these results do not demonstrate an association between use of SSRIs during pregnancy and autism. But do they convincingly show that there is no association? Not really. The 95% CI goes up to 1.6, which represents a 60% increase in the risk of autism. That is a lot of risk. The investigators therefore do not give a firm negative conclusion but instead conclude (p. 2406), "On the basis of the upper boundary of the CI, our study could not rule out a relative risk up to 1.6, and therefore the association warrants further study."

P Value and statistical significance

The authors presented only the CI and not a P value. Since the 95% CI included a relative risk of 1.0 (no association), we know the P value testing the hypothesis of no association must be greater than 0.05. Accordingly, if you wanted to use the rubric of statistical hypothesis testing, you'd conclude the association is not statistically significant. But to interpret that conclusion, you really need to look at the CI and see how high it goes.

Example 4: Does tight hyperglycemic control benefit children in intensive care?

CI

When hospitalized patients have hyperglycemia (high blood sugar), it is hard to know what to do. Administering insulin can maintain blood glucose at normal levels, but this creates a danger of hypoglycemia (too little glucose in the blood), which has negative consequences. Macrae and colleagues (2014) did a controlled study asking whether tight hyperglycemic control (more, or more frequent, insulin) benefited children in intensive care. They randomized 1,369 patients to receive either tight or conventional glucose control. The primary outcome they tabulated was the number of days alive and not on a mechanical ventilator. On average, the patients who received the tighter control of hypergly-cemia remained alive (and off a ventilator) 0.36 days longer than patients given standard glucose control.

To generalize their findings to the larger population of children in intensive care, they presented the 95% CI, which ranged from −0.42 to 1.14 days. The 95% CI includes zero, so we can say with 95% confidence there is no evidence that the tighter glucose control was beneficial. But is this CI narrow enough that these findings can be considered a solid negative result? The authors stated that they would have considered a difference of two days to be clinically relevant. The upper limit of the CI was only a bit more than one day. So the authors concluded that these results show no evidence for a difference between the two treatments, and if there is a difference it is very likely to not be clinically relevant. These are solid negative results. (The investigators analyzed the data in several ways and also looked at subgroups of patients. The final conclusions are a bit more nuanced than those presented here.)

P value and statistical significance

The authors presented only the CI and not a P value. Since the 95% CI included zero days (no difference), we know the P value testing the hypothesis of no difference must be greater than 0.05. Accordingly, if you wanted to use the rubric of statistical hypothesis testing, you'd conclude the difference is not statistically significant.

Q & A

If the 95% CI just barely reaches the value that defines the null hypothesis, what can you conclude about the P value?

> If the 95% CI includes the value that defines the null hypothesis, you can conclude that the P value is greater than 0.05. If the 95% CI excludes the null hypothesis value, you can conclude that the P value is less than 0.05. So if the 95% CI ends right at the value that defines the null hypothesis, then the P value must equal 0.05.

If the 95% CI is centered on the value that defines the null hypothesis, what can you conclude about the P value?

> The observed outcome equals the value that defines the null hypothesis. In this case, the two-tailed P value must equal 1.000.

The 99% CI includes the value that defines the null hypothesis, but the P value is reported to be < 0.05. How is this possible?

> If the 99% CI includes the value that defines the null hypothesis, then the P value must be greater than 0.01. But the P value was reported to be less than 0.05. You can conclude that the P value must be between 0.01 and 0.05.

The 99% CI includes the value that defines the null hypothesis, but the P value is reported to be < 0.01. How is this possible?

> It is inconsistent. Perhaps the CI or P value was calculated incorrectly. Or perhaps thedefinitionofthenullhypothesisincludedinthe99%CIisnotthesamedefinitionused to compute the P value.

Two of the examples came from papers that reported CIs, but not P values or conclusions about statistical significance. Isn't this incomplete reporting?

> No. In many cases, knowing the P value and a conclusion about statistical significance really adds nothing to understanding the data. Just the opposite. Conclusions about statistical significance often act to reduce careful thinking about the size of the effect.

CHAPTER SUMMARY

- CIs and statistical hypothesis testing are closely related.
- If the 95% CI includes the value stated in the null hypothesis, then the P value must be greater than 0.05.
- If the 95% CI does not include the value stated in the null hypothesis, then the P value must be less than 0.05.
- The same rule works for 99% CIs and P < 0.01, or 90% CIs and P < 0.10.

CHAPTER 18

Interpreting a Result That Is Statistically Significant

Facts do not "speak for themselves," they are read in the light of theory.

STEPHEN JAY GOULD

When you see a result that is "statistically significant," don't stop thinking. That phrase is often misunderstood, as this chapter explains. All it means is that the calculated P value is less than a preset threshold you set. This means that the results would be surprising (but not impossible) if the null hypothesis were true. A conclusion of statistical significance does not mean the difference is large enough to be interesting or worthy of follow-up, nor does it mean that the finding is scientifically or clinically significant.

SEVEN EXPLANATIONS FOR RESULTS THAT ARE "STATISTICALLY SIGNIFICANT"

Let's consider a simple scenario: comparing enzyme activity in cells incubated with a new drug to enzyme activity in control cells. Your scientific hypothesis is that the drug increases enzyme activity, so the null hypothesis is that there is no difference. You run the analysis, and, indeed, the enzyme activity is higher in the treated cells. You run an unpaired t test (explained in Chapter 30). Because the P value is less than 0.05, you conclude the result is statistically significant. The following discussion provides seven possible explanations for why this happened.

Explanation 1: The drug had a substantial effect

Scenario: The drug really does have a substantial effect to increase the activity of the enzyme you are measuring.

Discussion: This is what everyone thinks when they see a small P value in this situation. But it is only one of seven possible explanations.

If you are interpreting results you read about in a scientific journal, the published effect size is likely to be larger than the true effect. Why? If investigators did lots of experiments, some will have larger effects and some smaller, and the average is the true effect size. Some of those experiments will reach a conclusion that the result is "statistically significant" and others won't. The first set will, on average, show effect sizes larger than the true effect size, and these are the experiments that are most likely to be published. Other experiments that wound up with smaller effects may never get written up. Or if they are written up, they may not get accepted for publication. So published "statistically significant" results tend to exaggerate the effect being studied (Gelman, & Carlin, 2014).

Explanation 2: The drug had a trivial effect

Scenario: The drug actually has a very small effect on the enzyme you are studying even though your experiment yielded a P value of less than 0.05.

Discussion: The small P value just tells you that the difference is large enough to be surprising if the drug really didn't affect the enzyme at all. It does not imply that the drug has a large treatment effect. A small P value with a small effect will happen with some combination of a large sample size and low variability. It will also happen if the treated group has measurements a bit above their true average and the control group happens to have measurements a bit below their true average, causing the effect observed in your particular experiment to happen to be larger than the true effect.

Explanation 3: The result is a false positive, also called a Type I error or false discovery

Scenario: The drug really doesn't affect enzyme expression at all. Random sampling just happened to give you higher values in the cells treated with the drug and lower levels in the control cells. Accordingly, the P value was small.

Discussion: This is called a *Type I error, false positive,* or *false discovery.* If the drug really has no effect on the enzyme activity (if the null hypothesis is true) and you choose the traditional 0.05 significance level ($\alpha = 5\%$), you'll expect to make a Type I error 5% of the time. Therefore, it is easy to think that you'll make a Type I error 5% of the time when you conclude that a difference is statistically significant. However, this is not correct. You'll understand why when you read about the false positive report probability (FPRP; the fraction of "statistically significant" conclusions that are wrong) later in this chapter.

Explanation 4: There was a Type S error

Scenario: The drug (on average) decreases enzyme activity, which you would only know if you repeated the experiment many times. In this particular experiment, however, random sampling happened to give you high enzyme activity in the drug-treated cells and low activity in the control cells, and this difference was large enough and consistent enough to be statistically significant.

Discussion: Your conclusion is backward. You concluded that the drug (on average) significantly increases enzyme activity when in fact the drug decreases enzyme activity. You've correctly rejected the null hypothesis that the drug does not influence enzyme activity, so you have not made a Type I error. Instead, you have made a *Type S error,* because the *sign* (plus or minus, increase or decrease) of the actual overall effect is opposite to the results you happened to observe in one experiment. Type S errors occur rarely but are not impossible (Gelman, & Carlin, 2014). They are also referred to as *Type III errors*.

Explanation 5: The experimental design was poor

Scenario: The enzyme activity really did increase in the tubes with the added drug. However, the reason for the increase in enzyme activity had nothing to do with the drug but rather with the fact that the drug was dissolved in an acid (and the cells were poorly buffered) and the controls did not receive the acid (due to bad experimental design). So the increase in enzyme activity was actually due to acidifying the cells and had nothing to do with the drug itself. The statistical conclusion was correct—adding the drug did increase the enzyme activity—but the scientific conclusion was completely wrong.

Discussion: Statistical analyses are only a small part of good science. That is why it is so important to design experiments well, to randomize and blind when possible, to include necessary positive and negative controls, and to validate all methods.

Explanation 6: The results cannot be interpreted due to ad hoc multiple comparisons

Scenario: You ran this experiment many times, each time testing a different drug. The results were not statistically significant for the first 24 drugs tested but the results with the twenty-fifth drug were statistically significant.

Discussion: These results would be impossible to interpret, as you'd expect some small P values just by chance when you do many comparisons. If you know how many comparisons were made (or planned), you can correct for multiple comparisons. But here, the design was not planned, so a rigorous interpretation is impossible.

Explanation 7: The results cannot be interpreted due to dynamic sample size

Scenario: You first ran the experiment in triplicate, and the result (n = 3) was not statistically significant. Then you ran it again and pooled the results (so now n = 6), and the results were still not statistically significant. So you ran it again, and, finally, the pooled results (with n = 9) were statistically significant.

Discussion: The P value you obtain from this approach simply cannot be interpreted. P values can only be interpreted at face value when the sample size, the experimental protocol, and all the data manipulations and analyses are planned in advance. Otherwise, you are P-hacking, and the results cannot be interpreted (see Chapter 16.)

Bottom line: When a P value is small, consider all the possibilities

Be cautious when interpreting small P values. Don't make the mistake of instantly believing the first explanation without also considering the possibility that the true explanation is one of the other six possibilities listed.

HOW FREQUENTLY DO TYPE I ERRORS
(FALSE POSITIVES) OCCUR?

What is a Type I error?

Type I errors (defined in Chapter 16) are also called *false discoveries* or *false positives*. You run an experiment, do an analysis, find that the P value is less than 0.05 (or whatever threshold you choose), and so reject the null hypothesis and conclude that the difference is statistically significant. You've made a Type I error when the null hypothesis is really true and the tiny P value was simply due to random sampling. If comparing two means, for example, random sampling may result in larger values in one group and smaller values in the other, with a difference large enough (compared to the variability and accounting for sample size) to be statistically significant.

There are two ways to quantify how often Type I errors occur:

The significance level (α)

The *significance level* (defined in Chapter 16) answers these equivalent questions:

- If the null hypothesis is true, what is the probability that a particular experiment will collect data that generate a P value low enough to reject that null hypothesis?
- Of all experiments you could conduct when the null hypothesis is actually true, in what fraction will you reach a conclusion that the results are statistically significant?

The False Positive Report Probability

The *False Positive Report Probability* (abbreviated FPRP), also called the False Positive Rate (abbreviated FPR, Colquhoun 2017) is the answer to these equivalent questions:

- If a result is statistically significant, what is the probability that the null hypothesis is really true?
- Of all experiments that reach a statistically significant conclusion, what fraction are false positives (Type I errors)?

The FPRP and the significance level are not the same

The significance level and the FPRP are the answers to distinct questions, so the two are defined differently and their values are usually different. To understand this conclusion, look at Table 18.1 which shows the results of many hypothetical statistical analyses that each reach a decision to reject or not reject the null hypothesis. The top row tabulates results for experiments in which the null hypothesis is true. The second row tabulates results for experiments in which the null hypothesis is not true. When you analyze data, you don't know whether the null hypothesis is true, so you could never actually create this table from an actual series of experiments.

	DECISION: REJECT NULL HYPOTHESIS	DECISION: DO NOT REJECT NULL HYPOTHESIS	TOTAL
Null hypothesis is true	A (Type I error)	B	A + B
Null hypothesis is false	C	D (Type II error)	C + D
Total	A + C	B + D	A + B + C + D

Table 18.1. The results of many hypothetical statistical analyses to reach a decision to reject or not reject the null hypothesis.

In this table, A, B, C, and D are integers (not proportions) that count numbers of analyses (number of P values). The total number of analyses equals A + B + C + D. The significance level is defined to equal A/(A + B). The FPRP is defined to equal A/(A + C).

The FPRP only considers analyses that reject the null hypothesis so only deals with the left column of the table. Of all these experiments (A + C), the number in which the null hypothesis is true equals A. So the FPRP equals A/(A + C).

The significance level only considers analyses where the null hypothesis is true so only deals with the top row of the table. Of all these experiments (A + B), the number of times the null hypothesis is rejected equals A. So the significance level is expected to equal A/(A + B).

Since the values denoted by B and C in Table 18.1 are unlikely to be equal, the FPRP is usually different than the significance level. This makes sense because the two values answer different questions.

THE PRIOR PROBABILITY INFLUENCES THE FPRP (A BIT OF BAYES)

What is the FPRP? Its value depends, in part, on the significance level and power you choose. But it also depends, in part, upon the context of the experiment. This is a really important point that will be demonstrated via five examples summarized in Table 18.2.

EXAMPLE	PRIOR PROBABILITY (%)	FPRP DEFINING "DISCOVERY" AS P < 0.05 (%)	FPRP DEFINING "DISCOVERY" AS P BETWEEN 0.045 AND 0.050 (%)
1	0.0%	100.0%	100%
2	1.0%	86.0%	97%
3	10.0%	36.0%	78%
4	50.0%	5.9%	27%
5	100.0%	0.0%	0%

Table 18.2. The false positive report probability (FPRP) depends on the prior probability and P value.

Examples 2 to 4 assume an experiment with sufficient sample size to have a power of 80%. All examples assume that the significance level was set to the traditional 5%. The FDR is computed defining "discovery" in two ways: as all results where the P value is less than 0.05, and as only those results where the P value is between 0.045 and 0.050 (based on Colquhoun, 2014.)

Example 1: Prior probability = 0%

In a randomized clinical trial, each subject is randomly assigned to receive one of two (or more) treatments. Before any treatment is given, clinical investigators commonly compare variables such as the hematocrit (blood count), fasting glucose (blood sugar), blood pressure, weight, and so on.

Since these measurements are made before any treatments are given, you can be absolutely, positively sure that both samples are drawn from the same population. In other words, you know for sure that all the null hypotheses are true. The *prior probability* (the probability before collecting data or running a statistical test) that there is a true difference between populations equals zero. If one of the comparisons results in a small P value, you know for sure that you have made a Type I error. In Table 18.1, C and D must equal zero, so A/(A + C), the FPRP, equals 1.0, or 100%.

Does it even make sense to compare relevant variables at the beginning of a clinical trial? Sure. If a clinically important variable differs substantially between the two randomly chosen groups before any treatment or intervention, it will be impossible to interpret the study results. You won't be able to tell if any difference in the outcome variable is due to treatment or to unlucky randomization giving you very different kinds of people in the two groups. So it makes sense to compare the groups in a randomized trial before treatment and ask whether the differences in relevant variables prior to treatment are large enough to matter, defining "large enough" either physiologically or clinically (but not based on a P value).

Example 2: Prior probability = 1%

Imagine that you have planned a large sensible experiment. But you have a crazy hypothesis that you want to test. It doesn't really fit any theory or prior data, so you know the chance that your theory will end up being true is about 1%. But it is easy enough to add one more drug to your experiment, so testing your crazy hypothesis won't cost much time or money.

What can we expect to happen if you test 1,000 such drugs?

- Of the 1,000 drugs screened, we expect that 10 (1%) will really work.
- Of the 10 drugs that really work, we expect to obtain a statistically significant result in 8 (because our experimental design has 80% power).
- Of the 990 drugs that are really ineffective, we expect to obtain a statistically significant result in 5% (because we set α equal to 0.05). In other words, we expect 5% × 990, or 49, false positives.
- Of 1,000 tests of different drugs, we therefore expect to obtain a statistically significant difference in 8 + 49 = 57. The FPRP equals 49/57 = 86%.

So even if you obtain a P value less than 0.05, there is an overwhelming chance that the results it a false positive. This kind of experiment is not worth doing unless you use a much stricter value for α (say 0.1% instead of 5%).

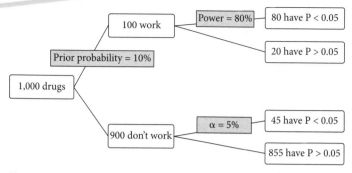

Figure 18.1. A probability tree of Example 3.
The FPRP equals 45/(80 + 45), or 36%.

Example 3: Prior probability = 10%

Imagine that you work at a drug company and are screening drugs as possible treatments for hypertension. You are interested in a mean decrease in blood pressure of 10 mmHg or more, and your samples are large enough that there is an 80% chance of finding a statistically significant difference (P < 0.05) if the true difference between population means is at least 10 mmHg. (You will learn how to calculate the sample size in Chapter 18.)

You test a drug that is known to weakly block angiotensin receptors, but the affinity is low and the drug is unstable. From your experience with such drugs, you estimate about a 10% chance that it will depress blood pressure. In other words, the prior probability that the drug works is 10%. What can we expect to happen if you test 1,000 such drugs? Figure 18.1 shows the probability tree.

- Of the 1,000 drugs screened, we expect that 100 (10%) will really work.
- Of the 100 drugs that really work, we expect to obtain a statistically significant result in 80 (because our experimental design has 80% power).
- Of the 900 drugs that are really ineffective, we expect to obtain a statistically significant result in 5% (because we set α equal to 0.05). In other words, we expect 5% × 900, or 45, false positives.
- Of 1,000 tests of different drugs, we therefore expect to obtain a statistically significant difference in 80 + 45 = 125. The FPRP equals 45/125 = 36%.

Example 4: Prior probability = 50%

In this example, the drug is much better characterized. The drug blocks the right kinds of receptors with reasonable affinity, and it is chemically stable. From your experience with such drugs, the prior probability that the drug is effective equals 50%. What can we expect to happen if you test 1,000 such drugs? Figure 18.2 shows the probability tree.

- Of those 1,000 drugs screened, we expect that 500 (50%) will really work.
- Of the 500 drugs that really work, we expect to obtain a statistically significant result in 400 (because our experimental design has 80% power).

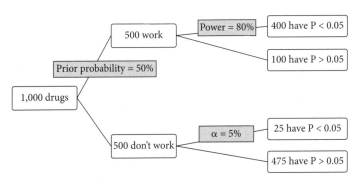

Figure 18.2. A probability tree of Example 4.
The FPRP equals 25/(400 + 25), or 5.9%.

- Of the 500 drugs that are really ineffective, we expect to obtain a statistically significant result in 5% (because we set α equal to 0.05). In other words, we expect 5% × 500, or 25, false positives.
- Of 1,000 tests of different drugs, therefore, we expect to obtain a statistically significant difference in 400 + 25 = 425. The FPRP equals 25/425 = 5.9%.

Example 5: Prior probability = 100%

You are running a screening assay to find a new drug that works on a particular pathway. You screen tens of thousands of compounds, and none (so far) work. To prove to yourself (and others) that your assay really will detect an active drug, you run a positive control every day that tests a drug already known to work on that pathway. This is important, because you want to make sure that all the negative results you obtain are the result of screening inactive drugs, not because the experimental system is simply not working on a given day (maybe you left out one ingredient).

The drug in the positive control always provokes a big response, so the P value will almost always less than 0.05. A false positive (Type I error) is impossible, because you are repeatedly testing a drug that is known to work. The FPRP is 0%.

The FPRP when the P value is just a tiny bit less than 0.05

So far, we have been defining a result to be a "discovery" when the P value is less than 0.05. That includes P values barely less than 0.05 as well as P values that are really tiny. Conclusions based on really tiny P values are less likely to be false positives than conclusions based on P = 0.049. The right column of Table 18.2 computes the FPRP defining a result to be a discovery when the P value is between 0.045 and 0.050 (based on Colquhoun, 2014). The FPRP values are much higher than they are when you define a discovery as being all P values less than 0.05.

Example 4 is for a situation where, before seeing any data, it is equally likely that there will be or will not be a real effect. The prior probability is 50%. If you observe a P value just barely less than 0.05 in this situation, the FDR is 27%. If you do a more exploratory experiment where the prior probability is only 10%, then the FPRP is 78%! These FPRPs are way higher than 5% that many people mistakenly expect them to be.

This is a really important point. A P value just barely less than 0.05 provides little evidence against the null hypothesis.

The FPRP is even higher when the power is low

Imagine that you are doing the experiments shown in Table 18.2, but you don't want to put so much effort into it. So you run the experiment with n = 3 in each of the two groups. You may not have bothered to do the calculation, but the power of this experimental design is only 15% (to detect a difference between population means equal to the SD, using a significance level of 5%). Let's do the calculations corresponding to the middle row of Table 18.2 with a prior probability of 10%.

- Of the 1,000 drugs, we expect that 100 (10%) will really work.
- Of the 100 drugs that really work, we expect to obtain a statistically significant result in 15 (because our experimental design with tiny sample size has only 15% power).
- Of the 900 drugs that are really ineffective, we expect to obtain a statistically significant result in 5% (because we set α equal to 0.05). In other words, we expect 5% × 900, or 45, false positives.
- Of 1,000 tests of different drugs, we therefore expect to obtain a statistically significant difference in 15 + 45 = 60. The FPRP equals 45/60 = 75%.

This analysis shows that if you have a scientific context with a low prior probability and an experimental design with low power, most "statistically significant" findings will be false positives.

What if you can't estimate the prior probability?

Estimating the prior probability can be difficult, maybe even impossible. For that reason, Colquhoun (2017) suggests an alternative approach.

1. Calculate the P value for a particular experiment or study.
2. Determine the power of that experimental design. For the examples below, I chose 80%.
3. Choose the highest value of the FPRP you would accept. For these examples, I choose 5% since many people incorrectly interpret P = 0.05 to mean FPRP = 5%.
4. Compute the prior probability required to have the desired FPRP with the actual P value and power.

Three examples computed using an online calculator linked from http://secure.graphpad.com/support/faqid/2069

- If P = 0.049 and you want the FPRP to be less than 5%, the prior probability must be greater than 87%. Yikes! To obtain a FPRP as low as 5% with a P value of 0.05, you must be almost sure (87% sure) that there is a non-zero effect before doing the experiment. This highlights that a P value close to 0.05 really provides little evidence against the null hypothesis.
- If P = 0.01 and you want the FPRP to be less than 5%, the prior probability must be greater than 55%.

- If P= 0.001 and you want the FPRP to be less than 5%, the prior probability must be greater than 16%.

Note these calculations were done for the exact P value specified, not for all possible P values less than the specified one.

BAYESIAN ANALYSIS

These calculations use a simplified form of a *Bayesian approach*, named after Thomas Bayes, who first published work on this problem in the mid-18th century. (Bayes was previously mentioned in Chapter 2.) The big idea of Bayesian analysis is to analyze data by accounting for prior probabilities.

There are some situations in which the prior probabilities are well defined, for example, in analyses of genetic linkage. The prior probability that two genetic loci are linked is known, so Bayesian statistics are routinely used in analysis of genetic linkage. There is nothing controversial about using Bayesian inference when the prior probabilities are known precisely.

In many situations, including the previous drug examples, the prior probabilities are little more than a subjective feeling. These feelings can be expressed as numbers (e.g., "99% sure" or "70% sure"), which are then treated as prior probabilities. Of course, the calculated results (the FPRPs) are no more accurate than are your estimates of the prior probabilities.

The examples in this chapter are a simplified form of Bayesian analysis. The rows of Table 18.2 show only two possibilities. Either the null hypothesis is true, or the null hypothesis is false and the true difference (or effect) is of a defined size. A full Bayesian analysis would consider a range of possible effect sizes, rather than only consider two possibilities.

ACCOUNTING FOR PRIOR PROBABILITY INFORMALLY

The examples in the previous section demonstrate that your interpretation of a statistically significant result depends on what you know before you collect the data—that is, the *prior probability* (see Chapter 2). Understanding the results requires combining the P value obtained from the experiment with the prior probability based on the context of the experiment or study.

Experienced scientists often account for the prior probability informally, without performing any additional calculations and without mentioning Bayes. When interpreting a result that is statistically significant, here are three scenarios that take into account prior probability:

- This study tested a hypothesis that is biologically sound and supported by previous data. The P value is 0.04. I have a choice of believing that the results occurred by a coincidence that will happen 1 time in 25 under the null hypothesis or of believing that the experimental hypothesis is true. Because the experimental hypothesis makes so much sense, I'll believe it. The null hypothesis is probably false.
- This study tested a hypothesis that makes no biological sense and has not been supported by any previous data. The P value is 0.04, which is lower

SAMPLE SIZE (PER GROUP)	P VALUE	RELATIVE RISK	95% CI
10	0.006	8.00	1.21 to 52.7
100	0.005	2.60	1.32 to 5.11
1,000	0.006	1.41	1.11 to 1.79
10,000	0.006	1.12	1.03 to 1.21

Table 18.3. Sample size matters when interpreting P values.

Four simulated experiments are tabulated, each with almost the same P value determined by comparing two proportions. In each experiment the proportion in the control group was 10%, which was compared to the proportion observed in the treated group (see Chapter 27). In addition to the sample size and P value, this table shows the relative risk (the ratio of the proportion in the treated group divided by the ratio in the control group) and its 95% CI. The P values for the top three rows were calculated by Fisher's exact test. The P value for the bottom row was calculated using a chi-square test. (Why is the P value 0.005 in the second row and not 0.006 like the others? Because with n = 100 and setting the control proportion to 10%, it is impossible to get a P value of 0.006.)

than the usual threshold of 0.05 but not by very much. I have a choice of believing that the results occurred by a coincidence that will happen 1 time in 25 under the null hypothesis or of believing that the experimental hypothesis is true. Because the experimental hypothesis is so unlikely to be correct, I think that the results are the result of coincidence. The null hypothesis is probably true.

• This study tested a hypothesis that makes no biological sense and has not been supported by any previous data. I'd be amazed if it turned out to be true. The P value is incredibly low (0.000001). I've looked through the details of the study and cannot identify any biases or flaws. These are reputable scientists, and I believe that they've reported their data honestly. I have a choice of believing that the results occurred by a coincidence that will happen 1 time in a million under the null hypothesis, or of believing that the experimental hypothesis is true. Although the hypothesis seems crazy to me, the data force me to believe it. The null hypothesis is probably false.

THE RELATIONSHIP BETWEEN SAMPLE SIZE AND P VALUES

A P value quantifies the probability of observing a difference or association as large (or larger) than actually observed if the null hypothesis were true. Table 18.3 demonstrates that the relationship between P value and the size of the observed effect depends on sample size. Each row in the table represents a simulated experiment that compares an observed proportion (the outcome tabulated by the experimenter) in control and treated groups. In each case, the observed proportion in the control group was 10%. The proportions in the expected group were chosen so that the P values in all the rows is about the same. In the top row, n = 10. The proportion in the treated group was 80%, so the ratio of the two proportions (tabulated as the relative risk; see Chapter 27) is 8.0. In the bottom row, n = 10,000 per group. The observed proportion in the treated group was only 11.2%, so the relative risk was 1.12. The P value was the same in both cases, 0.006.

Focus on the last row of Table 18.3. The effect is tiny, 11.2% versus 10.0% of the subjects had the outcome being studied. Even though the P value is small, it is unlikely that such a small effect would be worth pursuing. It depends on the scientific or clinical situation, but there are few fields where such a tiny association is important. With huge sample sizes, it takes only a tiny effect to produce such a small P value. In contrast, with tiny samples, it takes a huge effect to yield a small P value. It is essential when reviewing scientific results to look at more than just the P value.

COMMON MISTAKES

Mistake: Believing that a "statistically significant" result proves an effect is real
As this chapter points out, there are lots of reasons why a result can be statistically significant.

Mistake: Believing that if a difference is "statistically significant," it must have a large physiological or clinical impact
If the sample size is large, a difference can be statistically significant but also so tiny that it is physiologically or clinically trivial.

Mistake: Asking about the chance of a Type I error without any clarifying details
How often will you make Type I errors? This chapter has pointed out that are two answers because there really are two questions:

- If the null hypothesis is true, what is the chance of making a Type I error?
- If a finding is statistically significant, what is the chance you made a Type I error?

Mistake: Thinking that the FPRP equals the significance level
The FDR is not the same as the significance level. Mixing up the two is a common mistake. The significance level α answers this question: what fraction of experiments performed when the null hypothesis is true reach a conclusion that the effect is statistically significant? The FDR answers this question: in what fraction of experiments that conclude an effect is statistically significant is the null hypothesis really true?

Mistake: Not realizing that the FPRP depends on the scientific context
The FDR depends upon the scientific context, as quantified by the prior probability.

Mistake: Thinking that a P value just barely less than 0.05 is strong evidence against the null hypothesis
Table 18.2 tabulated the FPRP associated with P values just a tiny bit less than 0.05. When the prior probability is 50%, the FDR is 27%. If the prior probability is lower than 50% (as it often is in exploratory research), the FPRP is even higher.

Mistake: Focusing on one P value
Don't be bedazzled by one small P value. Look at all the data collected in the study.

Q & A

Is this chapter trying to tell me that it isn't enough to determine whether a result is, or is not, statistically significant? Instead, I actually have to think?
> Yes!

How is it possible for an effect to be statistically significant but not scientifically significant (important)?
> You reach the conclusion that an effect is statistically significant when the P value is less than 0.05. With large sample sizes, this can happen even when the effect is tiny and irrelevant. The small P value tells us that the effect would not often occur by chance but says nothing about whether the effect is large enough to care about.

When is an effect large enough to care about, to be scientifically significant (important)?
> It depends on what you are measuring and why you are measuring it. This question can only be answered by someone familiar with your particular field of science. It is a scientific question, not a statistical one.

Does the context of the experiment (the prior probability) come into play when deciding whether a result is statistically significant?
> Only if you take into account prior probability when deciding on a value for α. Once you have chosen α, the decision about when to call a result "statistically significant" depends only on the P value and not on the context of the experiment.

Is the False Positive Report Probability (FPRP) the same as the False Positive Risk (FPR) and the False Discovery Rate (FPRP)?
> Yes. The FPRP and FPR mean the same thing. The FDR is very similar, but usually is used in the context of multiple comparisons rather than interpreting a single P value.

Does the context of the experiment (the prior probability) come into play when calculating the FPRP?
> Yes. See Table 18.2.

Does your choice of a value for α influence the calculated value of the FPRP?
> Yes. Your decision for a value of α determines the chance that a result will end up in the first or second column of Table 18.1.

When a P value is less than 0.05 and thus the comparison is deemed to be statistically significant, can you be 95% sure that the effect is real?
> No! It depends on the situation, on the prior probability.

The FPRP answers the question: If a result is statistically significant, what is he chance that it is a false positive? What is the complement of the FPRP called, the value that answers the question: If a result is statistically significant, what is the chance that it is a true positive?
> One name is the *posterior probability*. That name goes along with prior probability, used earlier in this chapter. *Posterior* is the opposite of prior, so refers to a probability computed after collecting evidence. You estimate the prior probability based on existing data and theory. Then after collecting new evidence, you compute the posterior probability by combining the prior probability and the evidence. A synonym is the *positive predictive value*. The positive predictive value or posterior probability is computed as 1-FPRP (if the FPRP is expressed as a fraction), or 100%-FPRP% (if the FPRP is expressed as a percentage).

CHAPTER SUMMARY

- P values depend on the size of the difference or association as well as sample size. You'll see a small P value either with a large effect observed in tiny samples or with a tiny effect observed in large samples.
- A conclusion of statistical significance does not mean the difference is large enough to be interesting, does not mean the results are intriguing enough to be worthy of further investigation, and does not mean that the finding is scientifically or clinically significant.
- Don't mix up the significance level α (the threshold P value defining statistical significance) with the P value. You must choose α as part of the experimental design. The P value, in contrast, is computed from the data.
- Also, don't mix up the significance level α with the FPRP, which is the chance that a statistically significant finding is actually due to a coincidence of random sampling.
- The FPRP depends on the context of the study. In other words, it depends on the prior probability that your hypothesis is true (based on prior data and theory).
- Even if you don't do formal Bayesian calculations, you should consider prior knowledge and theory when interpreting data.
- A P value just barely less than 0.05 does not provide strong evidence against the null hypothesis.
- The conclusion that a result is "statistically significant" sounds very definitive. But in fact, there are many reasons for this designation.

TERMS INTRODUCED IN THIS CHAPTER

- Bayesian approach (p. 173)
- False Positive Report Probability (FPRP) (p. 168)
- False positive risk (FPR) (p. 168)
- Positive predictive value (p. 177)
- Posterior probability (p. 177)
- Prior probability (p. 169)

CHAPTER 19

Interpreting a Result That Is Not Statistically Significant

Extraordinary claims require extraordinary proof.

CARL SAGAN

When you see a result that is not statistically significant, don't stop thinking. "Not statistically significant" means only that the P value is larger than a preset threshold. Thus, a difference (or correlation or association) as large as what you observed would not be unusual as a result of random sampling if the null hypothesis were true. This does not prove that the null hypothesis is true. This chapter explains how to use confidence intervals to help interpret those findings that are not statistically significant.

FIVE EXPLANATIONS FOR "NOT STATISTICALLY SIGNIFICANT" RESULTS

Let's continue the simple scenario from the last chapter. You compare cells incubated with a new drug to control cells and measure the activity of an enzyme, and you find that the P value is large enough (greater than 0.05) for you to conclude that the result is not statistically significant. The following discussion offers five explanations to explain why this happened.

Explanation 1: The drug did not affect the enzyme you are studying

Scenario: The drug did not induce or activate the enzyme you are studying, so the enzyme's activity is the same (on average) in treated and control cells.

Discussion: This is, of course, the conclusion everyone jumps to when they see the phrase "not statistically significant." However, four other explanations are possible.

Explanation 2: The drug had a trivial effect

Scenario: The drug may actually affect the enzyme but by only a small amount.

Discussion: This explanation is often forgotten.

Explanation 3: There was a Type II error

Scenario: The drug really did substantially affect enzyme expression. Random sampling just happened to give you some low values in the cells treated with the drug and some high levels in the control cells. Accordingly, the P value was large, and you conclude that the result is not statistically significant.

Discussion: How likely are you to make this kind of Type II error? It depends on how large the actual (or hypothesized) difference is, on the sample size, and on the experimental variation. This topic is covered in Chapter 26.

Explanation 4: The experimental design was poor

Scenario: In this scenario, the drug really would increase the activity of the enzyme you are measuring. However, the drug was inactivated because it was dissolved in an acid. Since the cells were never exposed to the active drug, of course the enzyme activity didn't change.

Discussion: The statistical conclusion was correct—adding the drug did not increase the enzyme activity—but the scientific conclusion was completely wrong.

Explanation 5: The results cannot be interpreted due to dynamic sample size

Scenario: In this scenario, you hypothesized that the drug would not work, and you really want the experiment to validate your prediction (maybe you have made a bet on the outcome). You first ran the experiment three times, and the result (n = 3) was statistically significant. Then you ran it three more times, and the pooled results (now with n = 6) were still statistically significant. Then you ran it four more times, and finally the results (with n = 10) were not statistically significant. This n = 10 result (not statistically significant) is the one you present.

Discussion: The P value you obtain from this approach simply cannot be interpreted.

"NOT SIGNIFICANTLY DIFFERENT" DOES NOT MEAN "NO DIFFERENCE"

A large P value means that a difference (or correlation or association) as large as what you observed would happen frequently as a result of random sampling. But this does not necessarily mean that the null hypothesis of no difference is true or that the difference you observed is definitely the result of random sampling.

Vickers (2010) tells a great story that illustrates this point:

The other day I shot baskets with [the famous basketball player] Michael Jordan (remember that I am a statistician and never make things up). He shot 7 straight free throws; I hit 3 and missed 4 and then (being a statistician) rushed to the sideline, grabbed my laptop, and calculated a *P* value of .07 by Fisher's exact test. Now, you wouldn't take this *P* value to suggest that there is *no* difference between my basketball skills and those of Michael Jordan, you'd say that our experiment hadn't *proved* a difference.

A high P value does not prove the null hypothesis. Deciding not to reject the null hypothesis is not the same as believing that the null hypothesis is definitely true. The absence of evidence is not evidence of absence (Altman & Bland, 1995).

EXAMPLE: α_2-ADRENERGIC RECEPTORS ON PLATELETS

Epinephrine, acting through α_2-adrenergic receptors, makes blood platelets stickier and thus helps blood clot. We counted these receptors and compared people with normal and high blood pressure (Motulsky, O'Connor, & Insel, 1983). The idea was that the adrenergic signaling system might be abnormal in people with high blood pressure (hypertension). We were most interested in the effects on the heart, blood vessels, kidney, and brain but obviously couldn't access those tissues in people, so we counted receptors on platelets instead. Table 19.1 shows the results.

The results were analyzed using an unpaired t test (see Chapter 30). The average number of receptors per platelet was almost the same in both groups, so of course the P value was high, 0.81. If the two populations were Gaussian with identical means, you'd expect to see a difference as large or larger than that observed in this study in 81% of studies of this size.

Clearly, these data provide no evidence that the mean receptor number differs in the two groups. When my colleagues and I published this study decades ago, we stated that the results were not statistically significant and stopped there, implying that the high P value proves that the null hypothesis is true. But that was not a complete way to present the data. We should have interpreted the CI.

The 95% CI for the difference between group means extends from −45 to 57 receptors per platelet. To put this in perspective, you need to know that the average number of α_2-adrenergic receptors/platelet is about 260. We can therefore rewrite the CI as extending from −45/260 to 57/260, which is from −17.3% to 21.9%, or approximately plus or minus 20%.

It is only possible to properly interpret the CI in a scientific context. Here are two alternative ways to think about these results:

- A 20% change in receptor number could have a huge physiological impact. With such a wide CI, the data are inconclusive, because they are consistent with no difference, substantially more receptors on platelets from people with hypertension, or substantially fewer receptors on platelets of people with hypertension.

	CONTROLS	HYPERTENSION
Number of subjects	17	18
Mean receptor number (receptors/platelet)	263	257
SD	87	59

Table 19.1. Number of α_2-adrenergic receptors on the platelets of controls and people with hypertension.

• The CI convincingly shows that the true difference is unlikely to be more than 20% in either direction. This experiment counts receptors on a convenient tissue (blood cells) as a marker for other organs, and we know the number of receptors per platelet varies a lot from individual to individual. For these reasons, we'd only be intrigued by the results (and want to pursue this line of research) if the receptor number in the two groups differed by at least 50%. Here, the 95% CI extended about 20% in each direction. Therefore, we can reach a solid negative conclusion that either there is no change in receptor number in individuals with hypertension or that any such change is physiologically trivial and not worth pursuing.

The difference between these two perspectives is a matter of scientific judgment. Would a difference of 20% in receptor number be scientifically relevant? The answer depends on scientific (physiological) thinking. Statistical calculations have nothing to do with it. Statistical calculations are only a small part of interpreting data.

EXAMPLE: FETAL ULTRASOUNDS

Ewigman et al. (1993) investigated whether the routine use of prenatal ultrasounds would improve perinatal outcomes. The researchers randomly divided a large pool of pregnant women into two groups. One group received routine ultrasound exams (sonograms) twice during their pregnancy. The other group was administered sonograms only if there was a clinical reason to do so. The physicians caring for the women knew the results of the sonograms and cared for the women accordingly. The investigators looked at several outcomes. Table 19.2 shows the total number of adverse events, defined as fetal or neonatal deaths (mortality) or moderate to severe morbidity.

The null hypothesis is that the risk of adverse outcomes is identical in the two groups. In other words, the null hypothesis is that routine use of ultrasounds neither prevents nor causes perinatal mortality or morbidity, so the relative risk equals 1.00. Chapter 27 explains the concept of relative risk in more detail.

Table 19.2 shows that the relative risk is 1.02. That isn't far from the null hypothesis value of 1.00. The two-tailed P value is 0.86.

	ADVERSE OUTCOME	TOTAL	RISK (%)	RELATIVE RISK
Routine ultrasound	383	7,685	5.0	1.020
Only when indicated	373	7,596	4.9	
Total	756	15,281		

Table 19.2. Relationship between fetal ultrasounds and outcome.

The risks in Column 4 are computed by dividing the number of adverse outcomes by the total number of pregnancies. The relative risk is computed by dividing one risk by the other (see Chapter 27 for more details).

Interpreting the results requires knowing the 95% CI for the relative risk, which a computer program can calculate. For this example, the 95% CI ranges from 0.88 to 1.17.

Our data are certainly consistent with the null hypothesis, because the CI includes 1.0. This does not mean that the null hypothesis is true. Our CI tells us that the data are also consistent (within 95% confidence) with relative risks ranging from 0.88 to 1.17.

Here are three approaches to interpreting the results:

- The CI is centered on 1.0 (no difference) and is quite narrow. These data convincingly show that routine use of ultrasound is neither helpful nor harmful.
- The CI is narrow but not all that narrow. It certainly makes clinical sense that the extra information provided by an ultrasound will help obstetricians manage the pregnancy and might decrease the chance of a major problem. The CI goes down to 0.88, a risk reduction of 12%. If I were pregnant, I'd certainly want to use a risk-free technique that reduces the risk of a sick or dead baby by as much as 12% (from 5.0% to 4.4%)! The data certainly don't prove that a routine ultrasound is beneficial, but the study leaves open the possibility that routine ultrasound might reduce the rate of awful events by as much as 12%.
- The CI goes as high as 1.17. That is a 17% relative increase in problems (from 5.0% to 5.8%). Without data from a much bigger study, these data do not convince me that ultrasounds are helpful and make me worry that they might be harmful.

Statistics can't help to resolve the differences among these three mindsets. It all depends on how you interpret the relative risk of 0.88 and 1.17, how worried you are about the possible risks of an ultrasound, and how you combine the data in this study with data from other studies (I have no expertise in this field and have not looked at other studies).

In interpreting the results of this example, you also need to think about benefits and risks that don't show up as a reduction of adverse outcomes. The ultrasound picture helps reassure parents that their baby is developing normally and gives them a picture to bond with and show relatives. This can be valuable regardless of whether it reduces the chance of adverse outcomes. Although statistical analyses focus on one outcome at a time, you must consider all the outcomes when evaluating the results.

HOW TO GET NARROWER CIS

Both previous examples demonstrate the importance of interpreting the CI in the scientific context of the experiment. Different people will appropriately have different opinions about how large a difference (or relative risk) is scientifically or clinically important and will therefore interpret a not statistically significant result differently.

If the CI is wide enough to include values you consider clinically or scientifically important, then the study is inconclusive. In some cases, you might be able to narrow the CIs by improving the methodology and thereby reducing the SD. But in most cases, increasing the sample size is the only approach to narrowing the CI in a repeat study. This rule of thumb can help: if sample size is increased by a factor of four, the CI is expected to narrow by a factor of two. More generally, the width of a CI is inversely proportional to the square root of sample size.

WHAT IF THE P VALUE IS REALLY HIGH?

If you ran many experiments in which the null hypothesis was really true, you'd expect the P values to be uniformly distributed between 0.0 and 1.0. Half of the P values would have values greater than 0.5, and 10% would have values greater than 0.9. But what do you conclude when the P value is really high?

In 1866, Mendel published his famous paper on heredity in pea plants (Mendel, 1866). This was the first explanation of heredity and recessive traits and really founded the field of genetics. Mendel proposed a model of recessive inheritance, designed an experiment with peas to test the model, and showed that the data fit the model very well. Extremely well! Fisher (1936) reviewed these data and then pooled all of Mendel's published data to calculate a P value that answered the question: Assuming that Mendel's genetic theory is correct and every plant was classified correctly, what is the probability that the deviation between expected and observed would be as great or greater than actually observed?

The answer (the P value) is 0.99993. So there clearly is no evidence of deviation from the model. The deviation from theory is not statistically significant. That is where most people would stop. Fisher went further. If that null hypothesis were true, and you ran similar experiments many times, the P values would be uniformly distributed between zero and 1. So what is the chance of getting a P value of 0.99993 or higher? The answer is $1 - 0.99993$ or 0.00007 or 0.007%. That is pretty small. It could be a rare coincidence. Fisher concluded that since the data presented match the theory so well (with very little of the expected random deviation from the theory), Mendel (or his assistants) must have massaged the data a bit. The data presented are simply too good to be true.

Q & A

If a P value is greater than 0.05, can you conclude that you have disproven the null hypothesis?

> No.

If you conclude that a result is not statistically significant, it is possible that you are making a Type II error as a result of missing a real effect. What factors influence the chance of this happening?

> The probability of a Type II error depends on the significance level (α) you have chosen, the sample size, and the size of the true effect.

By how much do you need to increase the sample size to make a CI half as wide?

A general rule of thumb is that increasing the sample size by a factor of 4 will cut the expected width of the CI by a factor of 2. (Note that 2 is the square root of 4.)

What if I want to make the CI one-quarter as wide as it is?

Increasing the sample size by a factor of 16 will be expected to reduce the width of the CI by a factor of 4. (Note that 4 is the square root of 16.)

Can a study result can be consistent both with an effect existing and with it not existing?

Yes! Clouds are not only consistent with rain but also with no rain. Clouds, like noisy results, are inconclusive (Simonsohn, 2016).

CHAPTER SUMMARY

- If a statistical test computes a large P value, you should conclude that the findings would not have been unusual if the null hypothesis were true.
- You should *not* conclude that the null hypothesis of no difference (or association, etc.) has been proven.
- When interpreting a high P value, the first thing to do is look at the size of the effect.
- Also look at the CI of the effect.
- If the CI includes effect sizes that you would consider to be scientifically important, then the study is inconclusive.

CHAPTER 20

Statistical Power

There are two kinds of statistics, the kind you look up, and the kind you make up.

REX STOUT

The power of an experimental design answers this question: If the true effect is of a specified size and the experiment is repeated many times, what fraction of the results will be statistically significant? The concept of power can help when deciding how large a sample size to use (see Chapter 26) and can help to interpret results that are not statistically significant.

WHAT IS STATISTICAL POWER?

The definition of a P value begins with "If the null hypothesis is true . . ."

But what if the null hypothesis is false and the treatment really does affect the outcome? Even so, your data may reach the conclusion that the effect is not statistically significant. Just by chance, your data may yield a P value greater than 0.05 (or whatever significance level you chose). The following question therefore becomes relevant:

> If there really were a difference, relative risk, correlation, or so on (which we will collectively call an *effect*) of a specified value in the overall population, what is the chance of obtaining an effect that is statistically significant in one particular sample?

The answer, called the *power* of the experiment, depends on four values:

- The sample size
- The amount of scatter (if comparing values of a continuous variable) or starting proportion (if comparing proportions)
- The size of the effect you hypothesize exists
- The significance level you choose

Given these values, power is the fraction of experiments in which you would expect to find a statistically significant result. By tradition, power is usually expressed as a percentage rather than a fraction.

	DECISION: REJECT NULL HYPOTHESIS	DECISION: DO NOT REJECT NULL HYPOTHESIS	TOTAL
Null hypothesis is true	A	B	A + B
Null hypothesis is false	C	D	C + D

Table 20.1. Definition of power.
This table shows the results of A + B + C + D experiments. In some cases, the null hypothesis is true (top row), whereas in other cases it is not (bottom row). The first column tabulates results that are statistically significant (leading you to reject the null hypothesis), whereas the second column tabulates results that are not statistically significant. Power is the fraction of experiments that achieve a statistically significant result when the null hypothesis is false. Thus, power is defined as C/(C + D).

Table 20.1 (which repeats Table 18.2) shows the results of many hypothetical statistical analyses, each analyzed to reach a conclusion of "statistically significant" or "not statistically significant." Assuming the null hypothesis is not true, power is the fraction of experiments that reach a statistically significant conclusion. So power equals C/(C + D).

DISTINGUISHING POWER FROM BETA AND THE FALSE POSITIVE REPORT RATE

Beta (β) is defined to equal 1.0 minus power. Note the similarity in the definitions of α and β.

- If the null hypothesis is true, then α is the chance of making the wrong decision (rejecting the null hypothesis). In Table 20.1, α equals A/(A + B).
- If the null hypothesis is false (with a specified alternative hypothesis), β is the chance of making the wrong decision (not rejecting the null hypothesis). In Table 20.1 it equals D/(C + D).

The False Positive Report Rate (FPRP) was defined in Chapter 18. In Table 20.1, it equals C/(A + C). It answers the question: If you reject the null hypothesis, what is the chance you are wrong.

Note that statistical power and the FPRP answer different questions so have different values.

AN ANALOGY TO UNDERSTAND STATISTICAL POWER

Here is a silly analogy to help illustrate the concept of statistical power (Hartung, 2005). You send your child into the basement to find a tool. He comes back and says, "It isn't there." What do you conclude? Is the tool there or not?

There is no way to be sure, so the answer must be a probability. What you really want to know is the probability that the tool is in the basement. But that value can't really be calculated without knowing the prior probability and using Bayesian thinking (see Chapter 18). Instead, let's ask a different question: If the tool really is in the basement, what is the chance your child would have found it? The answer, of course, is: it depends. To estimate the probability, you'd want to know three things:

- How long did he spend looking? If he looked for a long time, he is more likely to have found the tool than if he looked for a short time. The time spent looking for the tool is analogous to sample size. An experiment with a large sample size has high power to find an effect, while an experiment with a small sample size has less power.
- How big is the tool? It is easier to find a snow shovel than the tiny screwdriver used to fix eyeglasses. The size of the tool is analogous to the size of the effect you are looking for. An experiment has more power to find a big effect than a small one.
- How messy is the basement? If the basement is a real mess, he was less likely to find the tool than if it is carefully organized. The messiness is analogous to experimental scatter. An experiment has more power when the data are very tight (little variation), and less power when the data are very scattered.

If the child spent a long time looking for a large tool in an organized basement, there is a high chance that he would have found the tool if it were there. So you could be quite confident of his conclusion that the tool isn't there. Similarly, an experiment has high power when you have a large sample size, are looking for a large effect, and have data with little scatter (small SD). In this situation, there is a high chance that you would have obtained a statistically significant effect if the effect existed.

If the child spent a short time looking for a small tool in a messy basement, his conclusion that the tool isn't there doesn't really mean very much. Even if the tool were there, he probably would not have found it. Similarly, an experiment has little power when you use a small sample size, are looking for a small effect, and the data have lots of scatter. In this situation, there is a high chance of obtaining a conclusion of "not statistically significant" even if the effect exists.

Table 20.2 summarizes this analogy and its connection to statistical analyses.

	TOOLS IN BASEMENT	STATISTICS
High power	• Spent a long time looking • Large tool • Organized basement	• Large sample size • Looking for large effect • Little scatter
Low power	• Spent a short time looking • Small tool • Messy basement	• Small sample size • Looking for small effect • Lots of scatter

Table 20.2. The analogy between searching for a tool and statistical power.

POWER OF THE TWO EXAMPLE STUDIES

The power of an experimental design can be computed to detect various hypothetical differences. The left side of Figure 20.1 shows the power of the platelet study (described in Chapter 19) to detect various hypothetical differences between means. The power was calculated using the program GraphPad StatMate from the sample size of the study, the SDs of the two groups, and the definition of significance (P < 0.05).

If there really is no difference between population means, there is a 5% chance of obtaining a statistically significant result. That is the definition of statistical significance (using the traditional 5% significance level). So the curve intersects the Y-axis at 5%. If there is a difference between population means, the power of this study depends on the size of that difference. If the difference is tiny (left side of the graph), then the power is low. If the difference is large, then so is the power, which approaches 100% with very large differences.

The right side of Figure 20.1 shows the power of the ultrasound study. If the relative risk is really 1.0 (no effect), there is a 5% chance of obtaining a statistically significant result, because that is the definition of statistical significance. So the curve shows 5% power with a relative risk of 1.00, and more power as the effect gets more pronounced (lower relative risk, showing that an ultrasound is beneficial).

The shapes in Figure 20.1 are universal. What varies between studies is the horizontal location of the curve.

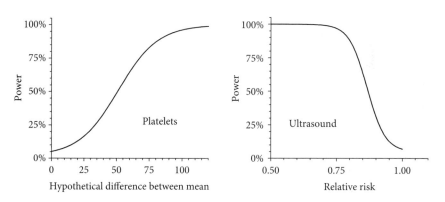

Figure 20.1. The power of two experiments.
The graph on the left summarizes the experiment in Table 19.1, which compared the α_2-receptor number in controls and people with hypertension. The graph on the right summarizes the data in Table 19.2, evaluating the advantages of a fetal ultrasound. For any hypothetical difference between means (X-axis), this graph shows the power of the study to find a statistically significant difference (P < 0.05). Note that an increasing effect for the platelet example (larger difference in receptor number) goes from left to right, whereas an increasing effect for the ultrasound example (smaller relative risk) goes from right to left.

WHEN DOES IT MAKE SENSE TO COMPUTE POWER?

It makes sense to compute statistical power in two situations:

- When deciding how many subjects (or experimental units) you need. Chapter 26 explains.
- When evaluating or critiquing completed experiments. It can be useful to ask how much power an experiment had to detect an effect of some specified size. That specified effect size must be chosen based on your scientific goals and not based on the effect actually observed in a particular experiment. Note, however, that it does not make sense to compute the power of the experiment to detect the effect actually observed. The next section explains why.

COMMON MISTAKES: POWER

Mistake: Believing that a study design has a single power

It is never possible to just ask, "What is the power of this experimental design?" That question is simply meaningless. Power can be computed for any proposed effect size. So there are a range of power values.

If you want to compute a single power value, you must ask, "What is the power of this experimental design to detect an effect of a specified size?" The effect size might be a difference between two means, a relative risk, or some other measure of treatment effect.

Which effect size should you calculate power for? This is not a statistical question but rather a scientific question. It only makes sense to do a power analysis when you think about the data scientifically. It makes sense to compute the power of a study design to detect an effect that is the smallest effect you would care about.

Mistake: Believing that calculating observed power (post hoc power) is helpful

When is the study is already completed, some programs augment their results by reporting the power to detect the effect size (or difference, relative risk, etc.) actually observed in that particular experiment. The result is sometimes called *observed power*, and the procedure is sometimes called a *post hoc power analysis* or *retrospective power analysis*.

Many (perhaps most) statisticians (and I agree) think that these computations are useless and misleading helpful (Hoenig & Heisey, 2001; Lenth, 2001; Levine & Ensom, 2001). If your study reached a conclusion that the difference is not statistically significant, then—by definition—its power to detect the observed effect is very low. You learn nothing new by such a calculation. It can be useful to compute the power of the study to detect a difference that would have been scientifically or clinically worth detecting. It is not worthwhile to compute the power of the study to detect the difference (or effect) actually observed.

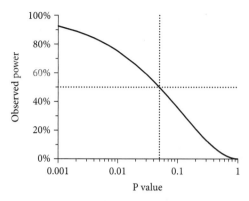

Figure 20.2. The relationship between P value and observed power when the significance level (alpha) is set to 0.05.

Hoenig and Helsey (2001) derived the equation use to create this curve. The dotted lines show that when the P value equals 0.05, the observed power equals 50%.

Hoenig and Helsey (2001) pointed out that the observed power can be computed from the observed P value as well as the value of α you choose (usually 0.05). The observed power conveys no new information. Figure 20.2 shows the relationship between P value and observed power, when α is set to 0.05.

Q & A

In Figure 20.2, why is the observed power 50% when the P value equals 0.05?

> If the P value is 0.05 in one experiment, that is your best guess for what it will be in repeat experiments. You expect half the P values to be higher and half to be lower. Only that last half will lead to the conclusion that the result is statistically significant, so the power is 50%.

If I want to use a program to compute power, what questions will it ask?

> To calculate power, you must enter the values for α, the expected SD, the planned sample size, and the size of the hypothetical difference (or effect) you are hoping to detect.

Why do I have to specify the effect size for which I am looking? I want to detect any size effect.

> All studies will have a small power to detect tiny effects and a large power to detect enormous effects. You can't calculate power without specifying the effect size for which you are looking.

When does it make sense to do calculations involving power?

> In two situations. When planning a study, you need to decide how large a sample size to use. Those calculations will require you to specify how much power you want to detect some hypothetical effect size. After completing a study that has results that are not statistically significant, it can make sense to ask how much power that study had to detect some specified effect size.

When calculating sample size, what value for desired power is traditionally used?

> Most sample size calculations are done for 80% power. Of course, there is nothing special about that value except tradition.

Can one do all sample size/power calculations using standard equations?

> Usually. But in some cases it is necessary to run computer simulations to compute the power of a proposed experimental design.

Power analyses traditionally set $\alpha = 0.05$ and $\beta = 0.20$ (because power is set to 80%, and 100% – 80% = 20% = 0.20). Using these traditional values implies that you accept four times the chance of a Type I error than a Type II error (because 0.20/0.05 = 4). Is there any justification for this ratio?

> No. Since the relative "costs" of Type I and Type II errors depend on the scientific context, so should the choices of α and β.

CHAPTER SUMMARY

- The *power* of an experimental design is the probability that you will obtain a statistically significant result assuming a certain effect size in the population.
- Beta (β) is defined to equal 1.0 minus power.
- The power of an experiment depends on sample size, variability, the choice of α to define statistical significance, and the hypothetical effect size.
- For any combination of sample size, variability, and α, there will be a high power to detect a huge effect and a small power to detect a tiny effect.
- Power should be computed based on the minimum effect size that would be scientifically worth detecting, not on an effect observed in a prior experiment.
- Once the study is complete, it is not very helpful to compute the power of the study to detect the effect size that was actually observed.

TERMS INTRODUCED IN THIS CHAPTER

- Observed power (p. 190)
- Post hoc power analysis (p. 190)
- Power (p. 186)
- Retrospective power analysis (p. 190)

CHAPTER 21

Testing for Equivalence
or Noninferiority

The problem is not what you don't know, but what you know that ain't so.

WILL ROGERS

In many scientific and clinical investigations, the goal is not to find out whether one treatment causes a substantially different effect than another treatment. Instead, the goal is to find out whether the effects of a new treatment are equivalent (or not inferior) to those of a standard treatment. This chapter explains how the usual approach of statistical hypothesis testing is not useful in this situation. A conclusion that the difference between two treatments is statistically significant does not answer the question about equivalence, and neither does a conclusion that the difference is not statistically significant.

EQUIVALENCE MUST BE DEFINED SCIENTIFICALLY, NOT STATISTICALLY

Imagine that you've created a generic drug and want to show that it works just as well as a standard drug. There are many aspects of drug action that can be compared, and we'll focus on one question: Is the peak blood plasma concentration of the two formulations of the drug equivalent? In other words, is the new drug absorbed as well as the standard drug? People are given one drug formulation and then (after a washout period) the other, randomizing the order. The peak concentration of each drug formulation is measured in blood samples taken from each person.

When you compare the two drug formulations, you will always see some difference in peak plasma concentration. It doesn't make sense to ask whether the two produce precisely the same outcome. When asking about *equivalence*, the question is whether the outcomes are close enough to be clinically or scientifically indistinguishable.

How close is close enough? "Close enough" must be defined as a range of treatment effects that are considered scientifically or clinically trivial. This requires thinking about the scientific or clinical context of your experiment.

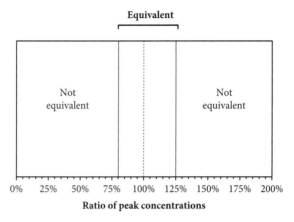

Figure 21.1. The zone of equivalence is defined using scientific criteria.
The FDA has said that for comparing two formulations of a drug, the results are deemed equivalent when the ratio of peak drug concentrations is between 80% and 125%.

Statistical calculations and conclusions about statistical significance are completely irrelevant in defining this range, called the *equivalence zone*, *equivalence margin*, or *region of practical equivalence* (ROPE).

When testing a generic drug, the US Food and Drug Administration (FDA) defines two drug formulations to be equivalent when the ratio of their peak concentrations in blood plasma, and the entire 90% CI of that ratio, is between 0.80 and 1.25 (FDA, 2012). This definition is based on clinical thinking about drug action. In other circumstances, equivalence might be defined differently.

Figure 21.1 shows this definition. Make sure you understand Figure 21.1, because it is the basis for Figures 21.2 and 21.3.

Note that the zone of equivalence as defined in Figure 21.1 does not appear to be symmetrical around 100%, but, in fact, it sort of is. It is fairly arbitrary whether you calculate the ratio of peak levels as the peak level of the new formulation divided by the peak level of the standard formulation or whether you define it as the peak level of the standard formulation divided by the peak level of the new formulation. The reciprocal of 80% is 125% (1/0.8 = 1.25), and the reciprocal of 125% is 80% (1/1.25 = 0.80). So the zone of equivalence is, in a practical sense, symmetrical. Another way to look at it: the zone of equivalence is symmetrical on a logarithmic axis. Zones of equivalence can be defined in various ways and can be asymmetrical even without thinking about reciprocals.

IF THE MEAN IS WITHIN THE EQUIVALENCE ZONE

Figure 21.2 shows data from three drugs for which the mean ratio of peak concentrations is within the zone of equivalence. The fact that the mean value is within the zone of equivalence does not prove equivalence. Two drugs are defined to be bioequivalent (by the FDA) only when the entire 90% CI for the ratio of peak concentrations lies within the equivalence zone.

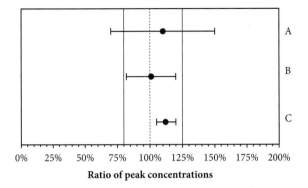

Figure 21.2. Three drug formulations for which the ratio of peak concentrations are within the equivalent zone.

For Drug A, the CI extends outside of the equivalent zone, so the results are inconclusive. The results are consistent with the two drugs being equivalent or not equivalent. In Drugs B and C, the 90% CIs lie completely within the equivalent zone, so the data demonstrate that the two drugs are equivalent to the standard drug.

This is the case for Drugs B and C. Those 90% CIs lie entirely within the equivalence zone, so the data demonstrate that Drugs B and C are equivalent to the standard drug with which they are being compared.

In contrast, the 90% CI for Drug A is partly in the equivalence zone and partly outside of it. The data are inconclusive.

IF THE MEAN IS OUTSIDE OF THE EQUIVALENCE ZONE

Figure 21.3 shows data from three drugs for which the mean ratio of peak concentrations is in the not-equivalent zone. This does not prove the drugs are not equivalent. You must look at the entire 90% CI to make such a determination.

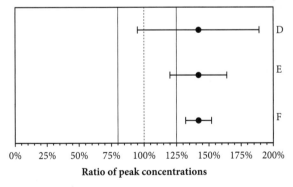

Figure 21.3. Results of three drugs for which the mean ratio of peak concentrations is in the not-equivalent zone.

With Drugs D and E, the CI includes both equivalent and not-equivalent zones, so the data are not conclusive. The 90% CI for Drug F is entirely outside the equivalence zone, proving that it is not equivalent to the standard drug.

The 90% CIs for Drugs D and E are partly within the equivalent zone and partly outside it. The data are not conclusive.

The 90% CI for Drug F lies totally outside the equivalence zone. These data prove that Drug F is not equivalent to the standard drug.

APPLYING THE USUAL APPROACH OF STATISTICAL HYPOTHESIS TESTING TO TESTING FOR EQUIVALENCE

It is possible to apply the ideas of statistical hypothesis testing to testing for equivalence, but it is tricky (Wellek, 2002; Lakens, 2017). The approach is to pose two distinct null hypotheses and define statistical significance using *two one-sided tests* (sometimes abbreviated TOST).

Let's continue the example of comparing peak drug concentrations for two drugs. The two drugs are shown to be equivalent when both of the following conditions are true (Figure 21.4).

- The mean value of the ratio is *greater* than 0.80 (the lower limit that defines equivalence), and this increase is statistically significant with a one-sided P value less than 0.05. The one-sided null hypothesis for this test is that the mean ratio is less than 0.80.
- The mean value of the ratio is *less* than 1.25 (the upper limit that defines equivalence), and this decrease is statistically significant with a one-sided P value less than 0.05. The one-sided null hypothesis for this test is that the mean ratio is greater than 1.25.

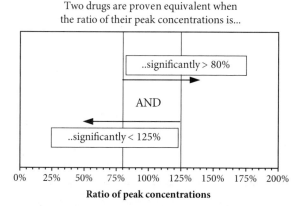

Figure 21.4. Applying the idea of statistical hypothesis testing to equivalence testing.
A conclusion of equivalence requires a statistically significant finding (P < 0.10) from two different tests of two different null hypotheses, shown by the vertical lines. Each null hypothesis is tested with a one-sided alternative hypothesis, shown by the arrows. Two drugs are considered equivalent when the ratio of peak concentrations is significantly greater than 80% and significantly less than 125%.

Juggling two null hypotheses and two P values (each one sided or one-tailed; the two terms are synonyms) is not for the statistical novice. The results are the same as those obtained using the CI approach described in the previous discussion. The CI approach is much easier to understand.

NONINFERIORITY TESTS

Equivalence trials attempt to prove that a new treatment or drug works about the same as the standard treatment. *Noninferiority* trials attempt to prove that a new treatment is not worse than the standard treatment.

To prove equivalence, all parts of the CI must be *within* the equivalence zone. To prove noninferiority, all parts of the CI must be to the right of the left (lower) border of the equivalence zone. The entire CI, therefore, is in a range that either shows the new drug is superior or shows that the new drug is slightly inferior but still in the zone defined to be practically equivalent. In Figure 21.2, Drugs B and C are noninferior. In Figure 21.3, all three drugs (D, E, and F) are noninferior. For all five of these drugs, the lower confidence limit exceeds 80%.

Table 21.1 (adapted from Walker and Nowacki, 2010) summarizes the differences between testing for differences, equivalence and noninferiority.

COMMON MISTAKES: TESTING FOR EQUIVALENCE

Mistake: Using the usual approach of statistical hypothesis testing when testing for equivalence

A common mistake is to simply compare the peak blood concentration of the two drugs using an unpaired t test (see Chapter 30). It seems as though the P value ought to be informative, but this approach is not helpful. If the results are not statistically significant, it is a mistake to conclude that the blood concentrations of the two drugs are equivalent. Similarly, if the results are statistically significant, it is a mistake to conclude that the two drugs are not equivalent.

TEST FOR...	NULL HYPOTHESIS	ALTERNATIVE HYPOTHESIS (CONCLUSION IF P VALUE IS SMALL)
Difference	No difference between treatments	Nonzero difference
Equivalence	A difference large enough to matter	Either no difference or a difference too small to matter
Noninferiority	Experimental treatment is worse than the standard treatment	Experimental treatment is either equivalent to the standard or better. It is not worse

Table 21.1. Distinguishing testing for difference, equivalence, and noninferiority.

Mistake: Testing for equivalence without being sure the standard treatment works

Snapinn (2000) and Kaul and Diamond (2006) reviewed the many issues one must consider when interpreting data that purport to demonstrate equivalence or noninferiority.

When comparing two drugs, the most important issue is this: you must be 100% sure the standard drug works. A conclusion that a new treatment is equivalent (or not inferior) to a standard treatment is only useful when you are positively sure that the standard treatment actually works better than a placebo. If the data from other studies leave any doubts about whether the standard treatment works, then it really doesn't make any sense to ask whether a new treatment is equivalent (or not inferior) to it.

Q & A

Why isn't a conclusion that a difference is not statistically significant enough to prove equivalence?

The P value from a standard statistical test, and thus the conclusion about whether an effect is statistically significant, is based entirely on analyzing the data. A conclusion about equivalence has to take the context into account. What is equivalent for one variable in one situation is not equivalent for another variable in another context.

Is it possible for a difference to be statistically significant but for the data to prove equivalence?

Surprisingly, yes. The conclusion that the difference is statistically significant just means the data convince you that the true difference is not zero. It doesn't tell you that the difference is large enough to care about. It is possible for the entire CI to include values that you consider to be equivalent. Look at Drug C in Figure 21.2.

Why 90% CIs instead of 95%?

Tests for equivalence use 90% CIs (I had this wrong in the prior edition). But the conclusions are for 95% confidence. The reason is complicated. Essentially you are looking at two one-sided CIs. So using a 90% CI yields a 95% confidence. Yes, this is confusing and not at all obvious.

How is testing for noninferiority different than testing for superiority?

Although it might initially appear so, the double negatives are not all that confusing. When testing for noninferiority (in the example presented in this chapter), you are asking if the data prove that a drug is not worse than a standard drug. You will conclude that Drug A is "not worse" than Drug B when the two drugs are equivalent or when Drug A is better.

CHAPTER SUMMARY

- In many scientific and clinical investigations, the goal is not to find out whether one treatment causes a substantially different effect than another treatment. Instead, the goal is to find out whether the effects of a new treatment are equivalent to (or not inferior to) that of a standard treatment.

- The usual approach of statistical hypothesis testing does not test for equivalence. Determining whether a difference between two treatments is statistically significant tells you nothing about whether the two treatments are equivalent.
- There is no point testing whether a new treatment is equivalent to a standard one unless you are really sure that the standard treatment works.
- Equivalence trials attempt to prove that a new treatment or drug works about the same as the standard treatment. Noninferiority trials attempt to prove that a new treatment is not worse than the standard treatment.

TERMS INTRODUCED IN THIS CHPATER

- Equivalence (p. 193)
- Equivalence margin (p. 194)
- Equivalence zone (p. 194)
- Noninferiority (p. 197)
- Region of practical equivalence (ROPE) (p. 194)
- Two one-sided tests (TOST) (p. 195)

PART E

Challenges in Statistics

CHAPTER 22

Multiple Comparisons Concepts

> If you torture your data long enough, they will tell you what-
> ever you want to hear.
>
> MILLS (1993)

Coping with multiple comparisons is one of the biggest challenges in data analysis. If you calculate many P values, some are likely to be small just by random chance. Therefore, it is impossible to interpret small P values without knowing how many comparisons were made. This chapter explains three approaches to coping with multiple comparisons. Chapter 23 will show that the problem of multiple comparisons is pervasive, and Chapter 40 will explain special strategies for dealing with multiple comparisons after ANOVA.

THE PROBLEM OF MULTIPLE COMPARISONS

The problem of multiple comparisons is easy to understand

If you make two independent comparisons, what is the chance that one or both comparisons will result in a statistically significant conclusion just by chance? It is easier to answer the opposite question. Assuming both null hypotheses are true, what is the chance that both comparisons will not be statistically significant? The answer is the chance that the first comparison will not be statistically significant (0.95) times the chance that the second one will not be statistically significant (also 0.95), which equals 0.9025, or about 90%. That leaves about a 10% chance of obtaining at least one statistically significant conclusion by chance.

It is easy to generalize that logic to more comparisons. With K independent comparisons (where K is some positive integer), the chance that none will be statistically significant is 0.95^K, so the chance that one or more comparisons will be statistically significant is $1.0 - 0.95^K$. Figure 22.1 plots this probability for various numbers of independent comparisons.

Consider the unlucky number 13. If you perform 13 independent comparisons (with the null hypothesis true in all cases), the chance is about 50% that one or more of these P values will be less than 0.05 and will thus lead to a conclusion that the effect is statistically significant. In other words, with 13 independent comparisons, there is about a 50:50 chance of obtaining at least one false positive finding of statistical significance.

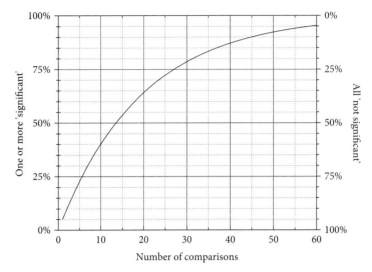

Figure 22.1. Probability of obtaining statistically significant results by chance.
The X-axis shows various numbers of statistical comparisons, each assumed to be independent of the others. The left Y-axis shows the chance of obtaining one or more statistically significant results (P < 0.05) by chance. The value on the left Y axis is computed as $1.0 - 0.95^X$, where X is the number of comparisons.

With more than 13 comparisons, it is more likely than not that one or more conclusions will be statistically significant just by chance. With 100 independent null hypotheses that are all true, the chance of obtaining at least one statistically significant P value is 99%.

A dramatic demonstration of the problem with multiple comparisons

Bennett and colleagues (2011) dramatically demonstrated the problem of multiple comparisons. They used functional magnetic resonance imaging (fMRI) to map blood flow in thousands of areas of a brain. Their experimental subject was shown a photograph of a person and asked to silently identify which emotion the person in the photograph was experiencing. The investigators measured blood flow to thousands of areas of the brain before and after presenting the photograph to their experimental subject. In two particular areas of the brain, blood flow increased substantially, and those differences were statistically significant (P < 0.001).

Sounds compelling. The investigators have identified regions of the brain involved with recognizing emotions. Right? Nope! The investigators could prove beyond any doubt that both findings were false positives resulting from noise in the instruments and were not caused by changes in blood flow. How could they be sure? Their experimental subject was a dead salmon! When they properly corrected for multiple comparisons to account for the thousands of brain regions at which they looked, neither finding was statistically significant.

CORRECTING FOR MULTIPLE COMPARISONS IS NOT ALWAYS NEEDED

Before considering two general approaches to correcting for multiple comparisons, let's pause to consider three scenarios in which corrections for multiple comparisons are not needed.

Corrections for multiple comparisons are not needed if the people reading the data take into account the number of comparisons

Some statisticians recommend that investigators never correct for multiple comparisons but instead report uncorrected P values and CIs with a clear explanation that no mathematical correction was made for multiple comparisons (Rothman, 1990). The people reading the results must then informally account for multiple comparisons. If all the null hypotheses are true, you'd expect 5% of the comparisons to have uncorrected P values of less than 0.05. Compare this number with the actual number of small P values.

This approach requires that all comparisons (or at least the *number* of comparisons) be reported. If the investigators only show the small P values, it is impossible to interpret the results.

Corrections for multiple comparisons are not essential when you have clearly defined one outcome as primary and the others as secondary

Many clinical trials clearly define, as part of the study protocol, that one outcome is primary. This is the key prespecified outcome on which the conclusion of the study is based. The study may do other secondary comparisons, but those are clearly labeled as secondary. Correction for multiple comparisons is often not used with a set of secondary comparisons.

Corrections for multiple comparisons may not be needed if you make only a few planned comparisons

Even if a study collects lots of data, you may want to focus on only a few scientifically sensible comparisons, rather than every possible comparison. The term *planned comparison* is used to describe this situation. These comparisons must be planned as part of the experimental design. It is cheating to look at the data and then decide which comparisons you wish to make. When you make only a few preplanned comparisons, many statisticians think it is OK to not correct for multiple comparisons.

Corrections for multiple comparisons are not essential when the results are complementary

Ridker and colleagues (2008) asked whether lowering LDL cholesterol would prevent heart disease in patients who did not have high LDL concentrations, did not have a prior history of heart disease, but did have an abnormal blood test suggesting the presence of some inflammatory disease. The study included almost

18,000 people. Half received a statin drug to lower LDL cholesterol and half received a placebo.

The investigators' primary goal (planned as part of the protocol) was to compare the number of end points that occurred in the two groups, including deaths from a heart attack or stroke, nonfatal heart attacks or strokes, and hospitalization for chest pain. These events happened about half as often to people treated with the drug compared with people taking the placebo. The drug worked.

The investigators also analyzed each of the end points separately. Those people taking the drug (compared with those taking the placebo) had fewer deaths, fewer heart attacks, fewer strokes, and fewer hospitalizations for chest pain.

The data from various demographic groups were then analyzed separately. Separate analyses were done for men and women, old and young, smokers and nonsmokers, people with hypertension and those without, people with a family history of heart disease and those without, and so on. In each of 25 subgroups, patients receiving the drug experienced fewer primary end points than those taking the placebo, and all of these effects were statistically significant.

The investigators made no correction for multiple comparisons for all these separate analyses of outcomes and subgroups, because these were planned as secondary analyses. The reader does not need to try to informally correct for multiple comparisons, because the results are so consistent. The multiple comparisons each ask the same basic question, and all the comparisons lead to the same conclusion—people taking the drug had fewer cardiovascular events than those taking the placebo. In contrast, correction for multiple comparisons would be essential if the results showed that the drug worked in a few subsets of patients but not in other subsets.

THE TRADITIONAL APPROACH TO CORRECTING FOR MULTIPLE COMPARISONS

The Familywise Error Rate

When each comparison is made individually without any correction for multiple comparisons, the traditional 5% significance level applies to each individual comparison. This is therefore known as *the per-comparison error rate*, and it is the chance that random sampling would lead *this particular comparison* to an incorrect conclusion that the difference is statistically significant when this particular null hypothesis is true.

When you correct for multiple comparisons, the significance level is redefined to be the chance of obtaining *one or more* statistically significant conclusions if *all* of the null hypotheses in the family are actually true. The idea is to make a stricter threshold for defining significance. If α is set to the usual value of 5% and all the null hypotheses are true, then the goal is to have a 95% chance of obtaining zero statistically significant results and a 5% chance of obtaining one or more statistically significant results. That 5% chance applies to the entire family of comparisons performed in the experiment, so it is called a *familywise error rate* or the *per-experiment error rate*.

What is a family of comparisons?

What exactly is a *family of related comparisons*? Usually, a family consists of all the comparisons in one experiment or all the comparisons in one major part of an experiment. That definition leaves lots of room for ambiguity. When reading about results corrected for multiple comparisons, ask about how the investigators defined the family of comparisons.

The Bonferroni correction

The simplest approach to achieving a familywise error rate is to divide the value of α (often 5%) by the number of comparisons. Then define a particular comparison as statistically significant only when its P value is less than that ratio. This is called a *Bonferroni correction*.

Imagine that an experiment makes 20 comparisons. If all 20 null hypotheses are true and there are no corrections for multiple comparisons, about 5% of these comparisons are expected to be statistically significant (using the usual definition of α). Table 22.1 and Figure 22.1 show that there is about a 65% chance of obtaining one (or more) statistically significant result.

If the Bonferroni correction is used, a result is only declared to be statistically significant when its P value is less than 0.05/20, or 0.0025. This ensures that if all the null hypotheses are true, there is about a 95% chance of seeing no statistically significant results among all 20 comparisons and only a 5% chance of seeing one (or more) statistically significant results. The 5% significance level applies to the entire family of comparisons rather than to each of the 20 individual comparisons.

Note a potential point of confusion. The value of α (usually 0.05) applies to the entire family of comparisons, but a particular comparison is declared to be statistically significant only when its P value is less than α/K (where K is the number of comparisons).

Example of a Bonferroni correction

Hunter and colleagues (1993) investigated whether vitamin supplementation could reduce the risk of breast cancer. The investigators sent dietary questionnaires

PROBABILITY OF OBSERVING SPECIFIED NUMBER OF SIGNIFICANT COMPARISONS

NUMBER OF SIGNIFICANT COMPARISONS	NO CORRECTION	BONFERRONI
Zero	35.8%	95.1%
One	37.7%	4.8%
Two or more	26.4%	0.1%

Table 22.1. How many significant results will you find in 20 comparisons?
This table assumes you are making 20 comparisons, that all 20 null hypotheses are true, and that α is set to its conventional value of 0.05. If there is no correction for multiple comparisons, there is only a 36% chance of observing no statistically significant findings. With the Bonferroni correction, this probability goes up to 95%.

to over 100,000 nurses in 1980. From the questionnaires, they determined the participants' intake of vitamins A, C, and E and divided the women into quintiles for each vitamin (i.e., the first quintile contains the 20% of the women who consumed the smallest amount). They then followed these women for eight years to determine the incidence rate of breast cancer. Using a test called the chi-square test for trend, the investigators calculated a P value to test the null hypothesis that there is no linear trend between vitamin-intake quintile and the incidence of breast cancer. There would be a linear trend if increasing vitamin intake was associated with increasing (or decreasing) incidence of breast cancer. There would not be a linear trend if (for example) the lowest and highest quintiles had a low incidence of breast cancer compared with the three middle quintiles. The authors determined a different P value for each vitamin. For Vitamin C, $P = 0.60$; for Vitamin E, $P = 0.07$; and for Vitamin A, $P = 0.001$.

Interpreting each P value is easy: if the null hypothesis is true, the P value is the chance that random selection of subjects will result in as large (or larger) a linear trend as was observed in this study. If the null hypothesis is true, there is a 5% chance of randomly selecting subjects such that the trend is statistically significant.

If no correction is made for multiple comparisons, there is a 14% chance of observing one or more significant P values, even if all three null hypotheses are true. The Bonferroni method sets a stricter significance threshold by dividing the significance level (0.05) by the number of comparisons (three), so a difference is declared statistically significant only when its P value is less than 0.050/3, or 0.017. According to this criterion, the relationship between Vitamin A intake and the incidence of breast cancer is statistically significant, but the intakes of Vitamins C and E are not significantly related to the incidence of breast cancer.

The terminology can be confusing. The significance level is still 5%, so α still equals 0.05. But now the significance level applies to the family of comparisons. The lower threshold (0.017) is used to decide whether each particular comparison is statistically significant, but α (now the familywise error rate) remains 0.05.

Example of Bonferroni corrections: Genome-wide association studies

Genome-wide association studies (GWAS) use the Bonferroni correction to cope with a huge number of comparisons. These studies look for associations between diseases (or traits) and genetic variations (usually single nucleotide polymorphisms). They compare the prevalence of many (up to 1 million) genetic variations between many (often tens of thousands) cases and controls, essentially combining about 1 million case-control studies (explained in Chapter 28). To correct for multiple comparisons, the threshold is set using the Bonferroni correction. Divide α (0.05) by the number of comparisons (1,000,000) to calculate the threshold, which is 0.00000005 (5×10^{-8}).

A tiny P value (less than is 0.00000005) in a GWAS is evidence that the prevalence of a particular genetic variation differs in the two groups you are studying. One of the groups is composed of people with a disease and the other group is composed of controls, so one explanation is that the disease is associated with that genetic variant. But there is another possible explanation. If the patients and controls are not well matched and have different ancestry (say the patients are largely of Italian ancestry and the controls are largely Jewish), you'd expect genetic differences between the two groups that have nothing to do with the disease being studied (see "The Challenge of Case-Control Studies" in Chapter 28).

CORRECTING FOR MULTIPLE COMPARISONS WITH THE FALSE DISCOVERY RATE

The false discovery rate (FDR) approach is an alternative approach to multiple comparisons that is especially useful when the number of simultaneous comparisons is large (Benjamini & Hochberg, 1995). To learn more about this approach, read the super clear nonmathematical review by Glickman, Rao and Schultz (2014).

Lingo: FDR

This approach does not use the term *statistically significant* but instead uses the term *discovery*. A finding is deemed to be a discovery when its P value is lower than a certain threshold. A discovery is false when the null hypothesis is actually true for that comparison. The FDR is the answer to these two equivalent questions (this definition actually defines the positive FDR, or pFDR, but the distinction between the pFDR and the FDR is subtle and won't be explained in this book):

- If a comparison is classified as a discovery, what is the chance that the null hypothesis is true?
- Of all discoveries, what fraction is expected to be false?

Controlling the FDR

When analyzing a set of P values, you can set the FDR to a desired value, abbreviated Q. If you set Q to 10%, then your goal is for at least 90% of the discoveries to be true and no more than 10% to be false discoveries (for which the null hypothesis is actually true). Of course, you can't know which are which.

A method developed by Benjamini and Hochberg (1995) sets the threshold values for deciding when a P value is low enough to be deemed a discovery. The method actually sets a different threshold value for each comparison. The threshold is tiny for the comparison with the smallest P value and much larger for the comparison with the largest P value. To understand why this makes sense, imagine that you computed 100 P values and all the null hypotheses were true. You'd expect the P values to be randomly distributed between 0.0 and 1.0. It would not be at all surprising for the smallest P value to equal 0.01. But it would be surprising (if all null hypotheses were true) for the median P value to be 0.01. You'd expect that value to be about 0.5. So it makes perfect sense to rank the P values from low to high and use that ranking when choosing the threshold that defines a discovery.

Here is a brief explanation of how those thresholds are determined. If all the null hypotheses were true, you'd expect the P values to be randomly scattered between zero and 1. Half would be less than 0.50, 10% would be less than 0.10, and so on. Let's imagine that you are making 100 comparisons and you have set

Q (the desired FDR) to 5%. If all the null hypotheses were true, you'd expect that the smallest P value would be about 1/100, or 1%. Multiply that value by Q. So you declare the smallest P value to be a discovery if its P value is less than 0.0005. You'd expect the second-smallest P value to be about 2/100, or 0.02. So you'd call that comparison a discovery if its P value is less than 0.0010. The threshold for the third-smallest P value is 0.0015, and so on. The comparison with the largest P value is called a discovery only if its value is less than 0.05. This description is a bit simplified but does provide the general idea behind the method. This is only one of several methods used to control the FDR.

If you set Q for the FDR method to equal alpha in the conventional method, note these similarities. For the smallest P value, the threshold used for the FDR method is α/k, which is the same threshold used by the Bonferroni method. For the largest P value, the threshold for the FDR method is α, which is equivalent to not correcting for multiple comparisons. P values between the smallest and largest are compared to a threshold that is a blend of the two methods.

Other ways to use the FDR

The previous section explained one approach: choose a desired FDR and use that to decide which results count as a discovery. In this approach, there is one FDR for the whole set of comparisons.

An alternative approach is to first decide on a threshold for defining discovery. For example, in a gene-chip assay, you might choose the 5% of genes whose expression changed the most. Given that definition, you would then compute the FDR. Again, there is one FDR for the entire set of comparisons.

A third alternative approach is to compute a distinct FDR for each comparison. For each comparison, define a threshold for discovery so that particular comparison just barely satisfies the definition. Using that definition, compute the overall FDR for all the comparisons. This value is called a q value (note the use of a lower-case letter). Repeat for each comparison. If you make 1,000 comparisons, you'll end up with 1,000 distinct q values.

COMPARING THE TWO METHODS OF CORRECTING FOR MULTIPLE COMPARISONS

Table 22.2 shows the results of many comparisons. You can't create this table with actual data, because the entries in the rows assume that you are Mother Nature and therefore know whether each null hypothesis is actually true. In fact, you never know that, so this table is conceptual.

The top row represents the results of comparisons for which the null hypothesis is in fact true—the treatment really doesn't work. The second row shows the results of comparisons for which there truly is a difference. The first column tabulates comparisons for which the P value was low enough to be deemed statistically significant (or a discovery, in the lingo of the FDR method discussed earlier). The second column tabulates comparisons for which the P value was high enough to be deemed not statistically significant (or not a discovery).

	DECISION: STATISTICALLY SIGNIFICANT, OR DISCOVERY	DECISION: NOT STATISTICALLY SIGNIFICANT OR NOT A DISCOVERY	TOTAL
Null hypothesis: True	A (false positive)	B	A + B
Null hypothesis: False	C	D (false negative)	C + D
Total	A + C	B+D	A + B+ C + D

Table 22.2. This table (identical to Table 18.2) shows the results of many statistical analyses, each analyzed to reach a decision to reject or not reject the null hypothesis.
The top row tabulates results for experiments for which the null hypothesis is really true. The second row tabulates experiments for which the null hypothesis is not true. When you analyze data, you don't know whether the null hypothesis is true, so you could never create this table from an actual series of experiments. A, B, C, and D are integers (not proportions) that count the number of analyses.

It would be nice if all comparisons ended up in Cells B or C, leaving A and D empty. This will rarely be the case. Even if the null hypothesis is true, random sampling will ensure that some comparisons will mistakenly yield a statistically significant conclusion and contribute to Cell A. And even if the null hypothesis is false, random sampling will ensure that some results will be not statistically significant and will contribute to Cell D.

A, B, C, and D each represent a number of comparisons, so the sum of A + B + C + D equals the total number of comparisons you are making.

If you make no correction for multiple comparisons

What happens if you make no correction for multiple comparisons and set α to its conventional value of 5%? Of all experiments done when the null hypothesis is true, you would expect 5% to be statistically significant just by chance. In other words, you would expect the ratio A/(A + B) to equal 5%. This 5% value applies to each comparison separately so is called a per-comparison error rate. In any particular set of comparisons, that ratio might be greater than 5% or less than 5%. But, on average, if you make many comparisons, that is the value you'd expect.

If you use the Bonferroni method to correct for multiple comparisons

If you use the Bonferroni method and set α to its conventional value of 5%, then there is a 5% chance of designating one or more comparisons (for which the null hypothesis is really true) as statistically significant. In other words, there is a 5% chance that the value A in Table 22.2 is greater than zero, and a 95% chance that A equals zero.

If you use the false discovery method to correct for multiple comparisons

In Table 22.2, the total number of discoveries equals A + C. If you set Q (the desired FDR) to 5%, then you would expect the ratio A/(A + C) to be no higher than 5% and the ratio C/(A + C) to exceed 95%.

Table 22.3 compares the three methods for dealing with multiple comparisons.

APPROACH	WHAT YOU CONTROL	FROM TABLE 22.1
No correction for multiple comparisons	α = if all null hypotheses are true, the fraction of all experiments for which the conclusion is statistically significant	$\alpha = A/(A + B)$
Bonferroni	α = if all null hypotheses are true, the chance of obtaining one or more statistically significant conclusions	α = probability that $A > 0$
False discovery rate	Q = the fraction of all discoveries for which the null hypothesis really is true	$Q = A/(A + C)$

Table 22.3. Three approaches to handling multiple comparisons.

Q & A

If you make 10 independent comparisons and all null hypotheses are true, what is the chance that none will be statistically significant?

If you use the usual 5% significance level, the probability that each test will be not statistically significant is 0.95. The chance that all 10 will be not significant is 0.95^{10}, or 59.9%.

Is the definition of a family of comparisons always clear?

No, it is a somewhat arbitrary definition, and different people seeing the same data could apply that definition differently.

If you make only a few planned comparisons, why don't you have to correct for those comparisons?

It makes sense to correct for all comparisons that were made, whether planned or not. But some texts say that if you only plan a few comparisons, you should get rewarded by not having to correct for multiple comparisons. This recommendation doesn't really make sense to me.

When using the Bonferroni correction, does the variable α refer to the overall family-wide significance level (usually 5%) or the threshold used to decide when a particular comparison is statistically significant (usually 0.05/K, where K is the number of comparisons)?

The significance level α refers to the overall familywide significance level (usually 5%).

Will you know when you make a false discovery?

No. You never know for sure whether the null hypothesis is really true, so you won't know when a discovery is false. Similarly, you won't know when you make Type I or Type II errors with statistical hypothesis testing.

Are there other ways to deal with multiple comparisons?

Yes. One way is to fit a multilevel hierarchical model, but this requires special expertise and is not commonly used, at least in biology (Gelman, 2012).

Physicists refer to the "look elsewhere effect." How does that relate to multiple comparisons?

It is another term for the same concept. When physicists search a tracing for a signal, they won't get too excited when they find a small one, knowing that they also looked elsewhere for a signal. They need to account for all the places they looked, which is the same as accounting for multiple comparisons.

CHAPTER SUMMARY

- The multiple comparisons problem is clear. If you make lots of comparisons (and make no special correction for the multiple comparisons), you are likely to find some statistically significant results just by chance.
- Coping with multiple comparisons is one of the biggest challenges in data analysis.
- One way to deal with multiple comparisons is to analyze the data as usual but to fully report the number of comparisons that were made and then let the reader account for the number of comparisons.
- If you make 13 independent comparisons with all null hypotheses true, there is about a 50:50 chance that one or more P values will be less than 0.05.
- The most common approach to multiple comparisons is to define the significance level to apply to an entire family of comparisons, rather than to each individual comparison.
- A newer approach to multiple comparisons is to control or define the FDR.

TERMS INTRODUCED IN THIS CHAPTER

- Bonferroni correction (p. 207)
- Discovery (p. 209)
- Family of comparisons (p. 206)
- Familywise error rate (p. 206)
- Genome-wide association studies (GWAS) (p. 208)
- Multiple comparisons (p. 203)
- Per-comparison error rate (p. 206)
- Per-experiment error rate (p. 206)
- Planned comparisons (p. 205)

CHAPTER 23

The Ubiquity of Multiple Comparisons

If the fishing expedition catches a boot, the fishermen should
throw it back, and not claim that they were fishing for boots.

JAMES L. MILLS

Chapter 22 explained the problem of multiple comparisons. This
chapter demonstrates how pervasive this problem is in many kinds
of data analysis.

OVERVIEW

Berry (2007, p. 155) concisely summarized the importance and ubiquity of multiple comparisons:

> Most scientists are oblivious to the problems of multiplicities. Yet they are everywhere. In one or more of its forms, multiplicities are present in every statistical application. They may be out in the open or hidden. And even if they are out in the open, recognizing them is but the first step in a difficult process of inference. Problems of multiplicities are the most difficult that we statisticians face. They threaten the validity of every statistical conclusion.

This chapter points out some of the ways in which the problem of multiple comparisons affects the interpretation of many statistical results.

MULTIPLE COMPARISONS IN MANY CONTEXTS

Multiple subgroups

Analyzing data divided into multiple subgroups is a common form of multiple comparisons. A simulated study by Lee and colleagues (1980) pointed out potential problems. They pretended to compare survival following two "treatments" for coronary artery disease. They studied a group of real patients with coronary artery disease randomly divided them into two groups, A and B. In a real study, they would give the two groups different treatments and compare survival after treatment. In this simulated study, they treated all subjects identically but analyzed the data as if the two random groups were actually given two distinct treatments. As expected, the survival of the two groups was indistinguishable.

They then divided the patients in each group into six subgroups depending on whether they had disease in one, two, or three coronary arteries and whether the heart ventricle contracted normally. Because these are variables that are expected to affect survival of the patients, it made sense to evaluate the response to "treatment" separately in each of the six subgroups. Whereas they found no substantial difference in five of the subgroups, they found a striking result among patients with three-vessel disease who also had impaired ventricular contraction. Of these patients, those assigned to Treatment B had much better survival than those assigned to Treatment A. The difference between the two survival curves was statistically significant, with a P value less than 0.025.

If this were an actual study comparing two alternative treatments, it would be tempting to conclude that Treatment B is superior for the sickest patients and to recommend Treatment B to those patients in the future. But in this study, the random assignment to Treatment A or Treatment B did not alter how the patients were actually treated. Because the two sets of patients were treated identically, we can be absolutely, positively sure that the survival difference was a coincidence.

It is not surprising that the authors found one low P value out of six comparisons. Figure 22.1 shows that there is a 26% chance that at least one of six independent comparisons will have a P value of less than 0.05, even if all null hypotheses are true.

If the difference between treatments had not been statistically significant in any of the six subgroups used in this example, the investigators probably would have kept looking. Perhaps they would have gone on to separately compare Treatments A and B only in patients who had previously had a myocardial infarction (heart attack), only in those with abnormal electrocardiogram findings, or only those who had previous coronary artery surgery. Since you don't know how many comparisons were possible, you cannot rigorously account for the multiple comparisons.

For analyses of subgroups to be useful, it is essential that the study design specify all subgroups analyses, including methods to be used to account for multiple comparisons.

Multiple sample sizes

If you perform an experiment and the results aren't quite statistically significant, it is tempting to run the experiment a few more times (or add a few more subjects) and then analyze the data again with the larger sample size. If the results still aren't significant, then you might do the experiment a few more times (or add more subjects) and reanalyze once again. The problem is you can't really interpret the results when you informally adjust sample size as you go.

If the null hypothesis of no difference is in fact true, the chance of obtaining a statistically significant result using that informal sequential approach is far higher than 5%. In fact, if you carry on that approach long enough, then every single experiment will eventually reach a statistically significant conclusion, even if the null hypothesis is true. Of course, "long enough" might be very long indeed, exceeding your budget or even your life span.

The problem with this approach is that you continue collecting data only when the result is not statistically significant and stop when the result is statistically significant. If the experiment were continued after reaching significance, adding more data might then result in a conclusion that the results are not statistically significant. But you'd never know this, because you would have stopped once significance was reached. If you keep running the experiment when you don't like the results but stop the experiment when you do like the results, you can't interpret the results at face value.

Statisticians have developed rigorous ways to allow you to dynamically adjust sample size. Look up *sequential data analysis*. These methods use more stringent criteria to define significance to make up for the multiple comparisons. Without these special methods, you can't interpret the results unless the sample size is set in advance.

Multiple end points: Outcome switching

When conducting clinical studies, investigators often collect data on multiple clinical outcomes. When planning the study, investigators should specify the outcome they care most about, called the *primary outcome*. That is the outcome that they most care about. The other outcomes are called *secondary outcomes*. By now you should realize why it is so important that the primary outcome be carefully defined in advance. If many outcomes are measured, you wouldn't be surprised if some show statistically significant differences by chance. The analysis of secondary outcomes can strengthen the scientific argument of a paper and lead to new hypotheses to study. But statistical analyses must be focused on the primary outcome chosen in advance. And the secondary outcomes too should be prespecified n the experimental protocol.

Every once in a while, it can make sense to change the definition of the primary outcome between study design and publication in response to a new understanding of the disease being studied. But this kind of change is appropriate only when fully documented in the paper and when the decision is made by people who have not yet seen any of the study data (Evans, 2007). Results will be misleading if the investigators first look at the data for all the outcomes, identify the outcome that has the biggest effect or the smallest P value, and then label that outcome as "primary." Yet this happens frequently:

- Chan and colleagues (2004) identified all protocols for randomized clinical trials approved in two cities over two years. A decade later, they looked at how the results of the studies were reported. In half the studies, they found that the primary outcome specified in the published paper was not the primary outcome defined in the original protocol.
- Vera-Badillo and colleagues (2013) found similar problems with the reporting of randomized clinical trials of various treatments of breast cancer.

They identified 30 studies that had posted the experimental protocol in advance. In seven of those studies, the primary end point reported in the publication differed from the primary end point planned in advance. They also found that when the results for the primary end point were not statistically significant, over half of the studies used "spin" to misrepresent the results, and many didn't even mention the primary outcome in the concluding sentence of the abstract.

• Goldacre and colleagues (2016) compared 67 clinical trials published in several major journals in 2015 with the protocol published in advance. Only nine of those studies followed the protocol exactly. On average, the publications omitted nearly half of the prespecified outcomes, and most reported several outcomes not specified in the protocol.

You can't draw proper inferences from clinical studies unless you can be sure that the investigators chose the primary outcome as part of the study plan and stuck with that decision when writing up the results and that the list of secondary outcomes was decided in advance and reported fully.

Multiple geographic areas

Five children in a particular school developed leukemia last year. Is that a coincidence? Or does the clustering of cases suggest the presence of an environmental toxin that caused the disease? These questions are very difficult to answer (Thun & Sinks, 2004). It is tempting to estimate the answer to a different question: What is the probability that five children in this particular school would all get leukemia this year? You could calculate (or at least estimate) the answer to that question if you knew the overall incidence rates of leukemia among children and the number of children enrolled in the school. The probability would be tiny. Everyone intuitively knows that and so is alarmed by the cluster of cases.

But you've asked the wrong question once you've already observed the cluster of cases. The school only came to your attention this year because of the cluster of cases, so you must consider all other schools and other years. You need to ask about the probability that five children in *any* school would develop leukemia in the same year. Clearly the answer to this question is much higher than the answer to the previous one. To calculate a precise answer, you would have to decide which geographical areas to include, which years to include, and which diagnoses to include.

About 1,000 cancer clusters are reported to US health officials each year. About three-quarters of these are not really clusters of similar kinds of cancers, but that still leaves several hundred cancer clusters each year. These are fully investigated to look for known toxins and to be alert to other findings that might suggest a common cause. But most disease clusters turn out to be coincidences. It is surprising to find a cluster of one particular disease in one particular place at any one particular time. But chance alone will result in many clusters of various diseases in various places at various times.

Multiple predictions or hypotheses

In 2000, the Intergovernmental Panel on Climate Change made predictions about future climate conditions. Pielke (2008) asked what seemed like a straightforward question: How accurate were those predictions over the next seven years? That's not long enough to seriously assess predictions of global warming, but it is a necessary first step. Answering this question proved impossible (Tierney, 2008). The problems were that the report contained numerous predictions and didn't specify which sources of climate data should be used. Did the predictions come true? The answer depends on the choice of which prediction to test and which data set to use for testing—"a feast for cherry pickers" (Pielke, 2008).

You can only evaluate the accuracy of a prediction or diagnosis when the timing, methodology, and data source are precisely defined. If there are multiple ways to assess the accuracy of predictions, you may get contradictory results.

It is also essential to record your predictions and not rely on memory alone. Psychologists have repeatedly demonstrated a phenomenon they call *hindsight bias* (Goodwin, 2010). People do a really bad job of remembering their predictions after the event occurs. Here's an example. College students were asked to state what they thought the chances were of a "not guilty" verdict in the O. J. Simpson murder trial (Bryant & Brockway, 1997). Becaue the defendant was so famous, the trial was heavily reported in the media and almost everyone had an opinion about the verdict. A week after Simpson was acquitted by the jury, the students were asked to recall their prediction of the probability of acquittal. These recollected probabilities were much higher than the recorded predictions, especially among people who originally predicted a small chance of acquittal. This effect is also called the *I-knew-it-all-along effect*. People tend to change their memory of their own predictions to match what actually ended up happening.

Kerr (1998) defined the acronym *HARK*, which means hypothesizing after the results are known. This occurs when a scientist first looks at the data, which includes many variables or subgroups, uses those data to pose a hypothesis, and then publishes the results so it appears that the hypothesis was stated before the data collection began. This can be viewed as another form of multiple comparisons, as complex data sets allow you to pose many hypotheses. If you search through the data to find a difference or association, that finding is likely to be a coincidence and so will not hold up in repeat experiments.

Multiple definitions of groups

When comparing two groups, the groups must be defined as part of the study design. If the groups are defined by the data, many comparisons are being made implicitly and the results cannot be interpreted.

Austin and Goldwasser (2008) demonstrated this problem. They looked at the incidence of hospitalization for heart failure in Ontario, Canada, in 12 groups of patients defined by their astrological sign (determined by their birthday). People born under the sign of Pisces happened to have the highest incidence of heart failure. After seeing this association, the investigators did a simple statistics test to compare the incidence of heart failure among people born under Pisces with the incidence of heart failure among all others (people born under all other 11 signs combined into one group). Taken at face value, this comparison showed that the people born under Pisces have a statistically significant higher incidence of heart failure than do people born under the other 11 signs (the P value was 0.026).

The problem is that the investigators didn't really test a single hypothesis; they implicitly tested 12. They only focused on Pisces after looking at the incidence of heart failure for people born under all 12 astrological signs. So it isn't fair to compare that one group against the others without considering the other 11 implicit comparisons.

Let's do the math. If you do 12 independent comparisons, what is the chance that every P value will be greater than 0.026? Since the probabilities of independent comparisons are independent of each other, we can multiply them. The probability of getting a P value greater than 0.026 in one comparison is $1.0 - 0.026$, or 0.974. The probability of two P values both being greater than 0.026 in two independent comparisons is $0.974 \times 0.974 = 0.949$. The chance that all 12 independent comparisons would have P values greater than 0.026 is $0.974^{12} = 0.729$. Subtract that number from 1.00 to find the probability that at least one P value of 12 will be less than 0.026. The answer is 0.271. In other words, there is a 27.1% chance that one or more of the 12 comparisons will have a P value less than or equal to 0.026 just by chance, even if all 12 null hypotheses are true. Once you realize this, the association between astrological sign and heart failure does not seem impressive.

Multiple ways to select variables in regression

Multiple regression (as well as logistic and proportional hazards regression) models predict an outcome as a function of multiple independent (input) variables. Some programs offer automated methods to choose which independent variables should be included in the model and which should be omitted. Chapters 37 and 38 discuss these techniques.

Freedman (1983) did simulations to show how these methods can lead to misleading results. His paper is reprinted as an appendix in a text by Good and Hardin (2006). He simulated a study with 100 subjects using data from 50 independent variables recorded for each subject. The simulations were performed so all the variation was random. The simulated outcome had no relationship at all to any of the simulated inputs. In other words, the data were all noise and no signal.

A multiple regression model (see Chapter 37) was used to find the equation that best predicted the outcome from all the inputs, and a P value was computed for each input variable. Each P value tested a null hypothesis that a particular input variable had no impact on predicting the outcome. The results were what you would expect: the P values were randomly spread between zero and 1. An overall P value, testing the null hypothesis that the overall model was not helpful in predicting the outcome, was high (0.53).

No surprises so far. Random variables went in, and a random assortment of P values came out. Anyone looking at the conclusion (and not knowing the data were randomly generated) would conclude that the 50 input variables have no ability to predict the outcome.

Freedman then selected the 15 input variables that happened to have the lowest P values (less than 0.25) and refit the multiple regression model using only those 15 input variables, rather than all 50 as before. Now the overall P value was tiny (0.0005). Of the 15 variables now in the model, 6 had an associated P value smaller than 0.05.

Anyone who saw only this analysis and didn't know the 15 input variables were selected from a larger group of 50 would conclude that the outcome variable is related to, and so can be predicted from, the 15 input variables. Because these are simulated data, we know that this conclusion would be wrong. There is, in fact, no consistent relationship between the input variables and the outcome variable.

By including so many variables and selecting a subset that happened to be predictive, the investigators were performing multiple comparisons. Freedman knew this, and that is the entire point of his paper. But the use of this kind of variable selection happens often in analyzing large studies. When multiple regression analyses include lots of opportunity for variable selection, it is easy to be misled by the results, especially if the investigator doesn't explain exactly what she has done.

Multiple ways to preprocess the data

As mentioned in Chapter 7, it is very common for scientists not to analyze the direct experimental measurements but instead to first process the data by transforming, normalizing, smoothing, adjusting (for other variables), or removing outliers (see Chapter 25). These kinds of manipulations can make lots of sense when done in a careful, preplanned way. However, you engage in multiple comparisons if you repeat the statistical analyses on data processed in multiple ways and then report analyses on the form that gives you the results for which you are hoping.

Multiple ways to dichotomize

When analyzing data, scientists sometimes divide the sample into two groups on the basis of a measurement. For example, one can use weight to designate people as obese or not obese, or can use age to categorize people as young or old. Dividing a group into two parts is called *dichotomizing*. It can make some analyses easier, but it throws away information and so is usually discouraged. But dichotomizing can be really misleading when the investigator tries different cut-off values and then chooses the cut-off that makes the resulting comparison have the smallest P value. This is a form of multiple comparisons.

Multiple ways to analyze the data

This story told by Vickers (2006) is only a slight exaggeration:

> STATISTICIAN: Oh, so you have already calculated the P value?
> SURGEON: Yes, I used multinomial logistic regression.
> STATISTICIAN: Really? How did you come up with that?
> SURGEON: Well, I tried each analysis on the SPSS drop-down menus, and that was the one that gave the smallest P value.

If you analyze your data many ways—perhaps first with a t test, then with a nonparametric Mann–Whitney test, and then with a two-way ANOVA (adjusting for another variable)—you are performing multiple comparisons.

Multiple ways to potentially analyze the data: The garden of forking paths

Investigators may not decide exactly how to analyze the data until they look at it. In the end, they only do one analysis. But had the data come out differently, they would have done a different analysis. Even if these choices follow a formal plan, the multiple potential analyses are enough to provoke a multiple comparisons issue. Gelman and Loken (2014) referred to this common problem as the *garden of forking paths*.

WHEN ARE MULTIPLE COMPARISONS DATA TORTURE OR P-HACKING?

This chapter lists the many forms of multiple comparisons. There is not much of a problem if all the comparisons are reported and the analyses adjusted appropriately. But there is a big problem when investigators try to wring statistical significance out of data by using all the methods listed in this chapter. Mills (1993) called this *data torture*. Simmons and colleagues (2012) call it *p-hacking* and lament the many *researcher degrees of freedom* (Simmons, Nelson, & Simonsohn, 2011).

Simmons and colleagues (2011) simulated a set of experiments that demonstrate how allowing researcher degrees of freedom can increase the chance of a false positive result. The simulation is for a simple situation in which the investigator compares two groups of subjects (n = 20 per group, each group being a mixture of men and women). The investigator first tests whether the mean value of a variable differs significantly between the two groups. If not, he tests a second variable, and if that doesn't "work," he computes the mean of the two variables and tests whether that mean differs between groups. If that doesn't yield a "statistically significant" result, he collects data from 10 more subjects in each group to increase the sample size from 20 to 30 per group and reanalyzes the data. If even that doesn't produce a "statistically significant" result, he runs a more sophisticated analysis that accounts for differences between men and women. This set of steps doesn't seem very different from what a lot of scientists do routinely.

The simulated data were all sampled from a single Gaussian distribution, so we can be absolutely sure that there is no difference between the two populations and that the null hypothesis is true. In this situation, using the SD of statistical significance, the investigators should reach an incorrect conclusion that the difference is statistically significant in 5% of the simulations. In fact, they found a "statistically significant" ($P < 0.05$) result in 31% of the simulated experiments! The use of researcher degrees of freedom as previously described (also called p-hacking) increased the chance of a false positive (Type I) error sixfold, from 5% to 31%. It would be easy to add even more researcher degrees of freedom by looking only at subgroups, removing outliers, analyzing the data after transforming to logarithms, switching to nonparametric tests, and so on. These methods would increase the false positive rate even further.

If investigators use even a few researcher degrees of freedom (i.e., allow themselves to p-hack), the chance of finding a bogus (false positive) "statistically significant" result increases. Statistical results can only be interpreted at face value if the analyses were planned in advance and the investigators don't try alternative analyses once the data are collected. When the number of possible analyses is not defined in advance (and is almost unlimited), results simply cannot be trusted.

HOW TO COPE WITH MULTIPLE COMPARISONS

This chapter shows you many ways in which multiple comparisons can impede progress in science. The best way to deal with multiple comparisons is to plan studies with focused hypotheses and defined methods to account for multiple comparisons in analyses. When evaluating results that include multiple comparisons, your best defense is the Bonferroni method discussed in Chapter 22.

When reviewing completed studies, you can correct for multiple comparisons if you know how many comparisons were made. Simply divide the usual threshold that defines statistical significance (often 0.05) by the number of comparisons. So if the study is making 100 comparisons, divide 0.05 by 100 and only declare a result to be statistically significant (at the 5% significance level with correction for multiple comparisons) when the P value is less than 0.0005.

Q & A

Why does this chapter only have only one Q&A and the rest have many?
Comparing chapters is a form of multiple comparisons you should avoid.

CHAPTER SUMMARY

- The problem of multiple comparisons shows up in many situations:
  - Multiple end points
  - Multiple time points
  - Multiple subgroups
  - Multiple geographical areas
  - Multiple predictions
  - Multiple geographical groups
  - Multiple ways to select variables in multiple regression
  - Multiple methods of preprocessing data

- ◆ Multiple ways to dichotomize
- ◆ Multiple statistical analyses
- ◆ Garden of forking paths
- You can only interpret statistical results when you know how many comparisons were made.
- All analyses should be planned, and all planned analyses should be conducted and reported. These simple guidelines are often violated.

TERMS INTRODUCED IN THIS CHAPTER

- Data torture (p. 221)
- Dichotomizing (p. 220)
- Garden of forking paths (p. 221)
- Hindsight bias (p. 218)
- Hypothesizing after the results are known (HARK) (p. 218)
- I-knew-it-all-along effect (p. 218)
- P-hacking (p. 221)
- Primary outcome (p. 216)
- Researcher degrees of freedom (p. 221)
- Secondary outcome (p. 216)

CHAPTER 24

Normality Tests

You need the subjunctive to explain statistics.

DAVID COLQUHOUN

Many statistical tests (including t tests, ANOVA, and regression) assume that the data were sampled from a Gaussian distribution (Chapter 10). When you analyze data or review analyses that others have done, you need to ask whether that assumption is reasonable. This sounds like a simple question, but it really isn't.

THE GAUSSIAN DISTRIBUTION IS AN UNREACHABLE IDEAL

In almost all cases, we can be 100% sure that the data were not sampled from an ideal Gaussian distribution. Why? Because an ideal Gaussian distribution extends infinitely in both directions to include both super-large values and extremely negative values. In most scientific situations, there are constraints on the possible values. Blood pressures, concentrations, weights, and many other variables cannot have negative values, so they can never be sampled from perfect Gaussian distributions. Other variables can be negative but have physical or physiological limits that don't allow extremely large values (or extremely low negative values). These variables also cannot follow a perfect Gaussian distribution. Micceri (1989) looked at 440 different variables that psychologists measure and concluded that every one deviated substantially from a Gaussian distribution. I suspect the same is true for most scientific fields.

Because almost no variables you measure follow an ideal Gaussian distribution, why use tests that rely on the Gaussian assumption? Plenty of studies with simulated data have shown that the statistical tests based on the Gaussian distribution are useful when data are sampled from a distribution that only approximates a Gaussian distribution. These tests are fairly robust to violations of the Gaussian assumption, especially if the sample sizes are large and equal. So the question that matters is not whether the data were sampled from an ideal Gaussian population but rather whether the distribution from which they were sampled is close enough to the Gaussian ideal that the results of the statistical tests are still useful.

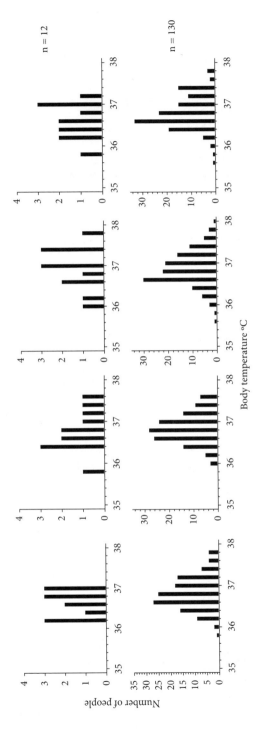

Figure 24.1. Gaussian distributions don't always look Gaussian.

These graphs show frequency distributions of values randomly chosen from a Gaussian distribution. The top four graphs are for samples with 12 values. The mean is 36.77, and the SD is 0.40. The bottom four graphs have a mean of 36.82 and an SD of 0.41, and n = 130. The variation from graph to graph (sample to sample) is simply the result of random variation when sampling from a Gaussian distribution. You rarely see smooth bell-shaped curves unless the sample size is enormous.

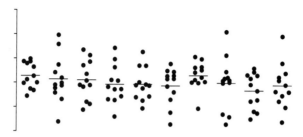

Figure 24.2. Samples from Gaussian distributions don't always look Gaussian.

All 10 samples were randomly sampled from Gaussian distributions. It is too easy to be fooled by random variation and think that the data are far from Gaussian.

WHAT A GAUSSIAN DISTRIBUTION REALLY LOOKS LIKE

The ideal Gaussian frequency distribution is shown in Figure 10.1. With huge data scts, this is what you expect a Gaussian distribution to look like. But what about smaller data sets?

Figure 24.1 shows the distribution of values randomly chosen from a Gaussian distribution. The top four distributions are four different samples with 12 values in each sample, and the bottom four graphs show samples with 130 values. Because of random sampling variation, none of these frequency distributions really looks completely bell shaped and symmetrical.

The bell-shaped Gaussian distribution is an ideal distribution of the population. Unless the samples are huge, actual frequency distributions tend to be less symmetrical and more jagged. This is just the nature of random sampling.

Figure 24.2 makes the same point by showing individual values rather than frequency distributions. Each sample was drawn from a Gaussian distribution. To many people, however, some of these distributions don't look Gaussian. Some people prefer to plot the cumulative frequency distributions, perhaps transforming the axes so that a Gaussian distribution becomes linear (called a QQ plot). But re-plotting the data doesn't change the issue. Unless the sample size is huge, random sampling can create distributions that don't really look Gaussian. It takes a lot of experience to interpret such plots.

QQ PLOTS

When investigators want to show that the data are compatible with the assumption of sampling from a Gaussian distribution, they sometimes show a QQ plot, like the ones in Figure 24.3. The Y axis plots the actual values in a sample of data. The X axis plots ideal predicted values assuming the data were sampled from a Gaussian distribution.

If the data are truly sampled from a Gaussian distribution, all the points will be close to the line of identity (shown in a dotted line in Figure 24.3). The data

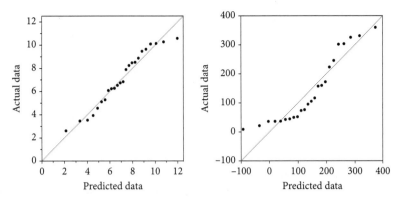

Figure 24.3. QQ Plots
The left panel shows a QQ plot made from data sampled from a Gaussian distribution.
The right panel shows data sampled from a lognormal distribution.

in the left panel of the figure are sampled from a Gaussian distribution, and the points lie pretty close to the line of identity.

Systematic deviation of points from the line of identity is evidence that the data are not sampled from a Gaussian distribution. But it is hard to say how much deviation is more than you'd expect to see by chance. It takes some experience to interpret a QQ plot. The data in the right panel of Figure 24.3 were not sampled from a Gaussian distribution (but rather from a lognormal distribution), and you can see that the points systematically deviate from the line of identity.

Here is a brief explanation of how a QQ plot is created. First, the program computes the percentile of each value in the data set. Next, for each percentile, it calculates how many SDs from the mean you need to go to reach that percentile on a Gaussian distribution. Finally, using the actual mean and SD computed from the data, it calculates the corresponding predicted value.

Why the strange name QQ? Q is an abbreviation for *quantile*, which is the same as a percentile but expressed as a fraction. The 95th percentile is equivalent to a quantile of 0.95. The 50th percentile (median) is the same as a quantile of 0.5. The name *QQ plot* stuck even though quantiles don't actually appear on most versions of QQ plots.

TESTING FOR NORMALITY

Statistical tests can be used to quantify how much a data set deviates from the expectations of a Gaussian distribution. Such tests are called *normality tests*.

How normality tests work
The first step in running a normality test is to quantify how far a set of values differs from the predictions of a Gaussian distribution. One of the more popular

normality tests is the *D'Agostino–Pearson omnibus K2 normality test*. This test first computes two values that quantify how far the distribution deviates from the Gaussian ideal:

- *Skewness* quantifies symmetry. A symmetrical distribution (such as the Gaussian) has a skewness of zero. An asymmetrical distribution with a heavy right tail (more large values) has a positive distribution, and one with a heavy left tail (more small numbers) has a negative skew.
- *Kurtosis* quantifies whether the tails of the data distribution matches the Gaussian distribution. A Gaussian distribution has a kurtosis of zero; a distribution with fewer values in the tails than a Gaussian distribution has a negative kurtosis; and a distribution with more values in the tails (or values further out in the tails) than a Gaussian distribution has a positive kurtosis. Although it is commonly thought to measure the shape of the peak, kurtosis actually tells you virtually nothing about the shape of the peak and only measures the presence of values in the tails—that is, outliers (Westfall, 2014).

The D'Agostino–Pearson omnibus K2 normality test combines the skewness and kurtosis into a single value that describes how far a distribution is from Gaussian. Other normality tests (such as the Shapiro–Wilk test, the Kolmogorov–Smirnov test, and the Darling–Anderson test) use other approaches to quantify the discrepancy between the observed distribution and the Gaussian distribution.

The meaning of a P value from a normality test

All normality tests compute a P value that answers the following question: If you randomly sample from a Gaussian population, what is the probability of obtaining a sample that deviates from a Gaussian distribution as much as (or more so than) this sample does?

Interpreting a high P value

If the P value from a normality test is large, all you can say is that the data are not inconsistent with a Gaussian distribution. Ugh! Statistics requires the use of double negatives.

A normality test cannot prove the data were sampled from a Gaussian distribution. All the normality test can do is demonstrate that the deviation from the Gaussian ideal is not more than you'd expect to see from chance alone.

Interpreting a small P value

If the P value from the normality test is small, you reject that null hypothesis and so accept the alternative hypothesis that the data are not sampled from a Gaussian population. But this is less useful than it sounds. As explained at the beginning of this chapter, few variables measured in scientific studies are completely Gaussian. So with large sample sizes, normality tests will almost always report a small P value, even if the distribution only mildly deviates from a Gaussian distribution.

ALTERNATIVES TO ASSUMING A GAUSSIAN DISTRIBUTION

If you don't wish to assume that your data were sampled from a Gaussian distribution, you have several choices:

- Identify another distribution from which the data were sampled. In many cases, you can then transform your values to create a Gaussian distribution. Most common, if the data come from a lognormal distribution (see Chapter 11), you can transform all values to their logarithms.
- Ignore small departures from the Gaussian ideal. Statistical tests tend to be quite robust to mild violations of the Gaussian assumption.
- Identify and remove outliers (see Chapter 25).
- Switch to a nonparametric test that doesn't assume a Gaussian distribution (see Chapter 41).

Don't make the mistake of jumping directly to the fourth option, using a nonparametric test. It is very hard to decide when to use a statistical test based on a Gaussian distribution and when to use a nonparametric test. This decision really *is* difficult, requiring thinking, perspective, and consistency. For that reason, the decision should not be automated.

COMMON MISTAKES: NORMALITY TESTS

Mistake: Using a normality test to automate the decision of when to use a nonparametric test

Nonparametric tests (explained in Chapter 41) do not assume a Gaussian distribution, so it seems logical to use the results of a normality test to decide whether to use a nonparametric test. In fact, the decision of when to use a nonparametric test is not straightforward, and reasonable people can disagree about when to use them. The results of a normality test are not very helpful in making that decision. The problem is that a normality test is designed to detect evidence that the distribution deviates from a Gaussian distribution, but it does not quantify whether that deviation is large enough to invalidate the usual tests that assume a Gaussian distribution. I delve into that subject in more detail in Chapter 41.

Mistake: Using a normality test with tiny samples

Normality tests try to determine the distribution (population) from which the data were sampled. That requires data. With tiny samples (three or four values), it really is impossible to make useful inferences about the distribution of the population.

Q & A

Does it make sense to ask whether a particular data set is Gaussian?

No. A common misconception is to think that the normality test asks whether a particular data set is Gaussian. But the term *Gaussian* refers to an entire distribution or population. It only makes sense to ask about the population or distribution from which your data were sampled. Normality tests ask whether the data are consistent with the assumption of sampling from a Gaussian distribution.

Is it ever possible to know for sure whether a particular data set was sampled from a Gaussian distribution?

Not unless the data were simulated. Otherwise, you can never know for sure the distribution from which the data were sampled.

Do normality tests determine whether the data are parametric or nonparametric?

No! The terms *parametric* and *nonparametric* refer to statistical tests, not data distributions.

Should a normality test be run as part of every experiment?

Not necessarily. You want to know whether a certain kind of data are consistent with sampling from a Gaussian distribution. The best way to find out is to run a special experiment just to ask about the distribution of data collected using a particular method. This experiment would need to generate plenty of data points but would not have to make any comparisons or ask any scientific questions. If analysis of many data points convinces you that a particular experimental protocol generates data that are consistent with a Gaussian distribution, there is no point in testing smaller data sets from individual runs of that experiment.

Will every normality test give the same result?

No. The different tests quantify deviations from normality in different ways, so they will return different results.

How many values are required to run a normality test?

It depends on which test you choose. Some work with as few as five values.

Does a normality test decide whether the data are far enough from Gaussian to require a nonparametric statistical test?

No. It is hard to define what "far enough" means, and the normality tests were not designed with this aim in mind.

How useful are normality tests?

Not very. It would be very rare for any set of scientific data to truly be sampled from an ideal Gaussian distribution. With small samples, the normality tests don't have much power to detect non-Gaussian distributions. With large samples, the tests have too much power, since they can detect minor deviations from the ideal that wouldn't affect the validity of a t test or ANOVA.

What is the difference between a Gaussian distribution and a Normal distribution?

No difference. Those two terms are synonyms.

CHAPTER SUMMARY

- The Gaussian distribution is an ideal that is rarely achieved. Few, if any, scientific variables completely follow a Gaussian distribution.
- Data sampled from Gaussian distributions don't look as Gaussian as many expect. Random sampling is more random than many appreciate.
- Normality tests are used to test for deviations from the Gaussian ideal.
- A small P value from a normality test only tells you that the deviations from the Gaussian ideal are more than you'd expect to see by chance. It tells you nothing about whether the deviations from the Gaussian ideal are large enough to affect conclusions from statistical tests that assume sampling from a Gaussian distribution.

TERMS INTRODUCED IN THIS CHAPTER

- D'Agostino–Pearson omnibus K2 normality test (p. 228)
- Kurtosis (p. 228)
- Normality test (p. 227)
- Quantile (p. 227)
- QQ plot (p. 227)
- Skewness (p. 228)

CHAPTER 25

Outliers

There are liars, outliers, and out-and-out liars.

ROBERT DAWSON

An outlier is a value that is so far from the others that it appears to have come from a different population. More informally, an outlier is a data point that is too extreme to fit your preconceptions. The presence of outliers can invalidate many statistical analyses. This chapter explains the challenges of identifying outliers.

HOW DO OUTLIERS ARISE?

Outliers—also called anomalous, spurious, rogue, wild, or contaminated observations—can occur for several reasons.

- Invalid data entry. The outlier may simply be the result of transposed digits or a shifted decimal point. If you suspect an outlier, the first thing to do is to make sure the data were entered correctly and that any calculations (changing units, normalizing, etc.) were done accurately.
- Biological diversity. If each value comes from a different person or animal, the outlier may be a correct value. It is an outlier because that individual may truly be different from the others. This may be the most exciting finding in your data!
- Random chance. In any distribution, some values are far from the others by chance.
- Experimental mistakes. Most experiments have many steps, and it is possible that a mistake was made.
- Invalid assumption. If you assume data are sampled from a Gaussian distribution, you may conclude that a large value is an outlier. But if the distribution is, in fact, lognormal, then large values are common and are not outliers. See Chapter 11.

THE NEED FOR OUTLIER TESTS

The presence of an outlier can spoil many analyses, either creating the appearance of differences (or associations, correlations, etc.) or blocking discovery of real differences (or associations, correlations, etc.).

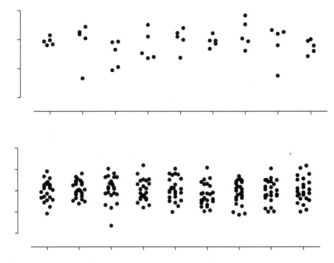

Figure 25.1. No outliers here. These data were sampled from Gaussian distributions.

All of these data sets were computer generated and sampled from a Gaussian distribution. But when you look at them, some points just seem too far from the rest to be part of the same distribution. They seem like real outliers—but they are not. The human brain is very good at seeing patterns and exceptions from patterns, but it is poor at recognizing random scatter.

It would seem that the presence of outliers would be obvious. If this were the case, we could deal with outliers informally. But identifying outliers is much harder than it seems.

Figure 25.1 shows the problem with attempting to identify outliers informally. It shows 18 data sets all sampled from a Gaussian distribution. Half of the samples have five values and half have 24 values. When you look at the graph, some points just seem to be too far from the rest. It seems obvious they are outliers. But, in fact, all of these values were sampled from the same Gaussian distribution.

One problem with ad hoc removal of outliers is our tendency to see too many outliers. Another problem is that the experimenter is almost always biased. Even if you try to be fair and objective, your decision about which outliers to remove will probably be influenced by the results you want to see.

FIVE QUESTIONS TO ASK BEFORE
TESTING FOR OUTLIERS

Before using an outlier test, ask yourself these five questions:

- Was there a mistake in data entry? If so, fix it. An outlier may simply be the result of transposed digits or a shifted decimal point.
- Is the extreme value really a code for missing values? In some programs, you simply leave a blank spot when a value is missing. In other programs, you might enter a value like 999. If you entered the wrong code for a

missing value (or configured the program incorrectly), that code might appear to be an outlier.

- Did you notice a problem during the experiment? If so, don't bother with outlier tests. Eliminate values if you noticed a problem with a value during the experiment.
- Could the extreme values be a result of biological variability? If each value comes from a different person or animal, the outlier may be a correct value. It is an outlier not because of an experimental mistake, but rather because that individual is different from the others. This may be the most exciting finding in your data!
- Is it possible the distribution is not Gaussian? If so, it may be possible to transform the values to make them Gaussian. Most outlier tests assume that the data (except the potential outliers) come from a Gaussian distribution.

OUTLIER TESTS

The question an outlier test answers
If you've answered no to all five of the previous questions, two possibilities remain:

- The extreme value came from the same distribution as the other values and just happened to be larger (or smaller) than the rest. In this case, the value should not be treated specially.
- The extreme value was the result of a mistake. This could be something like bad pipetting, a voltage spike, or holes in filters. Or it could be a mistake in recording a value. These kinds of mistakes can happen and are not always noticed during data collection. Because including an erroneous value in your analyses will give invalid results, you should remove it. In other words, the value comes from a different population than the other values and is misleading.

The problem, of course, is that you can never be sure which of these possibilities is correct. Mistake or chance? No mathematical calculation can tell you for sure whether the outlier came from the same or a different population than the others. An outlier test, however, can answer this question: If the values really were all sampled from a Gaussian distribution, what is the chance one value would be as far from the others as the extreme value you observed?

Interpreting a small P value
If this P value is small, conclude that the outlier is not from the same distribution as the other values. Assuming you answered no to all five questions previously listed, you have justification to exclude it from your analyses.

Interpreting a high P value
If the P value is high, you have no evidence that the extreme value came from a different distribution than the rest. This does not prove that the value came from the

same distribution as the others. All you can say is that there is no strong evidence that the value came from a different distribution.

How do outlier tests work?

Statisticians have devised several methods for detecting outliers. All of the methods first quantify how far the outlier is from the other values. This distance can be the difference between the extreme value and the mean of all values, the difference between the extreme value and the mean of the remaining values, or the difference between the extreme value and the next closest value. This value is then standardized by dividing by some measure of variability, such as the SD of all values, the SD of the remaining values, the distance to the closest value, or the range of the data. To determine whether the extreme value can be considered a statistically significant outlier, the calculated ratio is compared with a table of critical values.

One of the most popular outlier tests is the *Grubbs outlier test* (also called the *extreme studentized deviate test*). This test divides the difference between the extreme value and the mean of all values by the SD of all values.

IS IT LEGITIMATE TO REMOVE OUTLIERS?

Some people feel that removing outliers is cheating. It can be viewed that way when outliers are removed in an ad hoc manner, especially when you remove only those outliers that get in the way of obtaining results you like. But leaving outliers in the data you analyze can also be considered cheating, because it can lead to invalid results.

It is not cheating when the decision of whether to remove an outlier is based on rules and methods established before the data were collected and these rules (and the number of outliers removed) are reported when the data are published.

When your experiment has a value flagged as an outlier, it is possible that a coincidence occurred, the kind of coincidence that happens in 5% (or whatever level you pick) of experiments even if the entire scatter is Gaussian. It is also possible that the value is a "bad" point. Which possibility is more likely? The answer depends on your experimental system.

- If your experimental system generates one or more bad points in a small percentage of experiments, eliminate the value as an outlier. It is more likely to be the result of an experimental mistake than to come from the same distribution as the other points.
- If your system is very pure and controlled, such that bad values almost never occur, keep the value. It is more likely that the value comes from the same distribution as the other values than that it represents an experimental mistake.

AN ALTERNATIVE: ROBUST STATISTICAL TESTS

Rather than eliminate outliers, an alternative strategy is to use statistical methods that are designed to ensure that outliers have little effect on the results. Methods of data analysis that are not much affected by the presence of outliers are called *robust*. You don't need to decide when to eliminate an outlier, because the method is designed to accommodate them. Outliers just automatically fade away.

The simplest robust statistic is the median. If one value is very high or very low, the value of the median won't change, whereas the value of the mean will change a lot. The median is robust; the mean is not.

To learn more about robust statistics, start with the book by Huber (2003).

LINGO: OUTLIER

The term *outlier* is confusing because it gets used in three contexts:

- A value that appears to be far from the others.
- A value that is so far from the others that an outlier test has flagged it.
- A value that is so far from the others, that you decided to exclude it from an analysis.

When you see the term *outlier* used, make sure you understand whether the term is used informally as a value that seems a bit far from the rest, or is used formally as a point that has been identified by an outlier test.

COMMON MISTAKES: OUTLIER TESTS

Mistake: Not realizing that outliers are common in data sampled from a lognormal distribution

Most outlier tests are based on the assumption that all data—except the potential outlier(s)—are sampled from a Gaussian distribution. The results are misleading if the data were sampled from some other distribution.

Figure 25.2 shows that lognormal distributions can be especially misleading. These simulated values were all generated from a lognormal distribution. An outlier test, based on the assumption that most of the values were sampled from a Gaussian distribution, identified an outlier in three of the four data sets. But these values are not outliers. Extremely large values are expected in a lognormal distribution.

If you don't realize that the data came from a lognormal distribution, an outlier test would be very misleading. Excluding these high values as outliers would be a mistake and would lead to incorrect results. If you recognize that the data are lognormal, it is easy to analyze them properly. Transform all the values to their logarithms (see the right side of Figure 25.2), and those large values no longer look like outliers.

Mistake: Interpreting the significance level from an outlier test as if it applied to each point rather than to each data set

If you run an outlier test with a significance level of 5%, that means that if the data are all really sampled from a Gaussian distribution, there is a 5% chance that the

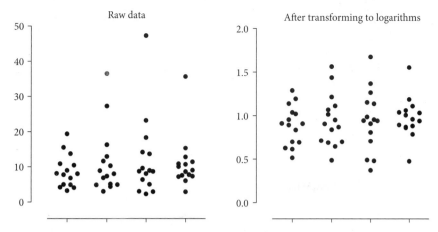

Figure 25.2. No outliers here. These data were sampled from lognormal distributions.
(Left) The four data sets were randomly sampled by a computer from a lognormal distribu-
tion. A Grubbs outlier test found a significant ($P < 0.05$) outlier in three of the four data sets.
(Right) Graph of the same values after being transformed to their logarithms. No outliers
were found. Most outlier tests are simply inappropriate when values are not sampled from a
Gaussian distribution.

outlier test would find an outlier in that set of data. Don't mistakenly interpret the
5% significance level to mean that each value has a 5% chance of being identified
as an outlier.

Mistake: Using a test designed to detect a single outlier when there are several outliers

Detecting multiple outliers is much harder than identifying a single outlier. The
problem is that the presence of the second outlier can *mask* the first one so that
neither is identified. Special tests are therefore needed. It is tempting to first
run an outlier test, then remove the extreme value, and then run the outlier test
again on the remaining values, but this method does a bad job of identifying
multiple outliers.

Mistake: Eliminating outliers only when you don't get the results you want

If you are going to remove outliers from your data, it must be done in a systematic
way, using criteria set up in advance. It is not OK to first analyze all the data,
then renanalyze after excluding outliers, and then finally decide to report only the
results that match your expectations.

Mistake: Truly eliminating outliers from your notebook

It can make sense to exclude outliers from graphs and analyses when you do so in
a systematic way that you report. But the outliers should still be recorded in your
notebook or computer files so that you have a full record of your work.

Q & A

What does it mean to remove or eliminate an outlier?

> When an outlier is eliminated, the analyses are performed as if that value were never collected. If the outliers are graphed, they are clearly marked. Removing an outlier from an analysis does not mean the outlier should be erased from the lab notebook. Instead, the value of the outlier and the reason it was excluded should be recorded.

Can outlier tests be used with linear or nonlinear regression?

> Yes. In fact, I published a method for doing so (Motulsky & Brown, 2006).

How should outlier removal be reported?

> A scientific paper should state how many values were removed as outliers, the criteria used to identify the outliers, and whether those criteria were chosen as part of the experimental design. When possible, report the results computed in two ways, with the outliers included and with the outliers excluded.

Can reasonable scientists disagree about how to deal with outliers?

> Yes!

Should outlier tests automatically be applied to all data sets?

> No! Before running any outlier test, ask yourself the five questions listed at the beginning of this chapter.

CHAPTER SUMMARY

- Outliers are values that lie very far from the other values in the data set.
- The presence of a true outlier can lead to misleading results.
- If you try to remove outliers manually, you are likely to be fooled. Random sampling from a Gaussian distribution creates values that often appear to be outliers but are not.
- Before using an outlier test, make sure the extreme value wasn't simply a mistake in data entry.
- Don't try to remove outliers if it is likely that the outlier reflects true biological variability.
- Beware of multiple outliers. Most outlier tests are designed to detect one outlier and do a bad job of detecting multiple outliers. In fact, the presence of a second outlier can cause the outlier test to miss the first outlier.
- Beware of lognormal distributions. Very high values are expected in lognormal distributions, but these can easily be mistaken for outliers.
- Rather than removing outliers, consider using robust statistical methods that are not much influenced by the presence of outliers.

TERMS INTRODUCED IN THIS CHAPTER

- Extreme studentized deviate test (p. 235)
- Grubbs outlier test (p. 235)
- Masking (p. 237)
- Outlier (p. 232)
- Robust statistical tests (p. 236)

CHAPTER 26

Choosing a Sample Size

There are lots of ways to mismanage experiments, but using the wrong sample size should not be one of them.

PAUL MATHEWS

Many experiments and clinical trials are run with a sample size that is too small, so even substantial treatment effects may go undetected. When planning a study, therefore, you must choose an appropriate sample size. This chapter explains how to understand a sample size calculation and how to compute sample size for simple experimental designs. Make sure you understand the concept of statistical power in Chapter 20 before reading this chapter. Sample size calculations for multiple and logistic regression are briefly covered in Chapters 37 and 38, respectively. Sample size for nonparametric tests is discussed in Chapter 41.

SAMPLE SIZE PRINCIPLES

The basic principles of sample size and power are quite straightforward. This section briefly explains these key ideas, the next section will interpret them in the context of a particular study, and later sections will discuss the complexities. To calculate necessary sample size, you must first answer four questions:

- What is the size of the effect you are looking for? The effect might be a difference between two means, a correlation coefficient, or a relative risk or some other ratio. The need to specify the size of the effect you are looking for is simple. The required sample size is inversely proportional to the square of the effect size. If you want to detect an effect size that is half as big, you need to quadruple (multiply by four) the sample size. If you want to detect an effect size that is one-third what you computed, you need to multiply the sample size by nine (Figure 26.1, right).
- How much power do you want? If there really is an effect (difference, relative risk, correlation, etc.) of a specified size in the overall population, how sure do you want to be that you will find a statistically significant result in your proposed study? The answer is called the power of the experiment (power is discussed in Chapter 20). If you want more power, you need a larger sample size (Figure 26.1, left). A standard for sample size

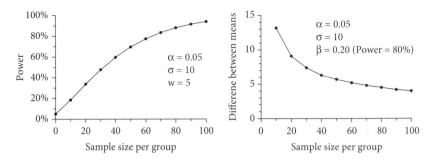

Figure 26.1. Graphing the trade-offs in choosing sample size.

(Left) Power increases as you increase sample size. These calculations were for an unpaired t test with the significance level set to 5%, the standard deviation within each group set to 10, and the effect size set to 5. (Right) Here we hold the power equal to 80% and plot the decrease in the effect size that we can detect as sample size increases.

calculations is to specify 80% power. Here are three approximate rules of thumb for changing to different values of power:

- If you are willing to accept 50% power, a sample size about half as large as required for 80% power will suffice.
- A sample size about 33% larger will take you from 80% to 90% power.
- A sample size about twice as large is needed to go from 80% to 99% power.
- What significance level do you want to use? The significance level, also called α, is the cutoff used to decide if a P value is low enough to label your results "statistically significant". It is usually set to 0.05 (sometimes expressed as a significance level of 5%), which means P values less than 0.05 earn the designation of "statistically significant". Although the 0.05 threshold is standard, it is arbitrary. If you choose to use a smaller (stricter) significance level, you'll need a larger sample size. For example, if you define statistical significance as P < 0.01 instead of P < 0.05, you'll need a sample size about 30% – 50% larger.
- How much variability do you expect? The signal you are looking for is obscured by the "noise" of random variation. If you expect lots of variation among the data, you'll need a larger sample size than if you expect little variation. If your outcome is binary, so is summarized as a proportion rather than a mean, what matters is how close to 50% you expect the proportion to be. It takes a larger sample size when the proportions are close to 50% than when the proportions are much higher or lower than 50%.

AN ALTERNATIVE WAY TO THINK ABOUT SAMPLE SIZE CALCULATIONS

Everything else being equal, a larger sample size produces a narrower CI. An alternative way to think about sample size determination is to state the maximum acceptable width of the CI, and then calculate the needed sample size. You can use the same calculations, programs, or tables as those used for computing sample size needed to achieve statistical significance. Here are the similarities and differences:

- Instead of entering the smallest difference you wish to detect, enter the largest acceptable margin of error, which equals half the width of the widest CI you'd find acceptable.
- Calculate α as 100% minus the confidence level you want. For 95% confidence intervals, enter α as 5% or 0.05.
- Estimate SD the same as you would for the usual methods.
- Enter power as 50%. But see the next paragraph to understand what this means.

When calculating sample size to achieve a specified margin of error, it is customary to do the calculations so the expected average margin of error will equal the value you specified. About half the time, the margin of error will be narrower than that specified value, and the other half of time the margin of error will be wider. Although the term *power* is often not used, these calculations are the same as those used to calculate sample size for 50% power. If you chose, say, 80% power instead, then you'd need a larger sample size and the margin of error would be expected to be narrower than the specified value 80% of the time (Metcalfe, 2011).

I prefer this approach to determining sample size, because it puts the focus on effect size rather than statistical significance. I think it is more natural to think about how precise you want to determine the effect size (the widest acceptable confidence interval), rather than the smallest effect size you can detect. But the two methods are equivalent.

INTERPRETING A SAMPLE SIZE STATEMENT

Reports of clinical trials (and other well-designed studies) often include a statement that reads something like this: "We chose to study 313 subjects in each group to have 80% power to detect a 33% relative reduction in the recurrence rate from a baseline rate of 30% with a significance level of 0.05 (two tailed)."

To understand this statement, you have to think about four questions (four which repeat the prior section).

What effect size?
Every sample size calculation requires you to decide the effect size for which you are looking, abbreviated Δ. Depending on the experimental design, the effect size might be a difference between means, a risk ratio, a correlation coefficient, an odds ratio, or something else.

Here, the investigators were looking for a 33% reduction in recurrence rate (from 30% down to 20%, a relative risk reduction of 33%, which is an absolute reduction of 10%). Entering this value is an essential part of sample size determination, because it takes more subjects to find a smaller effect than a larger effect..

What power?

Statistical power is explained in Chapter 20. The power of an experimental design is the answer to this question: If there truly is an effect of a specified size, what is the chance that your result will be statistically significant?

The example study chose a power of 80%. That means that if the effect the investigators wanted to see really exists, there is an 80% chance that the study would have resulted in a statistically significant result, leaving a 20% chance of missing the real difference.

If you want lots of statistical power, then you'll need a huge sample size. If moderate power is adequate, then the sample size can be smaller.

What significance level?

Recall that the significance level (α) of a study is the answer to the question, If the null hypothesis were actually true, what is the chance your result would be statistically significant? This study defined $\alpha = 0.05$, so P values less than 0.05 are defined to be statistically significant. If the investigators had chosen a stricter threshold (say, $\alpha = 0.01$), they would have needed a larger sample size.

What is the SD (or how large is the proportion)?

This example compares proportions (not means), so you don't need to enter the SD directly. Instead, you must estimate how far the proportion will be from 50%. If you expect the outcome to occur about half the time, the sample size must be larger than if you expect the proportion to be closer to 0% or 100%.

Here, the investigators estimated that the baseline recurrence rate (which they are hoping to reduce) is 30%.

The sample size statement restated

Now we can restate the sample size statement in simpler language:

The investigators assumed that the recurrence rate is normally 30% and asked whether the new treatment reduced the recurrence rate by one-third (or more), down to 20% (or less). They chose to use 313 subjects in each group so that if the true effect was a one-third reduction in recurrence rate, they would have an 80% chance of obtaining a statistically significant result (defined as P < 0.05).

LINGO: POWER

The word "power" tends to get emphasized when discussing sample size calculations. The calculations used to compute necessary sample size are commonly referred to as a *power analysis*. When discussing the effect size the experiment was designed to detect, some investigators say the experiment was *powered to detect* the specified effect size. That is just slang for choosing a large enough sample size for the specified effect size, significance level, and power. When the sample size is too small to detect a specified effect size, a study is sometimes referred to as being *underpowered*.

CALCULATING THE PREDICTED FPRP AS PART OF INTERPRETING A SAMPLE SIZE STATEMENT

The False Positive Report Probability (FPRP), also called the False Positive Rate (FPR) or False Discovery Rate (FDR) was defined in Chapter 18 in the context of interpreting a P value from a completed experiment. The FPRP can also be calculated for a proposed experiment from the choices of significance level and power, and an estimate of the prior odds that the null hypothesis is true. It is not common to compute a FPRP as part of sample size determination, but this section explains why it should be.

Table 26.1 shows the expected FPRP for a several choices of α and power and for several prior probabilities. The legend to the table shows how to easily compute FPRP for any values for α, power and prior probability you choose (the logic of the calculations is explained in Chapter 18).

The sample size statement in the example earlier in this chapter said that the investigators chose 80% power and $\alpha =0.05$. Those commonly used values correspond to the first row of Table 26.1. The columns show several values for the prior probability. You can never know the prior probability precisely, but can estimate it based on the scientific context. Let's consider two possible scenarios:

- The left column sets the prior probability to 1%. This would be a reasonable estimate of FPRP for a speculative experiment not founded on theory or

Power (%)	α	Prior Probability (%)			
		1%	10%	25%	50%
80%	0.05	86.1%	36.0%	15.8%	5.9%
90%	0.05	84.6%	33.3%	14.3%	5.3%
95%	0.05	83.9%	32.1%	13.6%	5.0%
99%	0.05	83.3%	31.3%	13.2%	4.8%
80%	0.01	55.3%	10.1%	3.6%	1.2%
90%	0.01	52.4%	9.1%	3.2%	1.1%
95%	0.01	51.0%	8.7%	3.1%	1.0%
99%	0.01	50.0%	8.3%	2.9%	1.0%
80%	0.001	11.0%	1.1%	0.4%	0.1%
90%	0.001	9.9%	1.0%	0.3%	0.1%
95%	0.001	9.4%	0.9%	0.3%	0.1%
99%	0.001	9.1%	0.9%	0.3%	0.1%

Table 26.1. Expected FPRP as a function of choices for α and power, and an estimate of the prior probability.

Each value was computed in Excel by this formula where *Prior* and *Power* are percentages and *Alpha* is a fraction:

=100((1-Prior/100)*Alpha)/ (((1-Prior/100)*Alpha)+((Prior/100)*(Power/100)))

prior data, but rather on a wild hunch. The corresponding FPRP is 86.1%. If such an experiment resulted in a P value less than 0.05, there would be an 86% chance that this is a false positive. The problem is that the almost 5% of the experiments will be false positives (5% of the 99% of the experiments where the null hypothesis is true) while fewer than 1% of the experiments will be true positives (80% of the 1% of the experiments where the experimental hypothesis is true. There is no point reading the rest of the paper in this case. The study design makes no sense. If the prior probability is low, α must be set to a lower value (and power to a higher value).

• The right column sets the prior probability is 50%. If the study was based on solid theory and/or prior data, this is a reasonable estimate of the prior probability and the FPRP is 5.9%. If such an experiment resulted in a P value less than 0.05, there would be only about a 6% chance that it is a false positive. That is not unreasonable. This is a well-designed study.

These two extremes demonstrate why experimental context (here quantified as prior probability and FPRP) must be considered when choosing a sample size for a study. Of course, one can never know the prior probability exactly, but it is essential to estimate it when choosing sample size or when interpreting a sample size statement. The usual choices of power (80%) and α (0.05) are reasonable for testing well-grounded hypotheses, but are really not helpful when testing unlikely hypotheses.

COMPLEXITIES WHEN COMPUTING SAMPLE SIZE

The first section of this chapter makes it seem very simple to calculate necessary sample size. This section explains some of the complexities.

How to choose a significance level

Think about prior probability and FPRP

Most investigators always choose to define statistical significance as $P < 0.05$ (so set α to 0.05%), simply because that is the traditional threshold. Don't do that. Instead, think about the context of your work and estimate the prior probability that your scientific hypothesis is true. Then use Table 26.1 to determine the predicted FPRP. If the predicted FPRP is high, there really is no point in performing the experiment or study. Some very rough rules of thumb from Table 26.1:

• If the prior probability is about 50% (because your experiment is based on solid theory or prior data) the usual choice of α =5% is reasonable.
• If the prior probability is about 10% (because your experiment is based on weak theory or prior data), you should set α to 1% or even lower.
• If the prior probability is about 1% (because your experiment is only vaguely based theory or prior data, but is mostly based on a hunch or wish), you should set α to 0.1% or even lower.

One-tail or two-tail P value?

If you plan to use a one-tail P value instead of a two-tail P value, required smaller sample size is reduced roughly 15–20% (depending on what power you choose).

Chapter 15 explains the difference, and explains what is involved in choosing a one-tail P value. You must predict the direction of the effect before collecting data, and firmly decide that if the difference goes in the other direction (no matter how large that effect is) you will conclude the difference is not statistically significant. It is not OK to choose to a one-tail P value when computing sample without also recording your prediction about which way the difference will go.

How to choose desired statistical power

In many fields of science, sample sizes are routinely computed to aim for a power of 80%. It isn't clear how that became the standard, but instead of choosing that standard value think carefully about what value of power to aim for.

One factor to consider is shown in Table 26.1 and discussed in the prior section. Choose a value of α and power that will give a reasonably low FPRP given your estimate of the prior probability.

Also think about the relative consequences of making a Type I error (false positive) and a Type II error (false negative). If you aim for 80% power to detect a specified effect size, that means that if the true effect size is the value you hypothesized, there is an 80% chance that your proposed experiment will result in a conclusion that the effect is "statistically significant" and a 20% chance that the result will conclude that the results are "not statistically significant". If you calculate sample size for 80% power and a 5% significance level, you are choosing to accept four times (20%/5%) the risk of a Type II error (missing a true effect of the specified size) compared to Type I error (falsely declaring a result is statistically significant, when really the populations are the same and the observed difference was due to random sampling). Is that a reasonable ratio? It depends on why you are doing the study.

How to choose the effect size you want to detect

To calculate sample size, you must specify the effect size you want to detect. The effect might be a difference between two means, a correlation coefficient, or a relative risk or some other ratio. You need to specify an effect size because a larger sample size is required to find a small effect than to find a large effect. If the true effect size is smaller than the one you specified, the experiment or study will have less power than you intended.

Sample size calculations should be based on the smallest effect size that is worth detecting scientifically (or clinically). It is not easy to specify the smallest effect size you'd care about. Sometimes the calculation is done in reverse. You specify the sample size you can afford, and find out what effect size can be detected with specified power. If you'd care about effects much smaller than that, then you may decide the experiment is not worth doing with the sample size you can afford.

Don't compute sample size based only on published effect sizes

Since it is hard to specify the smallest effect size you'd care about, it is common to base sample size determinations on the effect size seen in a prior published study or in a pilot study, and then calculate a sample size large enough to detect an effect that large. This approach presents two problems.

One problem is that you might care about effects that are smaller than the published (or pilot) effect size. You should compute the sample size required to detect the smallest effect worth detecting, not the sample size required to detect an effect size someone else has published (or you determined in a pilot study).

The second problem is that the published difference or effect is likely to inflate the true effect (Ioannidis, 2008; Zollner & Pritchard, 2007). If many studies are performed, the average of the effects detected in these studies should be close to the true effect. Some studies will happen to find larger effects, and some studies will happen to find smaller effects. The problem is that the studies that find small effects are much less likely to get published than the studies that find large effects. Therefore, published results tend to overestimate the actual effect size.

What happens if you use the published effect size to compute the necessary sample size for a repeat experiment? You'll have a large enough sample size to have 80% power (or whatever power you specify) to detect that published effect size. But if the real effect size is smaller, the power of the study to detect that real effect will be less than the power you chose.

Don't compute sample size based only on standard effect sizes

To avoid the difficulty of defining the minimum effect size they want to detect, many are tempted to pick a standard effect size instead. They use standard values for the significance level and power, so why not use a *standard effect size* too?

In a book that is almost considered the bible of sample size calculations in the social sciences, Cohen (1988) defines some standard effect sizes. He limits these recommendations to the behavioral sciences (his area of expertise) and warns that all general recommendations are more useful in some circumstances than others. Here are his guidelines for an unpaired t test:

- A "small" difference between means is equal to one-fifth the SD.
- A "medium" effect size is equal to one-half the SD.
- A "large" effect is equal to 0.8 times the SD.

If you choose standard definitions of α (0.05), power (80%), and effect size, then there is no need for any calculations. Choosing a standard effect size is really the same as picking a standard sample size. If you are comparing two groups (with a t test), you will need a sample size of 26 in each group to detect a large effect, 65 in each group to detect a medium effect, and 400 in each group to detect a small effect.

Lenth (2001) argues that you should avoid these "canned" effect sizes, and I agree. You should decide how large a difference you care to detect based on understanding the experimental system you are using and the scientific questions you are asking.

Don't try to compute sample size based on the effect size you expect to see

Sample size calculations should be based on the smallest effect size worth detecting, based on scientific or clinical considerations. Sample size calculations should not be based on the sample size you expect to see. If you knew the effect size to expect, it wouldn't really be an experiment! That is what you are trying to find out.

How to estimate the expected variability

Sample size programs require you to estimate the SD of your samples This is essential because the signal you are looking for is obscured by the "noise" of random variation. If you anticipate lots of variation among the values (large SD), you'll need a larger sample size than if you anticipate little variation.

Before planning a large study, try hard to lower the SD

The sample size you need is proportional to the square of the population standard deviation. If the variability is biological, there is nothing you can do to reduce it. But if the variation is entirely or mostly experimental error, consider each experimental step and try to reduce the variation. If you can improve the experimental methods to reduce the SD by a factor of two, you'll reduce needed sample size by a factor of four. Before planning a large study, it is worth spending a lot of effort and time to reduce the variation.

Standard deviations determined in small pilot experiments may too low

Some investigators run a small pilot experiment to determine the SD to enter when calculating sample size. But how accurate is that SD? Just by chance that pilot experiment may have happened to obtain data that are closely bunched together, making the SD of that sample lower than the true population SD. Or perhaps that experiment randomly obtained values that are far more scattered than the overall population, making the SD high. The SD of the published sample does not equal, and may be quite far from, the SD of the population. How far can the sample SD be from the population SD. It depends on sample size and can be quantified with a confidence interval.

Table 26.2 shows the 95% CI for the population SD, computed from a SD you compute from a sample randomly selected from a Gaussian distribution. With tiny

n	95% CI OF SD
2	$0.45 \times SD$ to $31.9 \times SD$
3	$0.52 \times SD$ to $6.29 \times SD$
5	$0.60 \times SD$ to $2.87 \times SD$
10	$0.69 \times SD$ to $1.83 \times SD$
25	$0.78 \times SD$ to $1.39 \times SD$
50	$0.84 \times SD$ to $1.25 \times SD$
100	$0.88 \times SD$ to $1.16 \times SD$
500	$0.94 \times SD$ to $1.07 \times SD$
1000	$0.96 \times SD$ to $1.05 \times SD$

Table 26.2. 95% CI of a SD.

Each row is for a different sample size, n. The table shows the 95% CI for the population SD expressed as multiples of the SD you computed from the sample. These calculations are based on the assumption that your sample is randomly sampled from a Gaussian population. The values were computed using an equation from Sheskin 2011, pp. 217–218) that are computed by these Excel formulae:

Lower limit: =SD*SQRT((n-1)/CHIINV((alpha/2), n-1))
Upper limit: =SD*SQRT((n-1)/CHIINV(1-(alpha/2), n-1))

samples, note how wide the confidence interval is. For example, with n = 3 the population SD can be (within 95% confidence) 6.29 times larger than the SD computed from the sample. If you compute a sample size with a SD that is lower than the true population SD, the computed sample size will also be too low. The required sample size is proportional to the square of the SD. So if the SD you obtain in a pilot experiment is half the true population SD, your computed sample size will be one quarter of what it should be. Conversely, your power will be a whole lot less than you think it is.

Standard deviations reported in publications may be too low

If similar experiments have been published, investigators can use the published SD when calculating sample size. But how accurate is that SD? There are two issues to consider.

The first issue is that the published SD could be too low or too large due to random sampling. With small samples, the published SD could be way too small. This is the same issue discussed in the previous section.

The second issue is that the investigator may have selected a "representative" experiment to publish, and the selection may be partly because the SD happened to be low, making the data look clean. If this were the case, the actual population SD is likely to be much higher than the value you used, so your calculated sample size will be too low.

Sometimes it is easier to estimate the ratio of effect size to SD

You don't actually need to separately state the smallest effect size you want to detect and also state the estimated standard deviation. The equations that compute sample size are based on the ratio. In some cases, you may find it easier to state the effect size you want to detect as a multiple of the SD. You can specify, for example, that you want to detect an effect size equal to half the SD. Or you could specify that you want to detect an effect size equal to 1.5 times the SD. You don't need to know the SD to enter the effect size this way.

How to plan for unequal sample sizes

Most studies are planned with the same sample size in all groups. However, it may make sense to plan studies with unequal sample sizes if one treatment is much more expensive, difficult, or risky than the other, or when it is difficult to find participants for one group (e.g., patients with a rare disease). The trade-off is that if you reduce the number of participants or animals in one group, you must increase the number in the other group even more, which means the total sample size must increase. These rules of thumb are helpful:

- You won't lose any power if you cut the sample size of one group by 25% if you also increase the size of the other group by 50%.
- Or you can cut the sample size of one group by 40% and triple the size of the other group.
- If you reduce the sample size in the one group to less than half of the number required had the sample sizes been equal, it is not possible to maintain statistical power by increasing the sample size of the other group.

How to plan for dropouts

Sample size calculations tell you the number of values you need to enter into your analyses. You'll need to start with a larger sample if you expect the sample size at the end of the study to be less than the same size you started with. This could be because some participants drop out of the study or because some of your repeated experiments can't be completed. If you expect about 20% of the participants to drop out (or 20% of your repeated experiments to fail), you should start with a 20% larger sample size than calculated.

EXAMPLES

Example 1: Comparing weights of men and women

You want to compare the average weight of adult men and women. How big a sample size do you need to test the null hypothesis that the men and women weigh the same on average? Of course, this is a silly question, as you know the answer already but going through this example may help clarify the concepts of sample size determination. You need to answer four questions:

- How big an effect are you looking for? For this example, the answer is already known. In an actual example, you have to specify the smallest difference you'd care about. Here we can use known values. In the US, the mean difference between the weight of men and women (for 30 year olds, but it doesn't vary that much with age) is 16 kg (Environmental Protection Agency, 2011, p. 36 NHANES-III data for 30-31-year-olds). An alternative way to look at this is to ask what is the widest that you are willing to accept for the confidence interval for the difference between the two mean weights. Here we are looking for a margin of error of 16 kg, which means the entire confidence interval extends that far in each direction so has a width of 32 kg.
- What power do you want? Let's choose the standard 80% power.
- What significance level? Let's stick with the usual definition of 5%.
- How much variability do you expect. The table with the average weights also states the SD of weights of 30-year-old men is 17 kg, and the SD for other ages is about the same.

Figure 26.2 shows how to use the program G*Power to compute sample size for this example. G*Power is free for Mac and Windows at www.gpower.hhu.de, and is described by Faul (2007). I used version 3.1.9.2. If you use a different version, the screen may not look quite the same. To use G*Power:

1. Choose a t test for the difference between two independent means.
2. Choose two-tail P values.
3. Enter the effect size, d, as 1.0. This is the ratio of the smallest difference between means you wish to detect divided by the expected SD.
4. Set alpha to 0.05 and power to 0.80. Enter both as fractions not percentages.
5. Enter 1.0 for the allocation ratio because you plan to use the same sample size for each group.

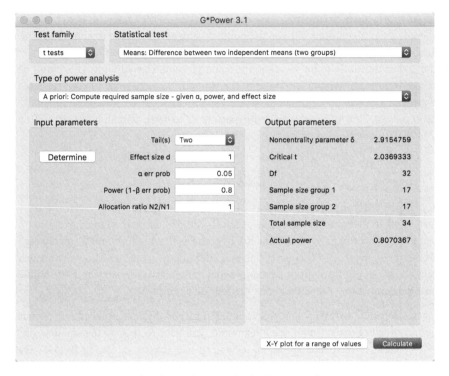

Figure 26.2. G*Power used to determine sample size for example 1.

6. Press Calculate at the bottom right.
7. Read that the required sample size for each group is 17. This value is reported as 16 per group if you use www.powerandsamplesize.com, which is based on a slightly different equation.

There are two ways to interpret these sample sizes:

1. Imagine that you collected weights from 17 men and 17 women in many experiments, that your estimate of the SD is close to correct, and that you computed the 95% CI for the difference between mean for each experiment. Since you chose 80% power, you expect that the margin of error of 80% of these intervals will be smaller than the effect size you specified, so the CI will extend less than 16 kg in each direction.
2. Imagine that you collected weights from 17 men and 17 women in many experiments, that your estimate of the SD is close to correct, and that you computed a P value testing the null hypothesis of no difference for each experiment. Since you chose 80% power, you'd expect to find a statistically significant difference (P < 0.05) in 80% of the experiments.

What if you want a much more definitive conclusion, defined by setting α to 1% and power to 99%? Change α and power and G*Power states that you need 50 people in each group.

These examples are silly in that you would never design an experiment to make this comparison. But it is a helpful example to appreciate the need for large samples. Look around. It is obvious that men, on average, weigh more than women. But to prove that (with only 1% chance of error in either direction) requires measuring the weight of about 100 people.

When you actually do studies, you usually can't look up the SD as we did here. And even if you can look it up, it may not be correct. The EPA reference actually lists two different tables of weights. The other one (Environmental Protection Agency, 2011, p. 38, NHANES-IV data for 30-31-year-olds) is quite different: the SD is 22 kg and the mean difference between men and women is 6 kg. If you based your sample size calculation on these values with an effect size of 6/22 = 0.27, the required sample size per group is 217 (α = 0.05; power = 80%) or 661 (α = 0.01; power = 99%). If you planned the study expecting the SD to be about 16 kg, but in fact it was about 22 kg, you would not have the statistical power you were expecting.

Example 2: Comparing two proportions

How many participants are needed to test the null hypothesis that a new surgical method is no different (in complication rate) than the standard procedure? You know the complications rate is about 5% and you want to detect an absolute change as small as 2%. you are looking for an absolute change of 2%, so a change from 5/100 down to 3/100. You define statistical significance as α = 0.05, and you wish to have 90% power.

Figure 26.3 shows how to determine the sample size using the free calculator at www.powerandsamplesize.com. This screenshot was taken in July 2017, and it may look different when you use it.

1. Choose the calculator to compare two proportions using two-sided P values.
2. Enter the predicted proportions for the two groups, 0.05 and 0.03. Enter fractions, not percentages.
3. Enter power as 0.90. Enter that fraction, not a percentage.
4. Enter the significance level as 5%. This calculator requires a percentage for this value, not a fraction.
5. Enter 1.0 as the sampling ratio. You want both groups to be the same size.
6. Click Calculate.
7. See that the required sample size per group is 2009. G*Power uses a slightly different equation and reports 2016 needed per group.

Let's review exactly what this means:

- Assume that the true difference between proportions is 0.02.
- Now imagine that you perform many experiments, with n = 2009 per group in each experiment.
- Because of random sampling, you expect to see a difference larger than 0.02 half the time and smaller than 0.02 the other half of the time.
- In 90% of those experiments, you expect to find a statistically significant result, with P < 0.05 (two-tailed). In the remaining 10% of the experiments,

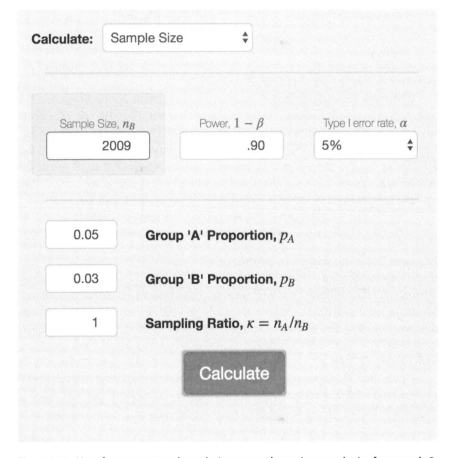

Figure 26.3. Use of www.powerandsamplesize.com to determine sample size for example 2.

you expect to observe a difference that is not statistically significant, so you will have made a Type II error.

- An alternative way of thinking about this: You'll expect that the margin of error to be less than 0.02 in 90% of the experiments (since you chose 90% power). This means you expect the 90% CI to extend less than 0.02 in each direction in 90% of these hypothetical experiments.

OTHER APPROACHES TO CHOOSING SAMPLE SIZE

Adaptive trials

The idea behind *adaptive trials* is simple: interim analyses, performed while the trial is proceeding, are used to decide the course of the study. This approach to designing large clinical trials is gaining acceptance (Chang and Balser, 2016; Bhatt and Mehta, 2016).

The simplest adaptive designs allow researchers to analyze the data only once at the middle of a trial and then decide to stop the trial if patients in one group are having substantially better outcomes than those in the other. This is a form of multiple comparisons, so the investigators need to take a statistical "hit" for peeking at the data while a trial is ongoing. For example, if you use a significance level of 0.01 for stopping the trial early, then the P value of the final analysis has to be less than about 0.049 to be deemed statistically significant at the 5% level.

Newer adaptive techniques are used to decide when to stop adding new patients to the study, when to stop the study, and what proportion of new subjects should be assigned to each alternative treatment. These decisions must be based on unambiguous protocols established not before collecting data but before analyzing data in a blinded study. Also, the interim analysis results must only be presented as a summary of each group and each group only designated by labels in a way that does not identify which treatment or experimental group is which, analyzing data to ensure that the statistical results are meaningful.

In some cases, the adaptive design will lead investigators to end a study early when it becomes clear that one treatment is much better than the other or when it becomes clear that the difference between the treatments is at best trivial. In other cases, the adaptive design will prolong a study (increase the sample size) until it reaches a crisp conclusion.

The methods used to adapt a study to interim findings are not straightforward, have some drawbacks (Fleming, 2006) and are not yet entirely standard. The investigators must choose the strategy of data analysis when planning the study and must lock in that strategy before looking at any data. Despite the name, you can't adapt your strategy after doing preliminary analyses. All adaptations must follow a preplanned strategy.

Simulations

Simulations provide a versatile approach to optimizing sample size for any kind of experimental design. Run multiple sets of computer simulations, each with various assumptions about variability, effect size, and sample sizes. Of course, you'll include a random factor so each simulation has different data. For each set of simulations, tabulate whatever outcome makes the most sense—often the fraction of the simulated results that have a P value less than 0.05 (the power). Then you can review all the simulations and see what advantages increasing sample size gives you. Knowing the cost and effort of the experiment, you can then choose a sample size that makes sense.

COMMON MISTAKES: SAMPLE SIZE

Mistake: Using standard values for significance level and power without thinking

Sample size calculations require you to specify the significance level you wish to use as well as the power. Many simply choose the standard values, a significance level (α) of 5% ($P < 0.05$ is significant) and a power of 80%. It is better to choose values based on the relative consequences of Type I and Type II errors.

Mistake: Ad hoc sequential sample size determination

The following approach will lead to invalid results. Collect and analyze some data. If the results are not statistically significant, collect some more data and reanalyze. Keep collecting more data until you obtain a statistically significant result or until you run out of money, time, or curiosity. This point is also discussed in Chapter 23.

Figure 26.4 is a simulation that demonstrates this point. The simulated data were sampled from populations with Gaussian distributions and identical means and standard deviations. An unpaired t test (see Chapter 30) was computed with n = 5 in each group, and the resulting P value (0.406) is plotted at the left of the graph. Note that the Y-axis is logarithmic to focus attention on the small P values. Then the t test was repeated with one more value added to each group. That P value was 0.825. Then one more value was added to each group (n = 7), and the P value was computed again (0.808). This was continued up to n = 75.

With these simulated data, we know for sure that the null hypothesis is true. If you ran t tests on many different samples of data, you'd expect to see a statistically significant result 5% of the time. But here, each sample included all the values in the prior sample, so the P values are not independent. Rather, they mimic what some scientists do: run the experiment a bunch of times, check the P value,

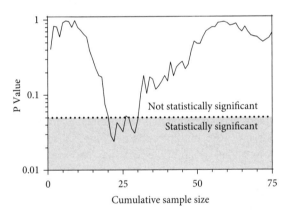

Figure 26.4. Beware of ad hoc sample size decisions.

This figure shows a simulation of a common mistake in data analysis. The (simulated) investigator started with n = 5 in each group and calculated a t test to determine a P value. The P value (plotted on the logarithmic Y-axis) was (much) greater than 0.05, so the investigator increased sample size to six in each group, and then seven, eight, nine, and so on. All the values analyzed were sampled from the same Gaussian distribution, so the null hypothesis is known to be true. The figure shows what would have happened had the study continued to include 75 in each group. However, the investigator actually would have stopped at 21 in each group, since the result happened to hit statistical significance then. Unless the analyses are done with special methods, P values simply cannot be interpreted when the sample size isn't determined in advance. Note that the values in this figure were calculated via simulation with random numbers. Another simulation would have looked different, but the main point would persist: the cumulative P value bounces around a lot, especially with small sample sizes.

increase sample size if the P value isn't as low as desired, calculate the P value again, and so on. If you could extend the sample size out to infinity, you'd always eventually see a P value less than 0.05, even if the null hypothesis were true. In practice, the investigator would eventually run out of money, time, or patience. Even so, the chance of hitting a P value less than 0.05 under the null hypothesis is way higher than 5%.

Someone viewing these results would have been very misled if the investigator had stopped collecting data the first time the P value dropped below 0.05 (at n = 21) and had simply reported that the P value was 0.0286 and the sample size was 21 in each group. It is impossible to interpret results if an ad hoc method is used to choose sample size.

Mistake: Not taking advantage of pairing when you can

The first example compute sample size for the unpaired t test. What if you compare measurements before and after intervention in the same subject, measurements made in two matched subjects, or a measurement made in the left eye (or ear, elbow, etc.) treated one way with the right eye (or ear, elbow, etc.) treated a different way. In all cases, the appropriate analysis is the paired t test (see Chapter 31).

Since each measurement has a matched control, you almost always gain some power over the same data analyzed by an unpaired t test. More important, you can design a study with a smaller sample size. How much lower depends on how well the two set of matched or paired results correlate. A rough rule of thumb: If the R^2 for the correlation between the before-treatment and after-treatment measurements (or the measurements on right side vs. the left side treated differently) is about 0.5, the sample size required will be less than half that required if there were no pairing.

Sample size programs report the number of pairs you need. When doing a study that does two measurements on each individual (maybe it compares left vs. right side or before vs. after treatment) you need the number of volunteers or animals suggested by the program. But if you collect data in matched pairs (two people from the same family, two animals from the same litter, etc.) then each animal or volunteer only provides one value. Therefore, the total sample size required is twice that reported by the sample size program.

Mistake: Not computing the anticipated false discovery rate (FPRP) as part of a sample size determination

Few sample size determinations are accompanied by a calculation of the anticipated FPRP, but I think they should be as shown in Table 26.1. If the resulting FPRP is too high, there really is no point in continuing with the study. Instead calculate required sample size using a stricter (lower) value for the significance level, and perhaps a higher value for power.

Mistake: Mixing up absolute and relative effect sizes

When using a sample size program, be careful about entering the effect size you are looking for. Is the program asking for the effect size as a difference, or as a ratio (relative effect size)? Example 2 earlier in this chapter was looking for a change in incidence from 5% down to 3%. This is an absolute change of 2% (5% − 3%) but a relative change of 40% ([5% − 3%]/5%). Different programs are different, and it depends on what kind of study you are planning. But you will end up with the wrong sample size if you enter an absolute effect size when the program is expecting a relative effect size or if you enter a relative effect size when the program expects an absolute effect size.

Mistake: Expecting there to be one sample size that is exactly right for your proposed study

Sample size calculations are often based on tradition (choices for α and power), a guess (SD), and an almost arbitrary decision (effect size). Essentially, the calculations use solid theory to compute sample size from nearly arbitrary inputs. Since the inputs are often arbitrary, the computed sample sizes are also somewhat arbitrary. Different knowledgeable people facing the same situation may come up with different estimates and decisions and so would compute different sample sizes. Sample size statements in journals tend to make sample size determination sound far more exact than it is.

Mistake: Rushing through a sample size calculations

Calculating sample size requires thought. You need to make decisions that have consequences in terms of the viability of your proposed project and the amount of effort and time it will take to complete. In some cases, the calculations can be tricky. It is worth the time to discuss the decisions with colleagues. Mathews (2010) suggests reducing the chances of making a mistake by performing the calculations with two different programs, or by hand and also with a program.

Mistake: Confusing the sample size needed for each group with the total needed sample size

Most programs compute the same size needed for each group, in which case the total sample size required is twice the value reported (assuming you are comparing two groups). If you mistakably think the program reported the total sample size when it actually reported the size for each group, you will use half the required sample size and your study will have much less power than you intended.

Q & A

Does 80% power mean that 80% of the participants would be improved by the treatment and 20% would not?

> No. Power refers to the probability that a proposed study would reach a conclusion that a given difference (or effect size) is statistically significant. Power has nothing to do with the fraction of participants who benefit from a treatment.

How much power do I need?

> Sample size calculations require you to choose how much power the study will have. If you want more power, you'll need a larger sample size. Often power is

set to a standard value of 80%. Ideally, the value should be chosen to match the experimental setting, the goals of the experiment, and the consequences of making a Type II error (see Chapter 16).

What are Type I and Type II errors?

See Table 16.2. A Type I error occurs when the null hypothesis is true but your experiment gives a statistically significant result. A Type II error occurs when the alternative hypothesis is true but your experiment yields a result that is not statistically significant.

My program asked me to enter β. What's that?

β is defined as 100% minus power. A power of 80% means that if the effect size is what you predicted, your experiment has an 80% chance of producing a statistically significant result and thus a 20% chance of obtaining a not-significant result. In other words, there is a 20% chance that you'll make a Type II error (missing a true effect of a specified size). β, therefore, equals 20%, or 0.20.

Why don't the equations that compute sample size to achieve a specified margin of error require you to enter a value for power?

The equations you'll find in other books for calculating sample size to obtain a specified margin of error (half-width of the CI) often don't require you to choose power. That is because they assume 50% power. If the assumptions are all true and you use the computed sample size, there is a 50% chance that the computed margin of error will be less than the desired value and a 50% chance that it will be larger.

Can a study ever have 100% power?

No.

Why does power decrease if you set a smaller value for α (without changing sample size)?

When you decide to make α smaller, you set a stricter criterion for finding a significant difference. The advantage of this decision is that it decreases the chance of making a Type I error. The disadvantage is that it will now be harder to declare a difference significant, even if the difference is real. By making α smaller, you increase the chance that a real difference will be declared not significant and thus decrease statistical power.

Why does power increase if you choose a larger n?

If you collect more evidence, conclusions will be more certain. Collecting data from a larger sample decreases the standard error and thus increases statistical power.

How can I do a sample size calculation if I have no idea what I expect the SD to be?

You can't. You'll need to perform a pilot study to find out.

How large a sample size do I need for that pilot study?

Julious (2005) suggests that 12 is a good number. But using the equation shown in the legend for Table 26.1, the 95% CI for the SD extends down to 0.7 times the actual population SD. Since sample size is proportional to the square of SD, this means the computed sample size (based on a SD from a pilot study with n = 12) could be half the size you actually need.

Is it useful to compute power of a completed study?

When a study reaches a conclusion that an effect is not statistically significant, it can be useful to ask what power that experimental design had to detect various effects. However, it is rarely helpful to compute the power to detect the effect actually observed. See Chapter 20.

Is determining sample size a calculation or a negotiation?

Although published statements about sample size always make it seem as though the investigator calculated sample size from α, β, and Δ, often the process is more of a negotiation. The investigators may have first said they were looking for a 10% chance in recurrence rates and then were horrified at the enormous number of participants that would have been required to produce such a result. They then may have changed their goal until *n* seemed "reasonable." They may also have fussed with the values of α and β. Such an approach is not cheating but rather is a smart way to decide on sample size. Parker and Berman (2003) pointed out that the goal often isn't to calculate the number of subjects needed but rather to ask what information can I learn from n subjects?

Is computing the sample size several times with different decisions and assumptions, a form of P-hacking that invalidates the results?

No. It if fine to "negotiate" with a sample size program when designing a study. This is not a form of P-hacking.

What should an investigator do when the required sample size is way larger than is feasible given constraints of budget and staff ?

Don't perform the study. It is far better to cancel a study in the planning stage than to waste time and money on a futile study that won't have sufficient power. If the experiment involves any clinical risk or expenditure of public money, performing such a study could even be considered unethical.

Is getting lots of statistical power the only reason to prefer a large sample size?

No. Here are some other reasons to prefer large sample sizes:

- Large samples let you better assess whether the data comply with the assumption of sampling from a Gaussian distribution).
- Large samples let you more easily detect outliers that you may want to consider removing before doing analyses (Chapter 25).
- Large samples let you assess whether the distribution is biphasic.
- Large samples offer a better opportunity to make informative analyses of subgoups.

Is there any sample size so small that it never makes sense to design an experiment that small, regardless of sample size calculations?

There is no clear consensus on this. I think most scientists and statisticians don't set any minimum. Kraemer and Blasey (2016) say they would never design or approve a study with less than 10 in each group, but don't really justify that opinion. Curtis and colleagues (2015) set the policy of the British Journal of Pharmacology that statistical analyses can only be published when the sample size is at least five per group, but do not justify this policy. de Winter (2013) uses simulations to support use of the t test with n = 3 per group.

CHAPTER SUMMARY

- Sample size calculations require that you specify the desired significance level and power, the minimum effect size for which you are looking, and the scatter (SD) of the data or the expected proportion.
- You need larger samples when you are looking for a small effect, when the SD is large, and when you desire lots of statistical power.

- It is often more useful to ask what you can learn from a proposed sample size than to compute a sample size from the four variables listed in the previous bullet.
- It is important to decide the magnitude of the effect for which you are looking. Using standard effect sizes or published effect sizes doesn't really help.
- Instead of determining sample size to detect a specified effect size with a specified power, you can compute the sample size needed to make the margin of error of a CI have a desired size.
- *Ad hoc sequential sample size determination* is commonly done but not recommended. You can't really interpret P values when the sample size was determined that way.
- Adaptive trials do not require you to stick with the sample size chosen at the beginning of the study. They allow you to modify sample size based on data collected during the study using algorithms chosen as part of the study protocol.
- If you can estimate the prior probability that your hypothesis is correct, you can compute the estimated FPRP of the study you are designing. If the FPRP is too high, it makes sense to design a larger study with a lower significance level and more power.

TERMS INTRODUCED IN THIS CHAPTER

- Adaptive trials (p. 252)
- Ad hoc sequential sample size determination (p. 259)
- Power analysis (p. 242)
- Powered to detect (p. 242)
- Standard effect sizes (p. 246)
- Underpowered (p. 242)

Statistical Tests

CHAPTER 27

Comparing Proportions

No one believes an hypothesis except its originator, but everyone believes an experiment except the experimenter.

W. I. B. BEVERIDGE

This chapter and the next one explain how to interpret results that compare two proportions. This chapter explains analyses of cross-sectional, prospective, and experimental studies in which the results are summarized as the difference or ratio of two incidence or prevalence rates.

EXAMPLE: APIXABAN FOR TREATMENT OF THROMBOEMBOLISM

Study design

Agnelli and colleagues (2012) tested whether apixaban would be beneficial in extended treatment of thromboembolism. They studied patients who had a venous thromboembolism—a clot in the leg vein—which can be dangerous if it breaks loose and goes to the lung. After all the patients completed the standard anticoagulation (blood thinner) therapy, the investigators randomly divided them into three groups that were randomly assigned to receive placebo or one of two doses of apixaban, which inhibits clotting factor Xa. We'll ignore the patients who got the higher dose and only discuss the lower dose here.

The investigators followed the patients for a year. The primary outcome they assessed was whether there was a recurrence of the thromboembolism. They measured other outcomes and analyzed these results too.

This study is called a randomized, double-blind prospective study.

It is a *randomized* study because the assignment of subjects to receive a drug or placebo was determined randomly. Patients or physicians could not request one treatment or the other.

It is *double blind* because neither patient nor investigator knew who was getting the drug and who was getting the placebo. This is sometimes called a *double-masked* study. Until the study was complete, the information about which patient got which drug was coded, and the code was not available to any of the participating subjects or investigators (except in case of a medical emergency).

TREATMENT	RECURRENT THROMBOEMBOLISM	NO RECURRENCE	TOTAL
Placebo	73	756	829
Apixaban (2.5 mg twice a day)	14	826	840
Total	87	1,582	1,669

Table 27.1. Data from the Agnelli et al. (2012) apixaban study. Each value is an actual number of patients.

It is *prospective* because the subjects were followed forward over time. Chapter 28 explains retrospective case-control studies, which look back in time.

Results of the study

The results are shown in Table 27.1, arranged in a customary manner with the rows representing alternative treatments and the columns representing alternative outcomes. Each subject was assigned to one row based on treatment and ended up in one column based on outcome. Therefore, each cell in the table is the actual number of people that were given one particular treatment and had one particular outcome.

Table 27.1 is called a *contingency table*. These tables show how the outcome is contingent on the treatment or exposure. By convention, a contingency table always shows the exact number of people (or some other experimental unit) with a certain treatment or exposure and a certain outcome. Tables of percentages or normalized rates are not called contingency tables.

The goal of the analyses is to generalize about the general population of patients who have had a thromboembolism by computing confidence intervals and a P value. The results (computed by GraphPad Prism) are shown in Table 27.2. Lots of programs can do these calculations, so this chapter only discusses interpretation of the results, not the details of the calculations.

P value (Fisher's test)	< 0.0001
P value summary	****
One-or two-sided	Two-sided
Statistically significant? ($\alpha = 0.05$)	Yes
Relative risk	5.283
95% CI of relative risk	3.034 to 9.229
Difference between fractions	0.07139
95% CI of difference	0.04986 to 0.09443
Number needed to treat	14.0
95% confidence interval of NNT	10.6 to 20.0

Table 27.2. Statistical results of the Agnelli et al. (2012) apixaban study.

CIs of the proportions

The disease progressed in 8.8% of the patients receiving the placebo (73/829), with the 95% CI ranging from 7.0% to 10.9% (Chapter 4 explains CIs of proportions). In contrast, the disease progressed in 1.7% of the people receiving apixaban, with the 95% CI ranging from 0.9% to 2.8%. These CIs present the findings clearly. The patients who received apixaban were a lot less likely to have a recurrent thromboembolism.

The attributable risk

One way to summarize the results is to calculate the difference between the two proportions (often expressed as percentages). Disease progressed in 8.8% of placebo-treated patients and in 1.7% of patients treated with apixaban, a difference of 7.1%. This difference between two incidence rates is called *attributable risk.*

The 95% CI of the attributable risk ranges from 5.0% to 9.4%. If we assume our subjects are representative of the larger population of adults with a thromboembolism, we are 95% confident that treatment with apixaban will reduce the incidence of disease progression by somewhere between 5.0% and 9.4%. Note that these calculations deal with the actual difference in incidence rates (subtraction), not the relative change (division).

Number needed to treat

Laupacis, Sackett, and Roberts (1988) have suggested reporting the reciprocal of the attributable risk, a value they call the *number needed to treat* (NNT). This value tells you how many patients would require treatment to reduce the expected number of cases of a defined end point by one. In this example, the reciprocal of 0.0714 is 14.0. This is easy to understand. For every 14 patients who receive the treatment for a year, you'd expect to prevent a recurrent thromboembolism in one. The other 13 would receive no benefit from the treatment (assuming the only benefit is to prevent thromboembolism). Of course, you don't know before treating a patient whether he or she will benefit.

The CI of the NNT is obtained by taking the reciprocal of each end of the CI of the attributable risk, so it extends from 10.6 to 20.0. In other words, you'd have to treat between 11 and 20 patients to prevent one thromboembolism.

There are two advantages to reporting results as the NNT. One advantage is that this approach avoids the need to think about small fractions. The other advantage is that it puts the results in a clinically relevant context. Is it worth treating 13 patients who won't benefit to help the one patient who will? Answering that question requires that you compare the benefit to that one patient (huge in this case) with the risks, costs, and inconvenience to the other 12. You also need to account for the availability of alternative treatments. Making that decision requires clinical judgment. Expressing the result as the NNT makes it easier to think about the trade-offs.

When the treatment or exposure causes harm, the term *NNT* doesn't fit, so this value is renamed the *number needed to harm.*

The relative risk

It is often more intuitive to think of the ratio of two proportions rather than their difference. This ratio is termed the *risk ratio* or *relative risk*. Disease progressed in 8.8% of placebo-treated patients and in 1.7% of patients treated with apixaban. The ratio is 8.8/1.7, or 5.2. In other words, subjects treated with the placebo were 5.2 times as likely as patients treated with apixaban to have a recurrent thromboembolism.

That seems a bit backwards. Let's flip the ratio to 1.7/8.8, which is 0.19. Patients receiving the drug had 19% the risk of a recurrent thromboembolism than did patients receiving the placebo.

The 95% CI of the relative risk extends from 0.11 to 0.33. The interpretation is now familiar. If we assume our subjects are representative of the larger population of adults with thromboembolism, we are 95% sure that treatment with apixaban will reduce the relative incidence of disease progression to between 0.11 to 0.33 times the risk.

Don't get confused by the two uses of percentages. In this example, the drug lowered the *absolute* risk of thromboembolism by 7.1% (the difference between 8.8% and 1.7%) and reduced the *relative* risk by 81% (which is 100% minus the ratio 1.7/8.8), down to a risk that is 19% that of placebo-treated patients.

In this example, the term *risk* is appropriate because it refers to disease recurrence. In other contexts, one alternative outcome may not be worse than the other, and the relative risk is therefore more appropriately termed the *relative probability* or *relative rate*.

P value

All P values test a null hypothesis, which in this case is that apixaban does not alter the risk of a recurrent thromboembolism. The P value answers the question, If the null hypothesis were true, what is the chance that random sampling of subjects would result in incidence rates as different (or more different) from what we observed?

The P value depends on sample size and on how far the relative risk is from 1.0. Two statistical methods can be used to compute the P value. Most would agree that the best test to use is *Fisher's exact test*. With large sample sizes (larger than used here), Fisher's test is mathematically unwieldy, so a chi-square test is used instead. With large samples, the two compute nearly identical P values.

The P value (calculated by a computer using either test) is tiny, less than 0.0001. Interpreting the P value is straightforward: if the null hypothesis is true, there is less than a 0.01% chance of randomly picking subjects with such a large (or larger) difference in incidence rates.

Other results

The results discussed above convincingly show that the lower dose (of two tested) of apixaban prevents recurrence of thromboembolism. Although statistical tests focus on one result at a time, you must integrate various results before reaching an overall conclusion. Here the investigators also showed the following:

- The results with the higher dose (5 mg twice a day) were very similar to the results summarized above for 2.5 mg twice a day.
- The main potential risk of an anticoagulant is bleeding, so it is reassuring that patients receiving apixaban (an anticoagulant) did not have more bleeding events than those treated with the placebo.
- The reduction in the number of recurrent thromboembolisms was similar in old and young, and was also similar in men and women.

ASSUMPTIONS

The ability to interpret the results of a prospective or experimental study depends on the following assumptions.

Assumption: Random (or representative) sample
The 95% CI is based on the assumption that your sample was randomly selected from the population. In many cases, this assumption is not true. You can still interpret the CI as long as you assume that your sample is representative of the population.

The patients in the example were certainly not randomly selected, but it is reasonable to think that they are representative of adults completing anticoagulant therapy following a venous thromboembolism.

Beware of the two uses of the term *random*. The subjects in the study were not randomly selected from the population of all people with venous thromboembolism. But each subject was randomly assigned to receive the drug or placebo.

Assumption: Independent observations
The 95% CI is only valid when all subjects are sampled from the same population and each has been selected independently of the others. Selecting one member of the population should not change the chance of selecting anyone else.

This assumption would be violated, for example, if the study included several people from one family.

Assumption: No difference between the two groups except treatment
In this study, subjects were randomly assigned to receive the drug or placebo, so there is no reason to think that the two groups differed. But it is possible that, just by chance, the two groups differed in important ways. The authors presented data showing that the two groups were indistinguishable in terms of age, weight, sex, kidney function, and other relevant variables.

COMPARING OBSERVED AND EXPECTED PROPORTIONS

Mendel's peas
Mendel pioneered the field of genetics. The chi-square test (developed after he died) can be used to compare actual data with the predictions of Mendel's model.

PHENOTYPE	OBSERVED NUMBER OF SEEDS IN THIS EXPERIMENT	EXPECTED PROPORTION	EXPECTED NUMBER IN THIS EXPERIMENT
Round and yellow	315	9/16	312.75
Round and green	108	3/16	104.25
Angular and yellow	101	3/16	104.25
Angular and green	32	1/16	34.75
Total	556	16/16	556.00

Table 27.3. Mendel's peas.
This shows one of Mendel's experiments (adapted from Cramer, 1999). The first column shows the actual distribution of traits in the seeds he collected. The last column shows the distribution expected by genetic theory.

Table 27.3 shows one of his experiments (from Cramer, 1999). In this experiment, he recorded whether peas were round or angular and also whether they were yellow or green. The yellow and round traits were dominant. Table 27.3 also shows the expected proportions of the four phenotypes and the expected number in an experiment with 556 peas. These expected numbers are not integers, but that is OK, because they are the average expectations if you did many experiments.

Is the discrepancy between the observed and expected distributions greater than we would expect to see by chance? The test that answers this is called the *chi-square test*. It computes a P value that answers the question, If the theory that generated the expected distribution is correct, what is the chance that random sampling would lead to a deviation from expected as large or larger as that observed in this experiment?

The P value is 0.93 (computed here using GraphPad Prism, but lots of programs can do the calculations). With such a high P value, there is no reason to doubt that the data follow the expected distribution. Note that the large P value does not prove the theory is correct, only that the deviations from that theory are small and consistent with random variation. (Many of Mendel's P values were high, which makes some wonder whether the data were fudged, as was briefly discussed in Chapter 19.)

How it works: Chi-square goodness-of-fit test

When running the chi-square test to compare observed and expected distributions, it is essential that you enter observed values that are the actual numbers of subjects in each category. Don't try to run the chi-square test with percentages or any kind of normalized value. The expected values are the number of subjects you expect to see in each category. These values do not need to be integers. Each expected value is the average number you'd expect in that category if the experiment were repeated many times. In any one experiment, the observed value must be an integer. But the expected value, averaged over many experiments, can be a fraction. The sum of all the observed values must equal the sum of all the expected values.

To quantify the discrepancy between the observed and expected distributions, the chi-square test does the following. For each category, it computes the difference between the observed and expected values and divides that difference by the expected value. It then adds up the square of all those ratios. The result is a chi-square value, pronounced *ki* (rhymes with *eye*) *square* and written as χ^2. Here is the value as an equation:

$$X^2 = \sum \frac{(\text{Observed} - \text{Expected})^2}{\text{Expected}}$$

For the Mendel example, $\chi^2 = 0.470$. The relationship between χ^2 and P value depends on how many categories there are. The number of degrees of freedom (df) is defined to equal the number of categories minus one. The example data have four categories and so have 3 df. This makes sense. Once you know the total number of peas and the number in three of the phenotype categories, you can figure out the number of peas in the remaining category. Knowing that $\chi^2 = 0.470$ and df = 3, the P value can be determined by computer or using a table.

The test is based on some approximations, which are only reasonably accurate when all the expected values are fairly large. If any expected value is less than 5, the results are dubious. This matters less when there are lots of categories (rows) and matters the most when there are only two categories (in which case the expected values should be 10 or higher).

Binomial test

The chi-square test described in the previous section is an approximation. When there are only two categories, the *binomial test* computes the exact P value, without any approximation or worry about sample size.

The coin-flipping example of Chapter 15 used the binomial test. To compute the binomial test, you must enter the total number of observations, the fraction that had one of the two outcomes, and the fraction expected (under the null hypothesis) of having that outcome. In the coin-flipping example, the expected proportion was 50%, but this is not always the case.

COMMON MISTAKES: COMPARING PROPORTIONS

Mistake: Mixing up the two uses of the chi-square test

The chi-square test is used in two distinct ways, and it is easy to mix them up.

The first part of the chapter explained how to compare two proportions. One way to compute a P value is with a chi-square test (although the Fisher's exact test is preferred). The data provide four observed values. The expected values are computed from those observed data (assuming the null hypothesis is true). The chi-square test then combines the discrepancies between observed and expected and computes a P value. To analyze the data using a program that performs the chi-square test, you would enter only the four values of the contingency table (see Table 27.1). The program would compute the expected values automatically.

The second part of this chapter explained how the chi-square test can be used to compare an observed distribution with a distribution expected by theory. To do this analysis using a computer program, you must enter both an observed and an expected count for each category. The expected counts must come from theory or external data and cannot be derived from the data being analyzed.

Mistake: Not entering actual counts into a contingency table

When analyzing a contingency table or when comparing observed and expected distributions, it is essential that you enter the actual number of observed counts. The results will be incorrect if you enter fractions or percentages or normalized rates (e.g., deaths per 1,000 people).

Mistake: Analyzing observed and expected data (like Mendel's peas) in a program designed to analyze a contingency table

If you want to compare observed and expected distributions, it is essential that you use a program designed for that purpose. Don't mistakenly use a program designed to analyze a contingency table. If you enter observed data into the first column and expected counts into the second column, you have *not* created a contingency table. If you analyze that table with a program designed to analyze a contingency table, the results will be meaningless.

Q & A

What if there are more than three groups or more than three outcomes?

It won't be possible to compute an attributable risk or relative risk, but it is possible to compute a P value. Fisher's test is limited to use on tables with two rows and two columns, but the chi-square test can analyze any size contingency table. Some programs offer exact tests for this situation.

If there are more than three rows or columns, does it matter in what order they are placed?

The usual chi-square test pays no attention to the order of rows or columns. If your table has two columns and three or more rows in which the order matters (e.g., doses or ages), the *chi-square test for trend* questions whether there is a significant trend between row number and the distribution of the outcomes.

What is Yates's correction?

When you use a program to analyze contingency tables, you might be asked about Yates's correction. The chi-square test used to analyze a contingency table can be computed in two ways. Yates's correction increases the resulting P value to adjust for a bias in the usual chi-square test, but it overcorrects.

Are special analyses available for paired data in which each subject is measured before and after an intervention?

Yes. McNemar's test is explained in Chapter 31.

Is it better to express results as the difference between two proportions or the ratio of the two proportions?

It is best, of course, to express the results in multiple ways. If you have to summarize results as a single value, I think the NNT is often the most useful way to do so. Imagine a vaccine that halves the risk of a particular infection so that the relative risk is 0.5. Consider two situations. In Situation A, the risk in unexposed people is 2 in 10 million, so halving the risk brings it down to 1 in 10 million. The difference is 0.0000001, so NNT = 10,000,000. In Situation B, the risk in unexposed people is 20%, so halving the risk would bring it down to 10%. The difference in risks is 0.10, so the NNT = 10. The relative risk is the same in the two situations, and the risk difference involves tiny fractions that are hard to say and remember. But the NNT is clear. In the first situation, you'd treat 10 million people to prevent one case of disease. In the second situation, you'd treat 10 people to prevent one case of disease. Expressed as NNT, the results are easier to understand and explain.

The NNT in the apixipan example is 14. Does that mean the drug doesn't work in 13 out of 14 patients who receive it?

No. The drug works as an anticoagulant, and presumably all the patients were anticoagulated. The NNT of 14 means that during the year's course of the study, you need to treat 14 patients to prevent one thromboembolism (the outcome measured in the study).

Can a relative risk be zero?

If none of the control group had the outcome, the relative risk will be zero.

Can a relative risk be negative?

No.

Is there a limit to how high a relative risk can be?

The maximum possible value depends on the risk in the control group. For example, if the risk in the control group is 25%, the relative risk cannot be greater than 4.0, because that would make the risk in the treated group be 100% (the maximum possible risk).

CHAPTER SUMMARY

- A contingency table displays the results of a study with a categorical outcome. Rows represent different treatments or exposures. Columns represent different outcomes. Each value is an actual number of subjects.
- When there are two treatments and two outcomes, the results can be summarized as the ratio (the relative risk), the difference between two incidence rates (the attributable risk), or the reciprocal of that difference (NNT).
- Evidence against the null hypothesis that the outcome is not related to the treatment is expressed as a P value, computed using either Fisher's exact test or the chi-square test.
- An observed distribution can be compared against a theoretical distribution using the chi-square test. The result is expressed as a P value testing the null hypothesis that the data were in fact sampled from a population that followed the theoretical distribution. When there are only two categories, the binomial test is preferred.

TERMS INTRODUCED IN THIS CHAPTER

- Attributable risk (p. 265)
- Binomial test (p. 269)
- Chi-square test (p. 268)
- Chi-square test for trend (p. 270)
- Contingency table (p. 264)
- Double-blind study (p. 263)
- Fisher's exact test (p. 266)
- Number needed to harm (p. 265)
- Number needed to treat (NNT) (p. 265)
- Prospective study (p. 263)
- Randomized study (p. 263)
- Relative risk (p. 266)
- Risk ratio (p. 266)

CHAPTER 28

Case-Control Studies

It is now proven beyond doubt that smoking is one of the leading causes of statistics.

FLETCHER KNEBEL

This chapter explains how to interpret results from a case-control
study. In this kind of study, the investigators select two groups of
subjects: incident cases with the disease or condition being studied and
controls selected from the same population that produced the cases. The
investigators then look back in time to compare the exposure of the two
groups to a possible risk factor (or treatment).

EXAMPLE: DOES A CHOLERA VACCINE WORK?

Study design
Cholera kills many people in Africa. Although vaccines were not very effective
against cholera in the past, newer vaccines hold great promise for preventing this
awful disease.

Lucas and colleagues (2005) investigated whether a cholera vaccine was effective. An ideal approach would be to recruit unvaccinated people, randomly
assign them to be vaccinated or not vaccinated, and follow both groups for many
years to compare the incidence of cholera. Such a study would take many years
to conduct, require a huge number of subjects, and require withholding a vaccine
from many. Instead, the investigators asked whether people with cholera are less
likely to have been vaccinated than those who have not gotten cholera. This kind
of study is called a *case-control study* because the investigators pick cases and
controls to study.

Table 28.1 shows the results of Lucas et al.'s study. Note the difference between
this contingency table and the one from the example in Chapter 27 (concerning
apixaban). In the apixaban example, the investigators set the row totals by choosing
how many subjects would receive each treatment and then observed the outcome
tabulated in the two columns. In this example, the investigators set the column
totals by choosing how many cases and controls to study and then determined
whether each person had been previously vaccinated.

	CASES (CHOLERA)	CONTROLS
Received vaccine	10	94
No vaccine	33	78
Total	43	172

Table 28.1. Case-control study to investigate association of cholera with lack of vaccination.

Each value is the number of people with the outcome tabulated in the columns and the vaccination status tabulated in the rows. *Source*: Lucas et al. (2005).

Computing the relative risk from case-control data is meaningless

It would be a mistake to look at Table 28.1 and divide 10 (number of people with cholera who were vaccinated) by 104 (the total number of vaccinated people in the study) to calculate a 10% risk that vaccinated people will get cholera. This is not a helpful calculation and is not a correct statement. The authors chose to have four times as many controls as cases. If they had chosen to use only twice as many controls as cases, that same calculation would have divided 10 into 52 and concluded that there is a 20% risk that vaccinated people will get cholera. Computing risks, and the relative risk, directly from a case-control study is invalid.

The odds ratio

Results from a case-control study are summarized as an *odds ratio*. Among the cases, the odds of being vaccinated are 10:33, or 0.303. The difference between odds and probability was explained in Chapter 2. Among the controls, the odds of being vaccinated are 94:78, or 1.205. The odds ratio therefore equals 0.303/1.205, or 0.25. This compares the odds of exposure (vaccination) in cases and controls. What we really want is a comparison of the odds of the outcome (cholera) in the exposed and unexposed. Fortunately, the odds ratio comparing exposure is the same as the odds ratio comparing outcome. Thus, we can conclude that the odds ratio of 0.25 quantifies the reduced odds of cholera among those who are vaccinated. A computer program can compute the 95% CI, which extends from 0.12 to 0.54.

For most study designs, the odds ratio calculated from a case-control study can be interpreted as a relative risk (see Q&A at the end of the chapter). For this example, the odds ratio is 0.25, so we can conclude that a vaccinated person has about 25% the chance of an unvaccinated person getting cholera.

In the case of vaccination studies, it makes sense to subtract the odds ratio from 1.0. Here that difference is 0.75, which means the vaccine is 75% effective in preventing cholera. (The numbers reported in the original paper are slightly different than these, because the researchers analyzed the data using a fancier method called *conditional logistic regression* to account for the matching of controls with each case.)

Interpreting a P value

When you analyze these data with a computer program, you will be asked which test to use to compute the P value. Most would agree that Fisher's exact test is preferred over a chi-square test. An alternative is to use a chi-square test (much easier if you are doing the calculations manually, but why would you?).

The P value for the example is 0.0003 (computed by Fisher's exact test). The null hypothesis is that there is no association between the vaccination and whether someone got cholera. The P value is easy to interpret. If the vaccination rates were really the same among patients and controls, there is only a 0.03% chance that random sampling in an experiment of this size would end up showing such a strong (or stronger) association between disease and vaccination. This is a two-tailed P value, so it includes the possibilities of seeing either more vaccination, or less vaccination, in people with cholera.

EXAMPLE: ISOTRETINOIN AND BOWEL DISEASE

Study design

Isotretinoin is a very effective drug for treating severe acne but there have been concerns about its toxicity. Several case reports have suggested that use of this drug might cause inflammatory bowel disease (IBD, which encompasses both ulcerative colitis and Crohn's disease). But the case reports really can't be interpreted except as vague warnings. Of course, some people treated with isotretinoin get IBD. The question is whether people who take the drug have a higher risk of getting the disease than do people who don't take the drug. Because IBD is fairly rare, it would be very hard to answer such a question in a prospective study, so Bernstein and colleagues (2009) performed a case-control study.

The cases consisted of every single case of IBD that was reported in a person less than 40 years old living in Manitoba, Canada, between 1995 and 2007. For each case, the researchers used 10 controls matched by age, sex, postal code, and medical insurance coverage.

Odds ratio

They analyzed the data for each disease (ulcerative colitis and Crohn's disease) separately and also analyzed the two diseases combined. The combined data are shown in Table 28.2. Of the 1,960 cases of people with IBD, 25 (1.28%) had taken isotretinoin prior to their diagnosis. Among cases, the odds of having received isotretinoin is 25/1935, or 0.0129. Of the 19,419 controls, 213 (1.10%) had taken isotretinoin. Among controls, the odds of having received isotretinoin are 213/19216, or 0.0111. The odds ratio equals the odds of the cases having been exposed to the possible risk factor divided by the odds of the controls having been exposed. In this example, the odds ratio is 0.0129/0.0111, or 1.17, with a 95% CI ranging from 0.77 to 1.75. Since that interval contains 1.0, the investigators concluded there was no evidence of an association between isotretinoin use and IBD. Even though the samples are huge, the CI is quite wide. This is because only a small fraction (about 1%) of either group took the drug.

	CASES (IBD)	CONTROLS (NO IBD)
Received isotretinoin	25	213
No isotretinoin	1,935	19,216
Total	1,960	19,419

Table 28.2. Case-control study to investigate association of inflammatory bowel disease (IBD) with isotretinoin.
Source: Bernstein et al. (2009).

P value

The researchers presented the results as an odds ratio with CI but did not also present a P value or statements of statistical significance. I think this was a wise decision. The odds ratio with its CI summarizes the results very clearly.

Even though the P value wasn't presented, we can infer something about it. Since the 95% CI contains 1.0 (the value that denotes no association), the P value must be greater than 0.05 (since 100% − 95% = 5% = 0.05). Since 1.0 is not very near either end of the interval, the P value must be quite a bit greater than 0.05. In fact, it is 0.43.

Generalizing the results

This study used data from every single case of IBD reported in Manitoba over a 12-year period. For these results to be generally useful, you need to assume that these patients in Manitoba during the period of 1995–2007 are representative of future patients all over the world. That seems like a reasonable assumption, since IBD (as far as I know) is not a disease that manifests itself differently in various locations or a disease that changes from year to year.

EXAMPLE: GENOME-WIDE ASSOCIATION STUDIES

As discussed in Chapter 22, genome-wide association studies (GWAS) look for associations between a disease or condition and genetic variations (usually single nucleotide polymorphisms). Many cases (often tens of thousands) and controls are tested, and a separate analysis is done for each variation. Essentially, a GWAS simultaneously runs many (perhaps a million) case-control studies at once. Because so many hypotheses are tested at once, analysis of the data must account for multiple comparisons, as explained in Chapter 22.

HOW ARE CONTROLS DEFINED?

Matched controls

Often investigators choose matched controls. For every case, they try to find one, or several, controls that match in defined ways. In the cholera example, the authors picked controls by visiting homes near the patient and identifying someone of the same gender and age. Identifying the cases and controls at about the same time, and matching gender and age is a strength of the study. But, there are some potential problems with this approach:

- The only way that controls became part of the study was to be home when the investigators visited. This method thus selected for people who stay home a lot. This selection criterion was not applied to the cases. This would cause bias if the incidence of cholera was lower in people who stay home a lot.
- The controls were chosen among people who live very near the cases. This could be a problem if the vaccination effort was more effective in some neighborhoods than others. If that were true, then a case who was vaccinated is more likely to get matched to a control who was a vaccinated, and a case that was not vaccinated would be more likely to be matched to a control that was not vaccinated. This would bias the results of the case-control study, because it would be biased to not find a link between vaccination and infection.

These kinds of problems are inherent in all case-control studies where the controls were chosen to match the cases. When you match for variables, you might inadvertently also match (indirectly) for exposure.

The investigators in this study were very aware of these potential problems. To test whether these biases affected their results, they ran a second case-control study. Here, the cases were patients with bloody diarrhea that turned out to not be caused by cholera. The controls were selected in the same way as in the first study. The second study, which had almost the same set of biases as the first, did not detect an association between diarrhea and cholera vaccination. The odds ratio was 0.64, with a 95% CI extending from 0.34 to 1.18. Because the CI spans 1.0, the study shows no association between cholera vaccination and bloody diarrhea not caused by cholera. The negative results of the second study suggest that the association reported in the main study really was caused by vaccination, rather than by any of the previously listed biases.

Population controls

Rather than matching controls to cases, some studies are designed so controls are chosen from the same population from which the cases are identified, with an effort to choose controls at about the same time that cases are identified. Many epidemiologists think this kind of design is much better, as it avoids the problems of matching. Sometimes population controls are selected with some matching to get controls of about the same age as the cases.

HOW ARE CASES DEFINED?

In most case-control studies, the cases are every single person with a defined disease or condition that develops in a defined population during a defined period of time. The key point is that someone becomes a case only when the disease occurs during a defined time period.

Interpreting the results is far more difficult when cases are people who already have the disease or condition. The problem is simple to understand. Prevalence depends both on the number of new cases and on survival. If an exposure is associated with greater prevalence, there are two wildly different explanations. One explanation is that the exposure caused a greater incidence of the disease. The other explanation is that there is no change in incidence but rather an increase in survival. Yikes! One explanation means the exposure is bad because it causes the disease. The other explanation means the exposure is good because it extends survival with the disease. Since this design makes it impossible (in most cases) to interpret the results clearly, most case-control studies are done by defining cases as people who develop the disease or condition (incidence of disease), not those who already have it (prevalence of disease).

EPIDEMIOLOGY LINGO

Bias
Bias refers to anything that causes the results to be incorrect in a systematic way.

Retrospective and prospective
Retrospective studies look back in time to gather information. *Prospective* studies collect information as it happens and into the future.

The term *retrospective study* is sometimes equated with case-control study, and the term *prospective study* is sometimes equated with cohort studies, but these equivalences are incorrect. In fact, some cohort studies are conducted retrospectively, and some case-control studies are conducted prospectively.

A retrospective cohort study defines two exposure groups (cohorts) by looking at old records and then ascertain the outcome based on current records or tests. A prospective case-control study identifies a population to study and then observes what happens (prospectively). As each case appear, appropriate information is collected. At the same time the same information is collected on a random selection of the population to serve as controls.

Incidence versus prevalence
Incidence is the rate of new cases of a disease. *Prevalence* is the fraction of the group that has the disease at some point of time.

Contingency table
Contingency tables (first introduced in Chapter 27) show how the outcome is contingent on the treatment or exposure. The rows represent exposure (or lack of exposure) to alternative treatments or possible risk factors. Each subject belongs to one row based on exposure or treatment. Columns denote alternative outcomes. Each subject also belongs to one column based on outcome. Therefore, each cell in the table is the number of people (or animals or lightbulbs…) that were in one particular exposure (or treatment) group and had one particular outcome.

Not all tables are contingency tables. Contingency tables always show the actual number of people (or some other experimental unit) in various categories.

Thus, each number must be a positive integer or zero. Tables of fractions, proportions, percentages, averages, changes, or durations are not contingency tables, nor are tables of rates, such as number of cases per 1,000 people. Table 27.3 compares observed and expected counts and therefore is not a contingency table.

COMMON MISTAKES: CASE-CONTROL STUDIES

Mistake: Picking an inappropriate control group

The results of a case-control study can only be taken at face value when the cases and controls come from the same population. *Selection bias* occurs when people selected as controls are not truly representative of the population that produced the cases.

To assess this, ask two simple questions:

- If any of the controls had developed the disease, would he or she have been part of the group of cases? Most case-control studies define as cases every single person who develops the condition or disease (in a certain period of time). Therefore, every control ought to be recognized as a case if he or she develops the disease. If this condition is not met, then the cases and controls came from different populations.
- If one of the cases had not developed the disease, would he or she have been in the group of people that might have been randomly chosen to be a control? The logic is the same as the previous point.

This is a problem in many GWAS studies. A tiny P value in a GWAS is evidence that the prevalence of that particular genetic variation differs in the two groups you are studying. One of the groups is composed of people with a disease and the other group is composed of controls, so one explanation is that the disease is associated with, and perhaps even caused by, that genetic variation. However, it may also be the case that the association is due to different ancestry of patients and controls and so has nothing to do with the disease being studied. For example, there may be more people of Jewish ancestry in one group and more of Italian ancestry in the other. Geneticists call this *population stratification*.

The huge size of a GWAS (thousands or tens of thousands of cases and controls) makes it very difficult to avoid population stratification but makes it possible to measure and (attempt to) correct for it (Price et al., 2010).

Mistake: Using a control group that is largely self-selected

The control group will not represent the population from which the cases came if many people decline to participate in the study, and that decision depends on either exposure to the possible risk factor or of family history of the disease. This is called *self-selection bias*. For example, if the study is of Alzheimer's disease, people may be more likely to agree to be a control for the study if one of their parents suffers from it. Accordingly, the controls that agree to participate are not a random sample of the relevant population.

Mistake: Collecting data in a way that is more accurate in cases than controls

After identifying cases and controls, the investigator looks back in time to ascertain exposure to the possible risk factor. Sometimes this is done by looking at existing records. But often it involves asking the cases and controls questions. But it is too easy to ask the questions differently for cases and controls or for controls and cases to have different memory of exposure (*recall bias*). In the cholera example:

- The subjects obviously knew they had cholera and so may have recalled their vaccination more vividly than the controls.
- The interviewers knew whether they were talking to a control subject or a case (someone who had cholera). Although they tried to ask the questions consistently, they may have inadvertently used different emphasis with the cases.
- The subjects had suffered from cholera and may have been motivated to help the researchers learn more about the disease. In contrast, the controls may have been more focused on ending the interview as quickly as possible. Thus, cases and controls may not have given equally detailed or accurate information.

Mistake: Using inconsistent or inaccurate diagnostic criteria

The odds ratio and its 95% CI can only be interpreted when the number of subjects in each category is tabulated correctly. This assumption would be violated in the cholera example if some of the patients didn't actually have cholera or if some of the controls did.

Mistake: Entering normalized data

When entering data into a contingency table, you must enter the actual number of subjects or events counted in each group. The results will be misleading if you enter percentages, normalized values, or any kind of rate. This mistake will probably not affect the computation of the odds ratio but will affect the calculation of its CI and the P value.

Mistake: Extrapolating the conclusions too far

The patients in the cholera example were certainly not a random selection of all those with cholera, but perhaps they are representative of adults with cholera in that particular city and perhaps all of Africa. It isn't so clear how far to generalize the results.

Similarly, the patients in the isotretinoin study were certainly not randomly selected. The investigators studied every single person in Manitoba who reported having an IBD during a 12-year period. Should the results be extrapolated to everyone in Manitoba in future years or to all patients with IBD? The results can only be interpreted if you assume that these people are similar to those who will exist elsewhere in future years.

Mistake: Thinking that a case-control study can prove cause and effect

An observational study can show an association between two variables. Depending on the situation, you may be able to make some tentative conclusions about causation. But, in general, association does not imply causation. Check out the cartoons in Chapter 45 if you are not convinced.

Q & A

What is the biggest advantage and challenge of case-control studies?

The biggest advantage of case-control studies is that they can be done relatively quickly with a relatively small sample size drawn from previously recorded data. The alternative, in most cases, would be huge, time-consuming studies. The biggest challenge of case-control studies is to pick the right controls.

Can an odds ratio be zero?

If none of the cases were exposed to the possible risk factor, the odds ratio will be zero. But see the next question.

How are odds ratios computed when one of the four values is zero?

If one of the four values is zero, then the computed odds ratio may be undefined as a result of division by zero (it depends on which value is zero). In this case, many investigators and programs add 0.5 to each value before calculating the odds ratio.

Can odds ratios be negative?

No.

Is there a limit to how high an odds ratio can be?

Not in theory, but it depends on sample size.

When is it reasonable to interpret an odds ratio as a relative risk?

The odds ratio from a case-control study is almost always a good estimate of the relative risk. Many books (including prior editions of this one) state that the odds ratio can be only interpreted as a relative risk when the disease is rare. But, in fact, odds ratios can be interpreted as relative risks in most, but not all, case control studies, even those of common diseases.

To understand the complexities of selecting appropriate controls and interpreting odds ratios from case-control studies of common diseases, read the appropriate chapters from the epidemiology text books by Rothman, Geenland, and Lash (2008) and by Grobbee and Hoes (2015). Also read the review papers by Vandenbroucke and Pearce (2012) and Greenland, Thomas, and Morgenstern (1986).

How large does an odds ratio need to be to be believable?

When a case-control study is not conducted within a clearly defined population, it is difficult to know how to interpret odds ratios. Is there really an association between risk factor and disease? Or is the study biased? Taubes (1995) quoted two epidemiologists who caution that when reviewing results from such a case-control study (especially one not based on clear theory or prior data), you should be very skeptical of results if the lower 95% confidence limit (or its reciprocal) is less than 3 (or even 4), even if the P value is tiny and the CI narrow.

Are case-control studies "quick and dirty" ways to advance medical knowledge?

It is easy to design and perform a poor-quality case-control study. It is far from easy or quick to design a high-quality case-control study. But once designed, a case-control study only requires data collection and analysis. In contrast, an experimental or prospective study also requires waiting for the results of the treatment or exposure to become apparent over time.

Is it possible for the same person to be both a case and a control?

Surprisingly, it is possible. Say you identify the controls at the beginning of the study and then collect cases as people in that population develop the disease or condition. It is possible that one of your controls then also becomes a case.

Since case-control studies are retrospective, is there any point in following a prespecified analysis plan?

Yes. If investigators didn't create and follow an analysis plan, people reading the study never know how many variations on the study they ran until they got a P value small enough to publish.

How is a case-control study analyzed when the investigator wants to adjust for other variables?

Using logistic regression as explained in Chapter 38.

What are the advantages and disadvantages of hospital-based case-control studies?

Sometimes case-control studies define cases as people coming to the hospital with one disease and controls as people coming to the same hospital with a different disease or condition. One advantage is that the logistics of this kind of study are easier to arrange than most other kinds of studies. Another advantage is that hospitalized controls tend to be more likely to agree to participate in a research study than are people who are not hospitalized. The disadvantage is that the cases and controls may come from very different populations. This is not much of an issue when the hospital is in a small city or is part of a health maintenance organization where all members go to the same hospital for all conditions. But in larger cities where people have a choice of hospitals, the hospitalized patients with one disease may have very different characteristics than patients who come to that hospital with another disease.

Do case-control studies always use an outcome variable with only two possible values?

Case-control studies are usually done with binary outcome variables, so are usually analyzed by calculation of an odds ratio using Fisher's test or logistic regression. But a case-control design could be used for a study that collects continuous data to be analyzed with a t test, or collects survival data to be analyzed by the log-rank test or Cox regression (see Chapter 29).

CHAPTER SUMMARY

- Case-control studies start by identifying people with and without a disease or condition and then compare exposure to a possible risk factor or to a treatment.
- The advantage of a case-control study compared to a cohort study is that a case-control study can be performed more quickly, can sometimes use existing data, and requires smaller sample sizes.

- The results of a case-control study are expressed as an odds ratio, along with its 95% CI. In most studies, the odds ratio from a case-control study can be interpreted as a relative risk.
- Ideally, case-control studies identify the cases and sample the controls from the same population. To know if this is the case, ask yourself if each of the controls, had they come down with the disease or condition, would have become a case. If the answer is no, the cases and controls came from different populations, which makes the odds ratio hard to interpret.
- Fisher's exact test (or the chi-square test) can compute a P value that tests the null hypothesis of no association between disease and exposure to the possible risk factor or treatment.

TERMS INTRODUCED IN THIS CHAPTER

- Bias (p. 278)
- Case-control study (p. 273)
- Contingency table (p. 273)
- Incidence (p. 278)
- Odds ratio (p. 274)
- Population stratification (p. 279)
- Prevalence (p. 278)
- Prospective (p. 278)
- Recall bias (p. 280)
- Retrospective (p. 278)
- Self-selection bias (p. 279)

CHAPTER 29

Comparing Survival Curves

A statistician is a person who draws a mathematically
precise line from an unwarranted assumption to a foregone
conclusion.

UNKNOWN AUTHOR

Chapter 5 explained how survival curves were generated and how
to interpret the CI of survival curves. This chapter explains how to
compare two survival curves.

EXAMPLE SURVIVAL DATA

In one study, patients with chronic active hepatitis were treated with prednisolone
or placebo, and their survival was compared (Kirk et al., 1980; raw data from
Altman & Bland, 1998). Some patients were still alive at the time the data were
collected, and these individuals are indicated in Table 29.1 with an asterisk. The
information that these patients lived at least up to the time of data collection is an
essential part of the data collected from the study. The values for these individuals
are entered into a computer program as censored values. In addition, one patient
taking prednisolone was lost to follow-up at 56 months. His data are also said to
be censored and are entered into a survival analysis program the same as the other
censored values. He lived at least 56 months, but we don't know what happened
after that.

Figure 29.1 shows the Kaplan-Meier graph of the survival data, as previously
explained in Chapter 5.

ASSUMPTIONS WHEN COMPARING SURVIVAL CURVES

The following list summarizes the assumptions, explained in detail in Chapter 5,
that must be accepted to interpret survival analyses.

- Random (or representative) sample
- Independent subjects
- Consistent entry criteria
- Consistent definition of the end point
- Clear definition of the starting point

PREDNISOLONE	CONTROL
2	2
6	3
12	4
54	7
56 (left study)	10
68	22
89	28
96	29
96	32
125*	37
128*	40
131*	41
140*	54
141*	61
143	63
145*	71
146	127*
148*	140*
162*	146*
168	158*
173*	167*
181*	182*

Table 29.1. Sample survival data from Kirk et al. (1980), with raw data from Altman and Bland (1998).

Patients with chronic active hepatitis were treated with prednisolone or placebo, and their survival was compared. The values are survival time in months. The asterisks denote patients still alive at the time the data were analyzed. These values are said to be *censored*.

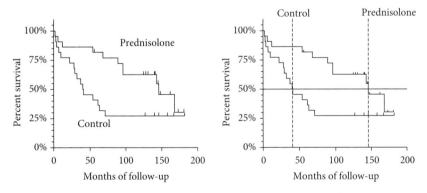

Figure 29.1. Kaplan-Meier survival curves created from the data in Table 29.1.

Each drop in the survival curve shows the time of one (or more) patient's death. The blips show the time of censoring. One subject was censored when he left the study. The data for the others were censored because they were alive when the data collection ended. The graph on the right shows how median survival times are determined.

- Time of censoring is unrelated to survival
- Average survival doesn't change during the study

When comparing survival curves, there are additional assumptions, as follows.

Assumption: Treatment groups were defined before data collection began

When you compare groups, the groups must be defined before data collection begins. It is not valid to divide a single group of patients (all treated the same way) into two groups based on whether they responded to treatment. It can be tempting, for example, to compare survival of subjects whose tumor got smaller as treatment progressed with those whose tumors stayed the same size, or to compare those whose lab tests improved with those whose lab tests didn't. Comparing survival of responders with nonresponders is invalid for two reasons.

The first reason is that a patient cannot be defined as a "responder" unless he or she survived long enough for you to measure the tumor. Any patient who died early in the study would certainly be defined as a member of the nonresponder group. In other words, survival influenced the assignment of a patient to a particular group. Therefore, you can't learn anything by comparing survival in the two groups.

The second reason is that the disease may be heterogeneous. The patients who responded may have had a different form of the disease than those who didn't respond. In this case, the responders may have survived longer even if they hadn't been treated.

You must define the groups you are comparing before starting the experimental phase of the study. Be very wary of studies that use data collected during the experimental phase of the study to divide patients into groups or to adjust the data.

Assumption: Groups are defined consistently as data are accrued

Some survival studies accrue patients over several years. In these studies, it is essential that the diagnostic groups be defined consistently.

Here is an example of when that assumption might be violated. Imagine you are comparing the survival of patients whose cancer has metastasized to the survival of patients whose cancer remains localized. Average survival is longer for the patients without metastases. Now imagine that a fancier scanner becomes available, making it possible to detect smaller tumors and so diagnose metastases earlier. What happens to the survival of patients in the two groups?

The group of patients without metastases will be smaller. The patients who are removed from the group will be those with small metastases that could not have been detected without the new technology. These patients will tend to die sooner than those patients without detectable metastases. By taking away these patients, the average survival of the patients remaining in the "no metastases" group will probably improve.

What about the other group? The group of patients with metastases will now be larger. The additional patients, however, will be those with small metastases. These patients will tend to live longer than patients with larger metastases. Thus, the average survival of all patients in the "with metastases" group will probably also improve.

Changing the diagnostic method paradoxically increases the average survival of both groups! Feinstein, Sosin, and Wells (1985) termed this paradox the *Will Rogers phenomenon* from a quote attributed to the humorist Will Rogers: "When the Okies left Oklahoma and moved to California, they raised the average intelligence in both states."

This phenomenon also causes a problem when making historical comparisons. The survival experience of people diagnosed with early stage breast cancer is much better now than it was prior to widespread use of screening mammography. But does that mean that screening mammography saves lives? That is a hard question to answer, since the use of screening mammography has nearly doubled the number of diagnoses of early breast cancer while barely changing the number of diagnoses of advanced breast cancer. Nearly a third of the people who are diagnosed with breast cancer by mammography have abnormalities that probably would never have caused illness (Bleyer & Welch, 2012). Comparing the survival curves of people diagnosed with early breast cancer then and now is pointless, as the diagnostic criteria have changed so much.

Assumption: Proportional hazards

Hazard is defined as the slope of the survival curve—a measure of how rapidly subjects are dying. The *hazard ratio* compares two treatments. If the hazard ratio is 2.0, then the rate of deaths in one treatment group is twice the rate in the other group. Figure 29.2 shows ideal curves with a consistent hazard ratio. Both curves

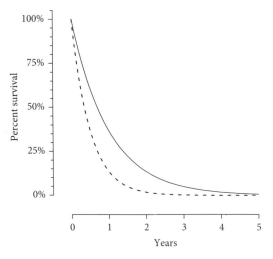

Figure 29.2. Definition of proportional hazards.
At every time point, the dotted curve is twice as steep as the solid curve. The hazard ratio is 2.0 (or 0.5, depending on how you compute it) at all time points.

start, of course, with 100% and end at 0%. At any time along the curve, the slope of one curve is twice the other. The hazards are proportional.

The assumption of *proportional hazards* means that the hazard ratio is consistent over time, and any differences are caused by random sampling. This assumption would be true if, at all times, patients from one treatment group are dying at about half the rate of patients from the other group. It would not be true if the death rate in one group is much higher at early times but lower at late times. This situation is common when comparing surgery (high initial risk, lower later risk) with medical therapy (less initial risk, higher later risk). In this case, a hazard ratio would essentially average together information that the treatment is worse at early time points and better at later time points, so would be meaningless.

If two survival curves cross, the assumption of proportional hazards is unlikely to be true. The exception would be when the curves only cross at late time points when few patients are still being followed.

The assumption of proportional hazards needs to be reasonable to interpret the hazard ratio (and its CI) and the CI for the ratio of median survival times.

COMPARING TWO SURVIVAL CURVES

Survival curves can be compared in several ways.

Hazard ratio

If the assumption of proportional hazards is reasonable, the two survival curves can be summarized with the hazard ratio, which is essentially the same as the relative risk (see Chapter 27). For the example data, the hazard ratio is 0.42, with a 95% CI ranging from 0.19 to 0.92. In other words, the treated patients are dying at 42% of the rate of control patients, and we have 95% confidence that the true ratio is between 19% and 92%. More specifically, this means that on any particular day or week, a treated patient has 42% of the chance of dying as a control patient.

If the two survival curves are identical, the hazard ratio would equal 1.0. Because the 95% CI of the sample data does not include 1.0, we can be at least 95% certain that the two populations have different survival experiences.

Ratio of median survival times

Median survival time is defined as the time it takes until half the subjects have died. The right panel of Figure 29.1 shows the median survival of the sample data. The horizontal line is at 50% survival. The time at which each survival curve crosses that line is the median survival. The median survival of the control patients is 40.5 months, whereas the median survival of the patients treated with prednisolone is 146 months, 3.61 times longer.

It is possible to compute a 95% CI for the ratio of median survivals. This calculation is based not only on the assumption of proportional hazards (see prior section) but also on an additional assumption: that the chance of dying in a small time interval is the same early in the study and late in the study. In other words, the survival curve follows the same pattern as an exponential decay curve.

This assumption does not seem unreasonable for the sample data. For these data, the ratio of median survivals is 3.61, with a 95% CI ranging from 3.14 to 4.07. In other words, we are 95% sure that treatment with prednisolone triples to quadruples the median survival time.

Confidence bands

Figure 29.3 shows the example survival curves with their 95% confidence bands. The curves overlap a bit, so plotting both curves together creates a very cluttered graph. Instead, the two are placed side by side. Both curves start at 100%, of course, so they overlap somewhat, but note that they don't overlap at all for many months. This graph should be enough to convince you that the prednisolone worked.

Note the difference between the confidence band, which is an area plotted on a graph, and a CI, which is a range of values.

P value

The comparison of two survival curves can be supplemented with a P value. All P values test a null hypothesis. When comparing two survival curves, the null hypothesis is that the survival curves of the two populations are identical and any observed discrepancy is the result of random sampling error. In other words, the null hypothesis is that the treatment did not change survival overall and that any difference observed was simply the result of chance.

The P value answers the question, If the null hypothesis is true, what is the probability of randomly selecting subjects whose survival curves are as different (or more different) than what was actually observed?

The calculation of the P value is best left to computer programs. The *log-rank method*, also known as the *Mantel–Cox method* (and nearly identical to the Mantel–Haenszel method), is used most frequently.

For the sample data, the two-tailed P value is 0.031. If the treatment was really ineffective, it is possible that the patients who were randomly selected to receive one treatment just happened to live longer than the patients who received

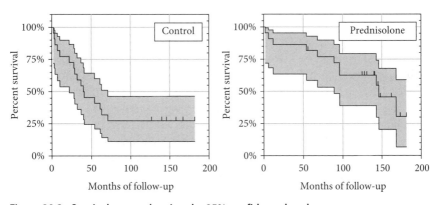

Figure 29.3. Survival curves showing the 95% confidence bands.

the other treatment. The P value tells us that the chance of this happening is only 3.1%. Because that percentage is less than the traditional significance cut-off of 5%, we can say that the increase in survival with treatment by prednisolone is statistically significant.

An alternative method to compute the P value is known as the *Gehan–Breslow–Wilcoxon method*. Unlike the log-rank test, which gives equal weight to all time points, the Gehan–Breslow–Wilcoxon method gives more weight to deaths at early time points. This often makes a lot of sense, but the results can be misleading when a large fraction of patients are censored at early time points. The Gehan–Breslow–Wilcoxon test does not require a consistent hazard ratio, but it does require that one group consistently have a higher risk than the other.

For the sample data, the P value computed by the Gehan–Breslow–Wilcoxon method is 0.011.

WHY NOT JUST COMPARE MEAN OR MEDIAN SURVIVAL TIMES OR FIVE-YEAR SURVIVAL?

Why compare entire survival curves? Why not just compare the mean or median time until death?

Why it is rarely helpful to compare mean time until death

There are three reasons why it is rarely helpful to compare the mean time until death:

- If anyone is still alive at the time you compile the data, it is not possible to compute the average time until death.
- If the data for any subject are censored, it is not possible to compute the average time until death.
- Even if every subject has died (and none were censored), the mean time until death may still not be meaningful. It is likely that the distribution of those values is not Gaussian. If so, summarizing the data with the mean may not be informative. Furthermore, calculations like the 95% CI of the mean won't be helpful if the distribution is far from Gaussian.

Median time until death

Why not just compare the median time until death? This would avoid the problem with mean times, since the median survival time can be calculated once half the subjects have died. The median time until death is a useful way to summarize survival data, but focusing on a single number (the median survival) can be misleading.

Figure 29.4 shows two simulated survival curves. The first parts of the curves are identical, so they appear superimposed. They have almost the same median survival (7.5 years). But after year 7, the two curves diverge drastically. If you had a choice between the two treatments, you'd be nuts not to choose the one with a survival curve that levels off with 42% of the patients still living after 20 years, rather than the treatment in which all subjects had died by year 9.

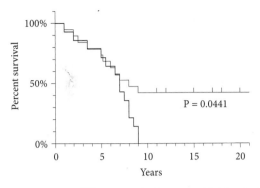

Figure 29.4. Two very different survival curves with identical five-year survival rates and identical median survival times.

Even though the median survival of the two curves is identical, as is the 5-year survival percentage, the two curves are very different. With one treatment, all patients are dead by 10 years. With the other treatment, 40% are still alive at 10 years.

Five-year survival

The success of cancer therapies is often expressed as a five-year survival. This tells you what fraction of the patients are still alive after five years. Five years is an arbitrary but standard length of time to tabulate survival. Figure 29.4 shows why focusing only on that one number can be misleading. The two curves in that figure have the same five-year survival (72%) but diverge a lot after five years. Even though the five-year survival is the same for both treatments, the 10-year survival is very different.

INTENTION TO TREAT

When comparing the survival of people randomly assigned to receive alternative treatments, you will often hit a snag. Some patients don't actually get the treatment they were randomized to receive. Others stop following the protocol.

It seems sensible to just exclude from analysis all data from people who didn't receive the full assigned treatment. This is called the *per protocol approach*, because it requires that you only compare survival of subjects who got the full treatment according to the study protocol. But this approach will lead to biased results if the noncompliance has any association with disease progression and treatment. Hollis and Campbell (1999) review an example of this. The study in question randomized patients with severe angina to receive a surgical or medical treatment. Some of the patients randomly assigned to the surgical group died soon after being randomized so never actually received the surgery. It would seem sensible to exclude them from the analysis, since they never received the treatment to which they were assigned. But that would exclude the earliest deaths from one assigned treatment (surgery) but not the other (medical). The results would then not be interpretable. The surgery group would have better survival, simply because people who died early were excluded from the analysis.

Dallal (2012) gives another example. Imagine a study of two alternative weight-loss treatments. One of the treatments is basically ineffective, but some people lose weight anyway. They stay in the study. Others don't lose weight and drop out of the study. If you only study those who remain in the study, you are selecting for those who lose weight, which would make a comparison between the two treatments invalid.

The preferred approach is called *intention to treat* (Hollis & Campbell, 1999; Montori & Guyatt, 2001). This approach is abbreviated ITT and is also called *analyze as randomized*. With ITT, data are analyzed based on the randomly assigned treatments even if:

- it later turns out that the patient didn't quite meet the entry criteria of the study;
- the treatment was not given; or
- they stopped the treatment for any reason.

Some studies analyze their data using both the per protocol and the ITT approaches. If the results are similar, then you know that the subjects who didn't get the assigned treatment (or didn't get a complete treatment) didn't affect the results much.

The ITT method applies to any analysis of clinical data, not just survival analysis. Some investigators use the intention-to-treat principle for analyzing efficacy but use the per protocol method for analyzing toxicity (e.g., Agnelli et al., 2012).

Q & A

Is the log-rank test the same as the Mantel–Haenszel test?

Almost. The two differ only in how they deal with multiple deaths at exactly the same time point. The results will be very similar.

Is the Gehan–Breslow-Wilcoxon test also the same?

No. The Gehan–Breslow–Wilcoxon method gives more weight to deaths at early time points, which makes a lot of sense. But the results can be misleading when a large fraction of patients are censored at early time points. In contrast, the log-rank test gives equal weight to all time points. The Gehan–Breslow–Wilcoxon test does not require a consistent hazard ratio, but it does require that one group consistently has a higher risk than the other. Of course, you should choose the test as part of the experimental design.

What if the two survival curves cross?

If two survival curves truly cross, then one group has a higher risk at early time points and the other group has a higher risk at later time points. If the two curves are based on plenty of data and the crossing point is near the middle of the time span, then the data violate the assumptions of both the log-rank and the Gehan–Breslow–Wilcoxon test, and you'd need to use more specialized methods that are beyond the scope of this book. In addition, the hazard ratio would be meaningless because it would be the average of bad news at early time points and good news at later time points.

When survival curves overlap only at later time points, it may be a matter of chance and mean nothing. At those late time points, fewer patients are followed, and the two curves may cross as the result of chance.

CHAPTER SUMMARY

- Survival analysis is used to analyze data for which the outcome is elapsed time until a one-time event (often death) occurs.
- Comparisons of survival curves can only be interpreted if you accept a list of assumptions.
- The hazard ratio is one way to summarize the comparison of two survival curves. If the hazard ratio is 2.0, then subjects given one treatment have died at twice the rate of subjects given the other treatment. Hazard ratios should be reported with a 95% CI.
- Another way to summarize the comparison of two survival curves is to compute the ratio of the median survival times, along with a CI for that ratio.
- A P value tests the null hypothesis that survival is identical in the two populations and the difference you observed is due to random assignment of patients.
- It is rarely helpful to compute the mean time until death for two reasons. One reason is that you can't compute the mean until the last person has died. The other reason is that survival times are rarely Gaussian.
- Computing the median survival is more useful than computing the mean survival, but that approach doesn't replace the need to view the survival curves.
- Most investigators analyze survival data from randomized trials using the ITT principle. In this approach, you analyze each person's data according to the treatment group to which they were assigned, even if they didn't actually get that treatment.

TERMS INTRODUCED IN THIS CHAPTER

- Analyze as randomized (p. 292)
- Five-year survival (p. 291)
- Gehan–Breslow–Wilcoxon method (p. 290)
- Hazard (p. 287)
- Hazard ratio (p. 287)
- Intention to treat (ITT) (p. 292)
- Log-rank method (p. 289)
- Mantel–Cox method (p. 289)
- Median survival (p. 288)
- Per protocol approach (p. 291)
- Proportional hazard (p. 288)
- Will Rogers phenomenon (p. 287)

CHAPTER 30

Comparing Two Means:
Unpaired t Test

Researchers often want Bioinformaticians to be Biomagicians, people who can make significant results out of non-significant data, or Biomorticians, people who can bury data that disagree with the researcher's prior hypothesis.

DAN MASYS

This chapter explains the unpaired t test which compares the means of two groups, assuming the data were sampled from a Gaussian population. Chapter 31 explains the paired t test, Chapter 39 explains the nonparametric Mann–Whitney test (and computer-intensive bootstrapping methods), and Chapter 35 shows how an unpaired t test can be viewed as a comparison of the fit of two models.

INTERPRETING RESULTS FROM
AN UNPAIRED t TEST

Example: Maximum relaxation of bladder muscles

Frazier, Schneider, and Michel (2006) measured how well the neurotransmitter norepinephrine relaxes bladder muscles. Chapter 11 looked at the concentrations required to get half-maximal relaxation as an example of lognormal data. Here we will look at the maximal relaxation that can be achieved by large doses of norepinephrine and then compare old with young rats.

The raw data are listed in Table 30.1 and plotted in Figure 30.1. The two means are fairly far apart, but there is a lot of variation and the two sets of data overlap considerably. Do young and old rats differ, or is the difference observed here just a matter of chance? An *unpaired t test*—as shown in Table 30.2—can help you answer that question.

CI for difference between means

The mean maximum response in old rats is 23.5 lower than the mean maximum response in young rats. The 95% CI for the difference ranges from 9.3 to 37.8 (see Figure 30.2). Because the CI does not include zero, we are 95% confident (given some assumptions) that the mean response in old animals is less than the mean response in young ones.

The width of the CI depends on three values:

- Variability. If the data are widely scattered (large SD), then the CI will be wider. If the data are very consistent (low SD), the CI will be narrower.
- Sample size. Everything else being equal, larger samples generate narrower CIs and smaller samples generate wider CIs.
- Degree of confidence. If you wish to have more confidence (i.e., 99% rather than 95%), the interval will be wider. If you are willing to accept less confidence (i.e., 90% confidence), the interval will be narrower.

OLD	YOUNG
20.8	45.5
2.8	55.0
50.0	60.7
33.3	61.5
29.4	61.1
38.9	65.5
29.4	42.9
52.6	37.5
14.3	

Table 30.1. Maximal relaxation of muscle strips of old and young rat bladders stimulated with high concentrations of norepinephrine (Frazier, Schneider, & Michel, 2006).

Larger values reflect more muscle relaxation. These values are graphed in Figure 30.1.

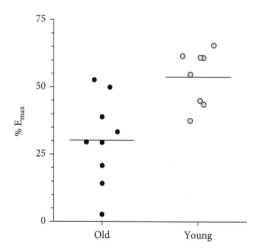

Figure 30.1. Maximal relaxation of muscle strips of old and young rat bladders stimulated with high concentrations of norepinephrine.

More relaxation is shown as larger numbers. Each symbol represents a measurement from one rat. The horizontal lines represent the means.

UNPAIRED T TEST

P value	0.0030
P value summary	**
Are means significantly different? (P < 0.05)	Yes
One-or two-tailed P value?	Two-tailed
t, df	t = 3.531, df = 15

HOW BIG IS THE DIFFERENCE?

Mean ± SEM of Column A (old)	30.17 ± 5.365, n = 9
Mean ± SEM of Column B (young)	53.71 ± 3.664, n = 8
Difference between means	23.55 ± 6.667
95% CI	9.335 to 37.76
R^2	0.4540

F TEST TO COMPARE VARIANCES

F, DFn, Dfd	2.412, 8, 7
P value	0.2631
P value summary	ns
Are variances significantly different?	No

Table 30.2. Results of unpaired t test from GraphPad Prism.

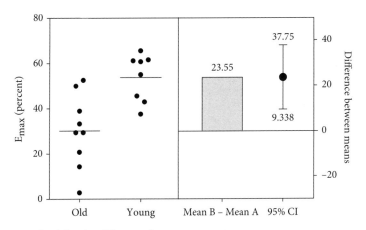

Figure 30.2. The CI for the difference between two means.
The first two columns show the actual data. The third column shows the difference between the two means, and the fourth column shows the 95% CI for that difference (plotted on the right axis).

P value

The null hypothesis is that both sets of data are randomly sampled from populations with identical means. The P value answers the question, If the null hypothesis were true, what is the chance of randomly observing a difference as large or larger than that observed in this experiment?

The two-tailed P value is 0.0030. Random sampling would rarely create a difference this big, so the difference between the two means is said to be statistically significant. The P value depends on the following:

- Difference between the means. Everything else being equal, the P value will be smaller when the means are farther apart.
- The SDs. Everything else being equal, the P value will be smaller when the data are more consistent (smaller SD).
- The sample size. Everything else being equal, the P value will be smaller when the sample sizes are larger.

R^2

Not all programs report R^2 with t test results, so you can ignore this value if it seems confusing. Chapter 35 will explain the concept in more detail. For these data, $R^2 = 0.45$. This means that a bit less than half (45%) of the variation among all the values is due to differences between the group means, and a bit more than half of the variation is due to variation within the groups.

t ratio

The P value is calculated from the *t ratio*, which is computed from the difference between the two sample means and the SD and sample size of each group. The t ratio has no units. In this example, t = 3.53. This number is part of the calculations of the t test, but its value is not very informative.

ASSUMPTIONS: UNPAIRED t TEST

When assessing any statistical results, it is essential to review the list of assumptions. The t test is based on a familiar list of assumptions (see Chapter 12).

- Random (or representative) samples
- Independent observations
- Accurate data
- Populations with Gaussian distributions (at least approximately)
- Equal SD of the two populations, even if their means are distinct (more about this assumption in the following discussion)

THE ASSUMPTION OF EQUAL VARIANCES

The unpaired t test depends on the assumption that the two data sets are sampled from populations that have identical SDs (and thus identical variances). Using statistical jargon, the assumption is of *homoscedasticity*.

Testing the assumption

In the sample of data collected in this experiment, you can see that there is more variability among the old animals than among the young. The SDs are

16.09 and 10.36. The old animals have an SD 1.55 times larger than that of the young animals. The square of that ratio (2.41) is called an *F ratio*.

The t test assumes that both groups are sampled from populations with equal variances. Are the data consistent with that assumption? To find out, let's compute another P value.

If that assumption were true, the distribution of the F ratio would be known, so a P value could be computed. This calculation depends on knowing the df for the numerator and denominator of the F ratio, abbreviated in Table 30.2 as DFn and DFd. Each df equals one of the sample sizes minus 1. The P value equals 0.26. If the assumption of equal variances were true, there would be a 26% chance that random sampling would result in this large a discrepancy between the two SD values as observed here (or larger still). The P value is large, so the data are consistent with the assumption.

Don't mix up this P value, which tests the null hypothesis that the two populations have the same SD, with the P value that tests the null hypothesis that the two populations have the same mean.

What if the assumption is violated?

What should you do if the P value (from the F test) is small, indicating that the assumption of equal variances is unlikely to be true? This is a difficult question, and there is no real consensus among statisticians. There are eight possible answers.

- Ignore the result. The t test is fairly robust to violations of the assumption of equal variances as long as the sample size isn't tiny and the two samples have an equal, or nearly equal, number of observations. This is probably the approach that most biologists use.
- Emphasize the result and conclude that the populations are different. If the SDs truly differ, then the populations are not the same. That conclusion is solid, whether the two means are close together or far apart.
- Transform the data (often to logarithms) in an attempt to equalize the variances and then run the t test on the transformed results.
- Instead of running the ordinary unpaired t test, run a modified t test that allows for unequal variance (the Welch modification to allow for different variances). If you always use this modified test, you don't have to think about the assumption of equal variances. The drawback is that this modified t test has less power to detect differences between means.
- Use the result of the test to compare variances to decide whether to run the usual unpaired t test or the modified t test that allows for unequal variance. Although this approach sounds sensible, it should not be used, because the results will be misleading (Moser & Stevens, 1992).
- Analyze the data using linear regression (as explained in Chapter 33) and weight by the reciprocal of each group's variability (as explained in Chapter 36).
- Use the result of the test that compares variances to decide whether to run a t test or the nonparametric Mann–Whitney test. This approach also sounds very reasonable but is not recommended. Chapter 41 explains why.
- Use your experience from other studies of the same kind of data to decide how to best analyze this specific data set.

OVERLAPPING ERROR BARS AND THE t TEST

Figure 30.1 showed the individual data points of the t test example. More often (for no good reason), you'll see data plotted as a bar graph showing mean and error bars. These error bars can show the SD or the SEM (Chapter 14 explained the difference between the two).

It is tempting to look at whether two error bars overlap and try to reach a conclusion about whether the difference between means is statistically significant. Resist that temptation, because you really can't learn much by asking whether two error bars overlap (Lanzante, 2005). Table 30.3 summarizes the rules of thumb for interpreting these error bars.

SD error bars

Figure 30.3 graphs the mean and SD of the sample rat data. The graph on the left is typical of graphs you'll see in journals, with the error bar only sticking up. But, of course, the scatter goes in both directions. The graph on the right shows this.

The two error bars overlap. The upper error bar for the data from the old rats is higher than the lower error bar for data from the young rats. What can you conclude from that overlap? Not much, because the t test also takes into account sample size (in this example, the number of rats studied). If the samples were larger (more rats), with the same means and same SDs, the P value would be much smaller. If the samples were smaller (fewer rats), with the same means and same SDs, the P value would be larger.

When the difference between two means is statistically significant (P 0.05), the two SD error bars may or may not overlap. Knowing whether SD error bars overlap, therefore, does not let you conclude whether the difference between the means is statistically significant.

SEM error bars

Figure 30.4 is similar to Figure 30.3 but plots the mean and SEM of the rat data, instead of the SD.

The two error bars do not overlap; the upper error bar for the data from the old rats is lower than the lower error bar for data from the young rats. What can you conclude from the lack of overlap?

TYPE OF ERROR BAR	CONCLUSION IF THEY OVERLAP	CONCLUSION IF THEY DON'T OVERLAP
SD	No conclusion	No conclusion
SEM	P > 0.05	No conclusion
95% CI	No conclusion	P < 0.05

Table 30.3. Rules of thumb for interpreting error bars that overlap or do not overlap.

These rules only apply when comparing two means with an unpaired t test and when the two sample sizes are equal or nearly equal.

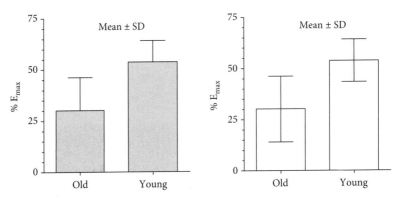

Figure 30.3. The same data plotted as mean and SD.

The left graph is similar to ones you'll often see in publications. The right graph shows the error bars going both up and down to emphasize that of course the scatter goes in both directions. To make the downward-pointing error bar obvious, the interior of the bar is kept white instead of shaded. The two error bars overlap. However, the fact that two SD error bars overlap tells you nothing about whether the difference between the two means is or is not statistically significant.

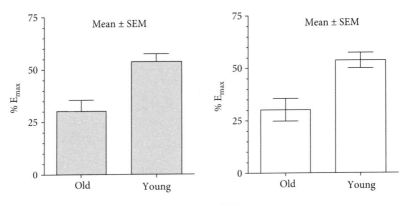

Figure 30.4. The same data plotted as mean and SEM.

Compare this figure with Figure 30.3. SEM error bars are always shorter than SD error bars. The left graph is similar to ones that you'll often see in publications. The right graph shows the error bars going both up and down to emphasize that, of course, the uncertainty goes in both directions. The two error bars do not overlap. However, the fact that two SEM error bars do not overlap tells you nothing about whether the difference between the two means is or is not statistically significant.

    The SEM error bars quantify how precisely you know the mean, taking into account both the SD and the sample size. Looking at whether the error bars overlap, therefore, lets you compare the difference between the mean with the precision of those means. This sounds promising. But, in fact, you don't learn much by looking at whether SEM error bars overlap. By taking into account sample size and considering how far apart two error bars are, Cumming, Fidler, and Vaux (2007) came up with some rules for deciding when a difference is significant. However, these rules are hard to remember and apply.

Here is a rule of thumb that can be used when the two sample sizes are equal or nearly equal. If two SEM error bars overlap, the P value is (much) greater than 0.05, so the difference is not statistically significant. The opposite rule does not apply. If two SEM error bars do not overlap, the P value could be less than or greater than 0.05.

CI error bars

The error bars in Figure 30.5 show the 95% CI of each mean. Because the two 95% CI error bars do not overlap and the sample sizes are nearly equal, the P value must be a lot less than 0.05—probably less than 0.005 (Payton, Greenstone, & Schenker, 2003; Knol, Pestman, & Grobbee, 2011). But it would be a mistake to look at overlapping 95% CI error bars and conclude that the P value is greater than 0.05. That relationship, although commonly believed, is just not true. When two 95% CIs overlap, the P value might be greater than 0.05, but it also might be less than 0.05. What you can conclude is that the P value is probably greater than 0.005.

HOW IT WORKS: UNPAIRED t TEST

Standard error of the difference between two means

Chapter 12 explained how the CI of a mean is computed from the standard error of the mean (SEM). Similarly, the CI of the difference between two means is computed from the standard error of the difference between the two means, which is computed by combining SEMs and both sample sizes. The standard error of the difference between two means will always be larger than either SEM but smaller than their sum. For the given example, the standard error of the difference between the two means (which is included in many programs' output) equals 6.67.

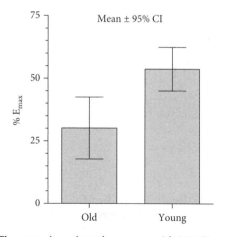

Figure 30.5. The same data plotted as means with 95% CIs.

Compare this figure with Figure 30.4. The 95% error bars are always longer than SEM error bars. The two error bars do not overlap, which means the P value must be much less than 0.05 and is probably less than 0.005.

CI

The CI for the difference between the two population means is centered on the difference between the means of the two samples. The margin of error, the distance the CI extends in each direction, equals the standard error of the difference (see the previous section) times a critical value from the t distribution (see Appendix D). For the sample data, the margin of error equals 6.67 (see previous section) times 2.1314 (the critical value from the t distribution for 95% CI and 15 df), or 14.22. The observed difference between means is 23.55, so the 95% CI extends from 23.55 − 14.22 to 23.55 + 14.22, or from 9.33 to 37.76.

t ratio

To determine the P value, first compute the t ratio, which equals the difference between the two sample means (23.5) divided by the standard error of that difference (6.67). In the rat example, $t = 3.53$. The numerator and denominator have the same units, so the t ratio has no units. Programs that compute the t test always report the t ratio, even though it is not particularly informative by itself.

Don't mix up the t ratio computed from the data with the critical value from the t distribution (see previous section) used to compute the CI.

P value

The P value is computed from the t ratio and the number of df, which equals the total number of values (in both groups) minus 2.

COMMON MISTAKES: UNPAIRED t TEST

Mistake: If the result is almost statistically significant, collect more data to increase the sample size and then recalculate the t test

You can only really interpret the results of any statistical calculation when the sample size was chosen in advance or you use special methods designed to handle sequential accumulation of data.

What's wrong with collecting more data when the P value is higher than 0.05? The problem is that it biases you toward getting small P values, even if the null hypothesis is true. If you stop when you get a P value you like but otherwise keep collecting more data, you can no longer interpret the P values. With this sequential approach, the chance of eventually reaching a conclusion that the result is statistically significant with P < 0.05 is much higher than 5%, even if the null hypothesis is true.

Mistake: If your experiment has three or more treatment groups, use the unpaired t test to compare two groups at a time

If you perform multiple t tests this way, it is too easy to reach misleading conclusions. Use one-way ANOVA, followed by multiple comparisons posttests, as explained in Chapters 39 and 40.

Mistake: If your experiment has three or more treatment groups, compare the largest mean with the smallest mean using a single unpaired t test

You won't know which group has the largest mean and which group has the smallest mean until after you have collected the data. So although you are formally just doing one t test to compare two groups, you are really comparing all the groups. Use one-way ANOVA.

Mistake: If the P value is larger than 0.05, try other tests to see whether they give a lower P value

If the P value from an unpaired t test is "too high," it is tempting to try a nonparametric test or to run the t test after removing potential outliers or transforming the values to logarithms. If you try multiple tests and choose to only report the results you like the most, those results will be misleading. You can't interpret the P value at face value when it was chosen from a menu of P values computed using different methods. You should pick one test before collecting data and use it.

Mistake: If the P value is small, concluding that the two distributions don't overlap very much

The t test is about separation of means. Figure 30.6 shows an experiment with n = 100 in each group. An unpaired t test results in a value of 0.0003. Yet the two distributions overlap considerably. The experiment provides pretty strong evidence that the two data sets are sampled from populations with different means. But the difference between means is tiny compared the variation within the two groups.

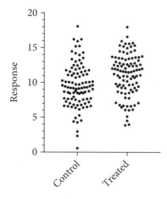

Figure 30.6. Tiny P value but extensive overlap.

The P value computed by unpaired t test is 0.0003, but the control and treated data sets overlap considerably. The 95% confidence interval for the difference between the two population means ranges from 0.73 to 2.44.

Q & A

Can a t test be computed from the mean, SD, and sample size of each group?

Yes. You don't need the raw data to compute an unpaired t test. It is enough to know the mean, SD or SEM, and sample size of each group.

Why is it called an unpaired test?

As its name suggests, this test is used when the values in the two groups are not matched. It should not be used when you compare two measurements in each subject (perhaps before and after an intervention) or if each measurement in one group is matched to a particular value in the other group. These kinds of data are analyzed using the paired t test, which is explained in Chapter 31.

Is the unpaired t test sometimes called Student's t test because it is the first statistical test inflicted on students of statistics?

No. W. S. Gossett developed the t test when he was employed by a brewery that required him to publish anonymously to keep as a trade secret the fact that they were using statistics. Gossett's pen name was "Student." He is no longer anonymous, but the name Student is still attached to the test.

How is the z test different than the t test?

The t test is computed using the t ratio, which is computed from the difference between the two sample means and the sample size and SD of the two groups. Some texts also discuss using the z test, based on the z ratio, which requires knowing the SD of the population from which the data are sampled. This population's SD on cannot be computed from the data at hand but rather must be known precisely from other data. Since this situation (knowing the population SD precisely) occurs rarely (if ever), the z test is rarely useful.

Does it matter whether or not the two groups have the same numbers of observations?

No. The t test does not require equal n. However, the t test is more robust to violations of the Gaussian assumption when the sample sizes are equal or nearly equal.

Why is the t ratio sometimes positive and sometimes negative?

The sign (positive or negative) of t depends on which group has the larger mean and in which order they are entered into the statistical program. Because the order in which you enter the groups into a program is arbitrary, the sign of the t ratio is irrelevant. In the example data, the t ratio might be reported as 3.53 or –3.53. It doesn't matter, because the CI and P value will be the same either way.

What would a one-tailed P value be in this example?

As explained in Chapter 15, the P value can be one- or two-sided. Here, the P value is two-sided (two-tailed). If the null hypothesis were true, the probability of observing a difference as large as that observed here with the young rats having the larger mean is 0.0015. That's a one-tailed P value. The chance of observing a difference as large as observed here with the old rats having the larger mean is also 0.0015. That's the other one-tailed P value. The sum of those two one-tailed P values equals the two-tailed P value (0.0030).

How is the t ratio reported with the t test different than the critical value of the t distribution used to calculate the CI?

The t ratio is computed from the data and used to determine the P value. The critical value of the t distribution, abbreviated t* in this book, is computed from the sample size and the confidence level you want (usually 95%) and is used to compute the CI. The value of t* does not depend on the data (except for sample size).

If you perform an unpaired t test with data that has very unequal variances, will the P value be too big or too small?

It can go either way depending on the data set (Delacre, Lakens, and Leys, 2017).

If you are not sure if the populations have equal variances or not, should you choose the regular t test because the variances might be the same, or should one choose Welch's t test because the variances might be different?

Ruxton (2006) and Delacre, Lakens, and Leys (2017) make a strong case to use the Welch's test routinely.

CHAPTER SUMMARY

- The unpaired t test compares the means of two unmatched groups.
- The results are expressed both as a CI for the difference between means and as a P value testing the null hypothesis that the two population means are identical.
- The P value can be tiny even if the two distributions overlap substantially. This is because the t test compares only the means.
- In addition to the usual assumptions, the unpaired t test assumes that the two populations follow Gaussian distributions with the same SD in each population.
- Looking at whether error bars overlap or not is not very helpful. The one rule of thumb worth remembering is that if two SEM error bars overlap, the P value is greater than 0.05.

TERMS INTRODUCED IN THIS CHAPTER

- F ratio (p. 298)
- Homoscedasticity (p. 297)
- t ratio (p. 297)
- Unpaired t test (p. 294)

CHAPTER 31

Comparing Two Paired Groups

To call in the statistician after the experiment is done may
be no more than asking him to perform a postmortem
examination: he may be able to say what the experiment
died of.

R. A. FISHER

This chapter explains the paired t test, which compares two matched or paired groups when the outcome is continuous, and the McNemar's test, which compares two matched or paired groups when the outcome is binomial. Chapter 30 explained the unpaired t test and Chapter 41 explains the nonparametric Wilcoxon test (and computer-intensive bootstrapping methods).

WHEN TO USE SPECIAL TESTS FOR PAIRED DATA

Often experiments are designed so that the same patients or experimental preparations are measured before and after an intervention. These data should not be analyzed with an unpaired t test or the nonparametric Mann–Whitney test. Unpaired tests do not distinguish variability among subjects that results from differences caused by treatment.

When subjects are matched or paired, you should use a special paired test instead. Paired analyses are appropriate in these types of protocols:

- A variable is measured in each subject before and after an intervention.
- Subjects are recruited as pairs and matched for variables such as age, postal code, or diagnosis. One of each pair receives one intervention, whereas the other receives an alternative treatment.
- Twins or siblings are recruited as pairs. Each receives a different treatment.
- Each run of a laboratory experiment has a control and treated preparation handled in parallel.
- A part of the body on one side is treated with a control treatment and the corresponding part of the body on the other side is treated with the experimental treatment (e.g., right and left eyes).

EXAMPLE OF PAIRED t TEST

Most of the examples in this book are from the fairly recent medical literature. For this example, however, we'll go back over a century to an example that was used by Ronald Fisher, one of the pioneers of statistics (Fisher, 1935). Fisher looked at an experiment by Charles Darwin that compared the growth of plants from seeds that were self-fertilized with the growth of plants from seeds that were cross-fertilized (Darwin, 1876).

Table 31.1 shows the data. Each row in the table represents one matched set, with the two kinds of seeds grown placed side by side. By designing the experiments this way, Darwin controlled for any changes in soil, temperature, or sunlight that would affect both kinds of seeds. Figure 31.1 graphs the data on a *before–after plot*. This kind of graph gets its name from its common use in experiments for which the same measurement is made for each subject before and after an experimental intervention. Here, the lines don't connect before and after measurements, but rather each of the 15 lines connects measurements of plants grown from matched seeds.

Figure 31.2 shows the same data as a bar graph, plotting the mean and SEM of each group. This kind of graph is commonly used, but it shows much less information than Figure 31.1. It shows you nothing about the pairing and only indirectly (via the SEM) gives you a sense of variation. When possible, avoid bar graphs and graph the actual data.

CROSS FERTILIZED	SELF-FERTILIZED	DIFFERENCE
23.500	17.375	6.125
12.000	20.375	–8.375
21.000	20.000	1.000
22.000	20.000	2.000
19.125	18.375	0.750
21.500	18.625	2.875
22.125	18.625	3.500
20.375	15.250	5.125
18.250	16.500	1.750
21.625	18.000	3.625
23.250	16.250	7.000
21.000	18.000	3.000
22.125	12.750	9.375
23.000	15.500	7.500
12.000	18.000	–6.000

Table 31.1. Sample data for paired t test.

These data were collected by Charles Darwin, who compared the growth of cross-fertilized and self-fertilized plants. The values are the height of the plants in inches (measured to the nearest one-eighth of an inch). Each row represents a matched pair of self- and cross-fertilized seeds, grown under the same conditions. The last column is the difference.

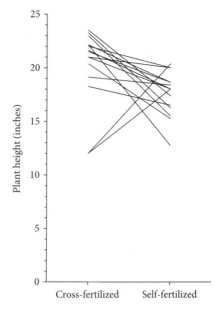

Figure 31.1. A before–after graph.
Usually, this kind of graph is used to plot values when the same measurement is made in each subject before and after an experimental intervention. Each of the 15 lines connects measurements from matched seeds.

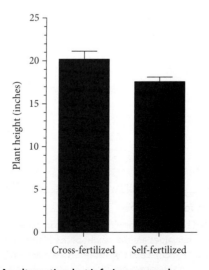

Figure 31.2. An alternative, but inferior, approach.
This graph plots the same data as Figure 31.1 but shows only the mean and SEM of each group without accounting for pairing.

PAIRED t TEST

P value	0.0497
P value summary	*
Are means significantly different? (P < 0.05)	Yes
One- or two-tailed P value?	Two-tailed
t, df	t = 2.148, df = 14
Number of pairs	15

HOW BIG IS THE DIFFERENCE?

Mean of differences	2.617
95% CI	0.003639 to 5.230

HOW EFFECTIVE WAS THE PAIRING?

Correlation coefficient (r)	−0.3348
P value (one-tailed)	0.1113
P value summary	ns
Was the pairing significantly effective?	No

Table 31.2. Results of a paired t test as computed by GraphPad Prism.

INTERPRETING RESULTS FROM A PAIRED t TEST

Table 31.2 shows the results of a paired t test.

CI

A paired t test looks at the difference in measurements between two matched subjects (as in our example; see Figure 31.3) or a measurement made before and after an experimental intervention. For the sample data, the average difference in plant length between cross- and self-fertilized seeds is 2.62 inches. To put this in perspective, the self-fertilized plants have an average height of 17.6 inches, so this difference is approximately 15%.

The 95% CI for the difference ranges from 0.003639 inches to 5.230 inches. Because the CI does not include zero, we can be 95% confident that the cross-fertilized seeds grow more than the self-fertilized seeds. But the CI extends from a tiny difference (especially considering that plant height was only measured to the nearest one-eighth of an inch) to a fairly large difference.

The width of the CI depends on three values:

- Variability. If the observed differences are widely scattered, with some pairs having a large difference and some pairs a small difference (or a difference in the opposite direction), then the CI will be wider. If the data are very consistent, the CI will be narrower.
- Sample size. Everything else being equal, a sample with more pairs will generate narrower CIs, and a sample with fewer pairs will generate wider CIs.
- Degree of confidence. If you wish to have more confidence (i.e., 99% rather than 95%), the interval will be wider. If you are willing to accept less confidence (i.e., 90% confidence), the interval will be narrower.

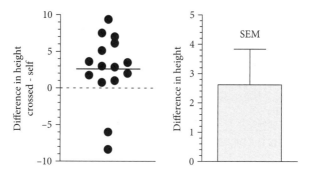

Figure 31.3. The difference between matched measurements.
Each of the circles on the left represents one paired experiment. The length of the self-fertilized plant is subtracted from the length of the cross-fertilized plant, and that difference is plotted. The cross-fertilized seed grew more than the self-fertilized seed in 13 of the 15 paired experiments, resulting in a positive difference. The other two experiments resulted in more growth in the cross-fertilized plant, and so the difference is negative. The graph on the right plots only the mean and SEM. This graph takes up the same amount of space but shows less information and isn't really easier to interpret.

P value

The null hypothesis is there is no difference in the height of the two kinds of plants. In other words, the differences that we measure are sampled from a population for which the average difference is zero. By now you should be able to easily state the question the P value answers.. If the null hypothesis were true, what is the chance of randomly observing a difference as large as or larger than the difference observed in this experiment?

The P value is 0.0497. If the null hypothesis were true, 5% of random samples of 15 pairs would have a difference this large or larger. Using the traditional definition, the difference is statistically significant, because the P value is less than 0.05.

As explained in Chapter 15, the P value can be one- or two-sided. Here, the P value is two-sided (two-tailed). Assuming the null hypothesis is true, the probability of observing a difference as large as that observed here with the self-fertilized plants growing more than the cross-fertilized plants is 0.0248. The chance of observing a difference as large as that observed here with the cross-fertilized plants growing more is also 0.0248. The sum of those two probabilities equals the two-sided P value (0.0497).

The size of the P value depends on the following:

- Mean difference. Everything else being equal, the P value will be smaller when the average of the differences is far from zero.
- Variability. If the observed differences are widely scattered, with some pairs having a large difference and some pairs having a small difference, then the P value will be higher. If the data are very consistent, the P value will be smaller.
- Sample size. Everything else being equal, the P value will be smaller when the sample has more data pairs.

How effective was the pairing?

Darwin's experiment worked with pairs of seeds. Each pair of seeds was exposed to the same conditions. But different pairs were measured at different times over several years. The *paired t test* was created to account for that pairing. If the seeds were grown under good conditions, both seeds would grow tall. If grown under worse conditions, both seeds would grow less. In either case, the difference between the two kinds of seeds would be consistent.

Figure 31.4 and Table 31.2 show that, in fact, there are no positive correlations among the data. Because the whole point of a paired t test is to use the internal controls to get more consistent data, the paired t test actually has a larger P value than would an unpaired t test conducted with these data (the P value from an unpaired t test for these data is 0.02). This situation is rare. Usually, if you design an experiment with matched samples, you'll see a strong positive correlation.

With these data, is it fair to switch to an unpaired t test? It certainly is not fair to "shop around" for the P value you want. You'd need to establish the protocol for switching before collecting any data.

Assumptions

When assessing any statistical results, it is essential to review the list of assumptions. The paired t test is based on a familiar list of assumptions (see Chapter 12).

- The paired values are randomly sampled from (or at least are representative of) a population of paired samples.
- In that population, the differences between the matched values follow a Gaussian distribution.
- Each pair is selected independently of the others.

How a paired t test works

For each set of before–after measurements or each set of matched pairs, a paired t test first computes the differences between the two values for each pair. Only this set of differences is used to compute the CI and P value.

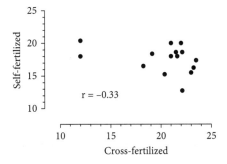

Figure 31.4. Testing the need for a paired t test.

The whole point of a paired t test is that you expect the paired values to be correlated— if one is higher, so is the other. This graph shows a negative correlation for sample data. This negative trend is far from convincing and is almost certainly a coincidence. But it is clear that there is no strong positive correlation. Chapter 32 explains correlation coefficients.

The CI for the mean difference is calculated exactly as explained in Chapter 12, except that the values used for the calculations are the set of differences.

The t ratio is computed by dividing the average difference (2.62 inches) by the SEM of those differences (1.22 inches). The numerator and denominator have the same units, so the t ratio is unitless. In the example, t = 2.15.

The P value is computed from the t ratio and the number of df, which equals the number of pairs minus 1.

THE RATIO PAIRED t TEST

Example data: Does a treatment change enzyme activity?

Table 31.3 shows example data that test whether treating cultured cells with a drug increases the activity of an enzyme. Five different clones of the cell were tested. With each clone, control cells and treated cells were tested side by side. In all five clones, the treatment increased the activity of the enzyme. Figure 31.5 plots the data as a before–after graph and also as a graph of the differences.

CONTROL	TREATED
24	52
6	11
16	28
5	8
2	4

Table 31.3. Sample data for ratio t test.
These data (which are not real) test whether treating cultured cells with a drug increases the activity of an enzyme.

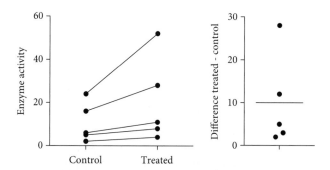

Figure 31.5. Enzyme activity in matched sets of control and treated cells.
The mean of the differences (treated minus control) is 10.0. The 95% CI for the average differences ranges from –3.4 to 23.4. Because that CI includes zero, you can't be sure (with 95% confidence) whether the treatment had any effect on the activity. The P value (two-tailed, from a paired t test) is 0.107, which is not low enough to be sure that the difference between control and treated is real; the difference could easily just be the result of random sampling. With only five pairs and inconsistent data, the data only lead to a very fuzzy conclusion.

Relative differences versus absolute differences

Look more carefully at the left panel of Figure 31.5. The difference between control and treated depends on where you start. The clones with higher enzyme activity in the control condition have a larger increment with treatment. The treatment multiplies the enzyme activity by a certain factor rather than adding a certain amount.

Figure 31.6 shows a simple way to align the multiplicative experimental model with the additive model used by the paired t test. This graph has a logarithmic Y-axis. Now, all five of the increases are about the same size. The treatment increases the logarithm of the outcome (enzyme activity) by a constant amount, because the treatment multiplies the activity by a constant factor rather than adding a constant factor. On a log scale, multiplication becomes addition.

If you are not comfortable working with logarithms, this process may seem a bit mysterious. But read on to see how logarithms are used practically to analyze these kinds of data.

Results from a ratio paired t test

To analyze these data with a paired t test, first transform all values to their logarithms (base 10) and then perform the paired t test on those results (logarithms and antilogarithms are reviewed in Appendix E). The mean of the difference between logarithms is 0.26. Transform that back to its original ratio scale: $10^{0.26} = 1.82$. On average, the treatment multiplies the enzyme activity by a factor of 1.82. In other words, the treatment increases activity by 82%.

The 95% CI for the difference of logarithms is 0.21 to 0.33. Transform each of those values to their antilogarithms to find the 95% CI for the ratio (treated/control), which is 1.62 to 2.14. This interval doesn't include 1.0 (a ratio of 1.0 means no change) and doesn't even come close to including 1.0. Thus, the increase in

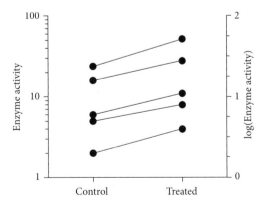

Figure 31.6. The same data as those shown in Figure 31.5 and Table 31.3, plotted on a logarithmic axis.

The right axis shows the logarithms. The left axis is labeled with the original values, logarithmically spaced. Both sets of data are plotted on both axes, which plot the same values but are expressed differently.

activity with treatment is very unlikely to be a coincidence. The P value is only 0.0003, so random variation would only rarely cause a treatment effect this large. Analyzed this way, the conclusion is very crisp.

I call this procedure (paired t test of data transformed to logarithms) a *ratio paired t test*, but that term is not widely used. However, the idea of transforming values to make the differences more consistent is entirely standard. Of course, you really should pick the analysis method before collecting data. It is not fair to analyze the data lots of different ways and then pick the P value you like the best.

McNEMAR'S TEST FOR A PAIRED CASE-CONTROL STUDY

McNemar's test analyzes matched data when the outcome is binomial (i.e., when there are two possible outcomes).

In a standard case-control study, the investigator compares a group of controls with a group of cases. As a group, the controls are chosen to be similar to the cases (except for the absence of disease and exposure to the putative risk factor). An alternative way to design a case-control study is to match each case with a specific control based on age, gender, occupation, location, or other relevant variables.

An example is shown in Table 31.4. The investigators studied 134 cases and 134 matched controls. Each entry in the table represents one pair (a case and a control). This is not a contingency table, and entering this table into a program designed to analyze contingency tables would lead to invalid results. Table 31.5 shows these data as a contingency table.

	CASES		
CONTROLS	RISK FACTOR +	RISK FACTOR –	TOTAL
Risk factor +	13	4	17
Risk factor –	25	92	117
Total	38	96	134

Table 31.4. **A matched pair case-control study.**
Each value in the table represents a matched case and control. The total number of the four values is 134, because the study used 134 pairs of people (so 268 people in all). This is not a contingency table, and analysis by programs designed for contingency tables would be invalid.

	RISK FACTOR +	RISK FACTOR –	TOTAL
Case	38	96	134
Control	17	117	134
Total	55	213	268

Table 31.5. **Data from Table 31.4 expressed as a contingency table.**
Each value in the table represents one person. The study had 134 patient-control pairs, so 268 people in total. This is a contingency table. This arrangement of the data obscures the fact that the study was designed with a matched control for each case. Table 31.4 is more informative.

The 13 pairs for which both cases and controls were exposed to the risk factor provide no information about the association between risk factor and disease. Similarly, the 92 pairs for which both cases and controls were not exposed to the risk factor provide no useful information. The association between risk factor and disease depends on the other two values in the table, and their ratio is the odds ratio.

For the example, the odds ratio is the number of pairs for which the case was exposed to the risk factor but the control was not (25), divided by the number of pairs for which the control was exposed to the risk factor but the case was not (4), which equals 6.25.

McNemar's test computes the 95% CI of the odds ratio and a P value from the two discordant values (25 and 4 for this example). Not all statistics programs calculate McNemar's test, but it can be computed using a free Web calculator in the QuickCalcs section of graphpad.com.

The 95% CI for the odds ratio ranges from 2.16 to 24.71. Assuming the disease is fairly rare, the odds ratio can be interpreted as a relative risk (see Chapter 28). These data show that exposure to the risk factor increases one's risk 6.25-fold and that there is 95% confidence that the range from 2.16 to 24.7 contains the true population odds ratio.

McNemar's test also computes a P value of 0.0002. To interpret a P value, the first step is to define the null hypothesis, which is that there really is no association between risk factor and disease, so the population odds ratio is 1.0. If that null hypothesis were true, the chance of randomly selecting 134 pairs of cases and controls with an odds ratio so far from 1.0 is only 0.02%.

COMMON MISTAKES: PAIRED t TEST

Mistake: If the result is almost statistically significant, collect more data to increase the sample size and then recalculate the test

You can only really interpret the results of any statistical calculation when the sample size is chosen in advance or you use special methods designed to handle sequential accumulation of data.

Mistake: If your experiment has three or more treatment groups, use the paired t test to compare two groups at a time

If you perform multiple t tests this way, it is too easy to reach misleading conclusions. Use repeated-measures one-way ANOVA, followed by multiple comparisons posttests, as explained in Chapters 39 and 40.

Mistake: If your experiment has three or more treatment groups, compare the largest mean with the smallest mean using a single paired t test

You won't know which group has the largest mean and which has the smallest mean until after you have collected the data. So, although you are formally just doing one t test to compare two groups, you are really comparing all the groups. Use repeated-measures one-way ANOVA.

Mistake: Using the paired t test when the ratios, not the differences, are a more consistent measure of treatment effects

The ratio t test, as explained in the previous discussion, is not commonly used, but it is often the best test for many situations.

Mistake: Analyzing the absolute value of the differences, rather than the actual differences

When computing the paired t test, the first step is to compute the difference for each pair. This must be done in a consistent way and must allow for negative values. If you are comparing measurements before and after an intervention, then you should compute the differences so that positive values mean the measurement went up and negative values mean the measurement went down. If you analyze absolute values (so that all differences are positive), the results will be meaningless.

Mistake: Deciding on the pairing only after seeing the data

Pairing or matching must be done as part of the experimental design and setup. It cannot be based on the values you are comparing with the t test.

Q & A

Can I decide how the subjects should be matched after collecting the data?

No. Pairing must be part of the experimental protocol that is decided before the data are collected. The decision about pairing is a question of experimental design and should be made long before the data are analyzed.

When computing the difference for each pair, in which order is the subtraction done?

It doesn't matter, so long as you are consistent. For each pair in the example, the length of the self-fertilized plant was subtracted from the length of the cross-fertilized plant. If the calculation were done the other way (if the length of the cross-fertilized plant was subtracted from the length of the self-fertilized plant), all the differences would have the opposite sign, as would the t ratio. The P value would be the same. It is very important, of course, that the subtraction be done the same way for every pair. It is also essential that the program doing the calculations doesn't lose track of which differences are positive and which are negative.

What if one of the values for a pair is missing?

The paired t test analyzes the differences between each set of pairs. If one value for a pair is missing, that pair can contribute no information at all, so it must be excluded from the analysis. A paired t test program will ignore pairs that lack one of the two values.

Can a paired t test be computed if you know only the mean and SD of the two treatments (or two time points) and the number of pairs?

No. These summary data tell you nothing about the pairing.

Can a paired t test be computed if you know only the mean and SD of the set of differences (as well as the number of pairs)? What about the mean and SEM?

Yes. All you need to compute a paired t test is the mean of the differences, the number of pairs, and the SD or SEM of the differences. You don't need the raw data.

Does it matter whether the two groups are sampled from populations that are not Gaussian?

> No. The paired t test only analyzes the set of paired differences, which it assumes is sampled from a Gaussian distribution. This doesn't mean that the two individual sets of values need to be Gaussian. If you are going to run a normality test on paired t test data, it only makes sense to test the list of differences (one value per pair). It does not make sense to test the two sets of data separately.

Will a paired t test always compute a lower P value than an unpaired test?

> No. With the Darwin example, an unpaired test computes a lower P value than a paired test. When the pairing is effective (i.e., when the set of differences is more consistent than either set of values), the paired test will usually compute a lower P value.

CHAPTER SUMMARY

- The paired t test compares two matched or paired groups when the outcome is continuous.
- The paired t test only analyzes the set of differences. It assumes that this set of differences was sampled from a Gaussian population of differences.
- The paired t test is used when the *difference* between the paired measurements is a consistent measure of change. Use the ratio t test when the *ratio* of the paired measurements is a more consistent measure of the change or effect.
- McNemar's test compares two matched or paired groups when the outcome is binomial (categorical with two possible outcomes).

TERMS INTRODUCED IN THIS CHAPTER

- Before–after plot (p. 307)
- McNemar's test (p. 314)
- Paired t test (p. 311)
- Ratio paired t test (p. 314)

CHAPTER 32

Correlation

> The invalid assumption that correlation implies cause is prob-
> ably among the two or three most serious and common errors
> of human reasoning.
>
> STEPHEN JAY GOULD

The association between two continuous variables can be quantified by the correlation coefficient, r. This chapter explains Pearson correlation. Chapter 41 explains nonparametric Spearman correlation.

INTRODUCING THE CORRELATION COEFFICIENT

An example: Lipids and insulin sensitivity

Borkman and colleagues (1993) wanted to understand why insulin sensitivity varies so much among individuals. They hypothesized that the lipid composition of the cell membranes of skeletal muscle affects the sensitivity of the muscle for insulin.

In their experiment, they determined the insulin sensitivity of 13 healthy men by infusing insulin at a standard rate (adjusting for size differences) and quantifying how much glucose they needed to infuse to maintain a constant blood glucose level. Insulin causes the muscles to take up glucose and thus causes the level of glucose in the blood to fall, so people with high insulin sensitivity will require a larger glucose infusion.

They also took a small muscle biopsy from each subject and measured its fatty acid composition. We'll focus on the fraction of polyunsaturated fatty acids that have between 20 and 22 carbon atoms (%C20–22).

Table 32.1 shows the data (interpolated from the author's published graph), which are graphed in Figure 32.1. Note that both variables are scattered. The mean of the insulin sensitivity index is 284, and the SD is 114 mg/m²/min. The CV equals the SD divided by the mean, so 114/284, or 40.1%. This is quite high. The authors knew that there would be a great deal of variability, and that is why they explored the causes of the variability. The fatty acid content is less variable, with a CV of 11.6%. If you don't look at the graph carefully, you could be misled. The X-axis does not start at zero, so you might get the impression that the variability is greater than it actually is.

INSULIN SENSITIVITY (mg/m²/min)	%C20–22 POLYUNSATURATED FATTY ACIDS
250	17.9
220	18.3
145	18.3
115	18.4
230	18.4
200	20.2
330	20.3
400	21.8
370	21.9
260	22.1
270	23.1
530	24.2
375	24.4

Table 32.1. Correlation between %C20–22 and insulin sensitivity.

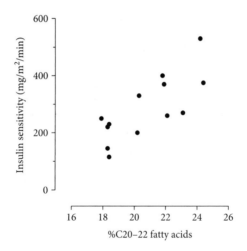

Figure 32.1. Correlation between %C20–22 and insulin sensitivity.

The graph shows a clear relationship between the two variables. Individuals whose muscles have more C20–22 polyunsaturated fatty acids tend to have greater sensitivity to insulin. The two variables vary together—statisticians say that there is a lot of *covariation* or *correlation*.

Correlation results

The direction and magnitude of the linear correlation can be quantified with a *correlation coefficient*, abbreviated r. Its value can range from –1 to 1. If the correlation coefficient is zero, then the two variables do not vary together at all.

CORRELATION	
r	0.77
95% CI	0.3803 to 0.9275
r^2	0.5929
P (two-tailed)	0.0021
Number of XY pairs	13

Table 32.2. Correlation results.

If the correlation coefficient is positive, the two variables tend to increase or decrease together. If the correlation coefficient is negative, the two variables are inversely related—that is, as one variable tends to decrease, the other one tends to increase. If the correlation coefficient is 1 or −1, the two variables vary together completely—that is, a graph of the data points forms a straight line.

The results of a computer program (GraphPad Prism 6) are shown in Table 32.2.

Correlation coefficient and its CI

In the example, the two variables increase together, so the correlation coefficient must be positive. But there is some scatter, so the correlation coefficient must be less than 1.0. In fact, the correlation coefficient equals 0.77.

It almost always makes sense to report any statistical result with a CI. Here the 95% CI for the correlation coefficient ranges from 0.38 to 0.93. Assuming these data were randomly sampled from a larger population (and that other assumptions listed later in this chapter are true), there is a 95% chance that this range includes the population correlation coefficient.

Note that the CI is not symmetrical. In this example, it extends much further below the correlation coefficient than it extends above it. This makes sense. The correlation coefficient can never be larger than 1.0 or smaller than −1.0, and so the CI is usually asymmetrical. The asymmetry is especially noticeable when r is far from zero and when the sample size is small.

r^2

The square of the correlation coefficient is an easier value to interpret than r. For the example, $r = 0.77$, so r^2 (referred to as "r squared") $= 0.59$. Because r is always between −1 and 1, r^2 is always between zero and 1. Note that when you square a positive fraction, the result is a smaller fraction.

If you can accept the assumptions listed in the next section, r^2 is the fraction of the variance shared between the two variables. In the example, 59% of the variability in insulin tolerance is associated with variability in lipid content. Knowing the lipid content of the membranes lets you explain 59% of the variance in the insulin sensitivity. That leaves 41% of the variance to be explained by other factors or by measurement error. X and Y are symmetrical in correlation analysis, so you can also say that 59% of the variability in lipid content is associated with variability in insulin tolerance.

P value

The P value is 0.0021. To interpret any P value, first define the null hypothesis. Here, the null hypothesis is that there is no correlation between insulin sensitivity and the lipid composition of membranes in the overall population. The two-tailed P value answers the question, If the null hypothesis were true, what is the chance that 13 randomly picked subjects would have an r greater than 0.77 or less than –0.77?

Because the P value is so low, the data provide evidence to reject the null hypothesis.

ASSUMPTIONS: CORRELATION

You can calculate the correlation coefficient for any set of data, and it may be a useful descriptor of the data. However, interpreting the CI and P value depends on the following assumptions.

Assumption: Random sample

Like all statistical analyses, you must assume that subjects are randomly selected from, or at least representative of, a larger population.

Assumption: Paired samples

Each subject (or each experimental unit) must have both X and Y values.

Assumption: Sampling from one population

Correlation assumes that all the points are randomly sampled from one population. If you sample some subjects from one population and some from another, the correlation coefficient and P value will be misleading.

Assumption: Independent observations

Correlation assumes that any random factor affects only one subject and not others. The relationship between all the subjects should be the same. In this example, the assumption of independence would be violated if some of the subjects were related (e.g., siblings). It would also be violated if the investigator purposely chose some people with diabetes and some without or if the investigator measured each subject on two occasions and treated the values as two separate data points.

Assumption: X values are not used to compute Y values

The correlation calculations are not meaningful if the values of X and Y are not measured separately. For example, it would not be meaningful to calculate the correlation between a midterm exam score and the overall course score, because the midterm exam is one of the components of the course score.

Assumption: X values are not experimentally controlled

If you systematically control the X variable (e.g., concentration, dose, or time), you should calculate linear regression rather than correlation (see Chapter 33).

You will get the same value for r^2 and the P value. However, the CI of r cannot be interpreted if the experimenter controlled the value of X.

Assumption: Both variables follow a Gaussian distribution

The X and Y values must each be sampled from populations that follow a Gaussian distribution, at least approximately. If this assumption isn't true, the P value cannot be interpreted at face value.

Assumption: All covariation is linear

The correlation coefficient would not be meaningful if, for example, Y increases as X increases up to a certain point but then Y decreases as X increases further. Curved relationships are common but are not quantified with a correlation coefficient.

Assumption: No outliers

Calculation of the correlation coefficient can be heavily influenced by one outlying point. Change or exclude that single point and the results may be quite different. Outliers can influence all statistical calculations, but especially when determining correlation. You should look at graphs of the data before reaching any conclusion from correlation coefficients. Don't instantly dismiss outliers as bad points that mess up the analysis. It is possible that the outliers are the most interesting observations in the study!

LINGO: CORRELATION

Correlation

When you encounter the word *correlation*, distinguish its strict statistical meaning from its more general usage. As used by statistics texts and programs, correlation quantifies the association between two continuous (interval or ratio) variables. However, the word *correlation* is often used much more generally to describe the association of any two variables, even if either (or both) of the variables are not continuous variables. It is not possible to compute a correlation coefficient to help you figure out whether survival times are correlated with choice of drug or whether antibody levels are correlated with gender.

Coefficient of determination

Coefficient of determination is just a fancy term for r^2. Most scientists and statisticians just call it *r square* or *r squared*.

HOW IT WORKS: CALCULATING THE CORRELATION COEFFICIENT

The calculation of the correlation coefficient is built into computer programs, so there is no reason to do it manually. The following explanation is provided to give you a feel for what the correlation coefficient means.

1. Calculate the average of all X values. Also calculate the average of all Y values. These two averages define a point at "the center of gravity" of the data.

2. Compare the position of each point with respect to that center. Subtract the average X value from each X value. The result will be positive for points to the right of the center and negative for points to the left. Similarly, subtract the average Y value from each Y value. The result will be positive for points above the center and negative for points below.

3. Standardize those X distances by dividing by the SD of all the X values. Similarly, divide each Y distance by the SD of all Y values. Dividing a distance by the SD cancels out the units, so these ratios are fractions without units.

4. Multiply the two standardized distances for each data point. The product will be positive for points that are northeast (product of two positive numbers) or southwest (product of two negative numbers) of the center. The product will be negative for points that are to the northwest or southeast (product of a negative and a positive number).

5. Add up all the products computed in Step 4.

6. Account for sample size by dividing the sum by $(n - 1)$, where n is the number of XY pairs.

If X and Y are not correlated, then the positive products in Step 4 will approximately balance out the negative ones, and the correlation coefficient will be close to zero. If X and Y are correlated, the positive and negative products will not balance each other out, and the correlation coefficient will be far from zero.

The nonparametric Spearman correlation (see Chapter 41) adds one step to the beginning of this process. First, you would separately rank the X and Y values, with the smallest value getting a rank of 1. Then you would calculate the correlation coefficient for the X ranks and the Y ranks, as explained in the previous steps.

COMMON MISTAKES: CORRELATION

Mistake: Believing that correlation proves causation

Why do the two variables in this example correlate so well? There are five possible explanations:

- The lipid content of the membranes determines insulin sensitivity.
- The insulin sensitivity of the membranes somehow affects lipid content.
- Both insulin sensitivity and lipid content are under the control of some other factor, such as a hormone.
- Lipid content, insulin sensitivity, and other factors are all part of a complex molecular/biochemical/physiological network, perhaps with positive and/or negative feedback components. In this case, the observed correlation would just be a peek at a much more complicated set of relationships.
- The two variables don't correlate in the population at all, and the observed correlation in our sample is a coincidence.

You can never rule out the last possibility, but the P value tells you how unlikely the coincidence would be. In this example, you would observe a correlation that strong (or stronger) in 0.21% (the P value) of experiments if there was no correlation in the overall population.

You cannot decide which of the first four possibilities is correct by analyzing only these data. The only way to figure out which is true is to perform additional experiments in which you manipulate the variables. Remember that this study simply measured both values in a set of subjects. Nothing was experimentally manipulated.

The authors, of course, want to believe the first possibility based on their knowledge of physiology. That does not mean they believe that the lipid composition is the *only* factor that determines insulin sensitivity, only that it is one factor.

Most people immediately think of the first two possibilities but ignore the rest. Correlation does not necessarily prove simple causality, however. Two variables can be correlated because both are influenced by the same third variable. Infant mortality in various countries is probably negatively correlated with the number of telephones per capita, but buying telephones will not make kids live longer. Instead, increased wealth (and thus increased purchases of telephones) is related to better plumbing, better nutrition, less crowded living conditions, more vaccinations, and so on.

Mistake: Focusing on the P value rather than the correlation coefficient

Arden and colleagues (2008) asked whether there is a correlation between intelligence and sperm count. Really!

They obtained data from a study of US Vietnam War–era veterans that measured both intelligence and the number and motility of sperm in ejaculate. The variation in sperm count (measured as sperm per milliliter, or sperm per ejaculate) among individuals was very skewed, so the researchers transformed all the values to logarithms. This approach makes perfect sense, because the values approximated a lognormal distribution (see Chapter 11).

With data from 425 men, Arden and colleagues computed the correlation coefficient for a measure of intelligence and three assessments of sperm quality (logarithm of sperm count per volume, logarithm of sperm count per ejaculate, and the percentage of those sperm that were motile). The results are shown in Table 32.3, which demonstrates four points:

- The r values are all positive. This tells you that the trend is positive—that is, that increased intelligence is associated with increased sperm count and motility.
- No corrections were made for multiple comparisons (see Chapters 22 and 23). This is appropriate because the three variables that assess sperm are all somewhat related. These aren't three independent measurements. It makes perfect sense to report all three P values without correction for multiple comparisons (but it wouldn't be fair to report only the lowest P values and not mention the other two). The fact that these values all are

	r	r^2	P VALUE
Log (sperm count per ml)	0.15	0.023	0.0019
Log (sperm count per ejaculate)	0.19	0.036	< 0.0001
Fraction of sperm that are motile (%)	0.14	0.020	0.0038

Table 32.3. Correlation between a measure of intelligence and three measures of sperm quality.
Data from Arden et al. (2008), with n = 425.

tiny and that all correlations go in the same direction helps make the data convincing.
- The P values are all quite small. If there really was no relationship between sperm and intelligence, there would only be a tiny chance of obtaining a correlation this strong.
- The r values are all small. It is easier to interpret the r^2 values, which show that only 2% to 3% of the variation in intelligence in this study is associated with variation in sperm count and motility.

The last two points illustrate a common situation. With large samples, data can show small effects (in this case, r^2 values of 2%–3%) and still have tiny P values. The fact that the P values are small tells you that the effect is unlikely to be a coincidence of random sampling. But the size of the P value does not tell you how large the effect is. The effect size is quantified by r and r^2. Here, the effect size is tiny.

Is an r^2 value of 2% to 3% large enough to be considered interesting and worth pursuing? That is a scientific question, not a statistical one. The investigators of this study seemed to think so. Other investigators might conclude that such a tiny effect is not worthy of much attention. I doubt that many newspaper reporters who wrote about the supposed link between sperm count and intelligence understood how weak the reported relationship really is.

Mistake: Interpreting the correlation coefficient without first looking at a graph

Don't try to interpret a correlation coefficient (or the corresponding P value) until you have looked at a graph of the data. Figure 32.2 shows why. This graph shows four artificial data sets designed by Anscombe (1973). The correlation coefficients are identical (0.816), as are the P values (0.0022), but the data are very different.

Mistake: Computing a correlation between two variables when one is computed from the other

Don't compute the correlation between a midterm exam score and the total course grade, because the midterm score is part of the grade. And don't compute the correlation between a variable and the difference between that variable and another.

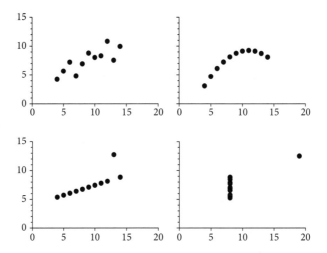

Figure 32.2. Four graphs with the same correlation coefficient.
In all four cases, the correlation coefficient (r) is 0.8164. Only the first graph seems to meet the assumptions of correlation.

If two variables A and B are completely independent with zero correlation, you'd expect the correlation coefficient of A versus A-B to equal about 0.7. See Figure 33.5 in the next chapter on linear regression to see why this is a problem.

Mistake: Mixing up correlation and linear regression

Correlation and linear regression are related but distinct. The next chapter explains linear regression and explains how it differs than correlation.

Q & A

Does it matter which variable is called X and which is called Y?
> No. X and Y are totally symmetrical in correlation calculations. But note this is not true for linear regression (see Chapter 33). When analyzing data with linear regression, you must carefully choose which variable is X and which is Y.

Do X and Y have to be measured in the same units to calculate a correlation coefficient? Can they be measured in the same units?
> X and Y do not have to be measured in the same units, but they can be.

In what units is r expressed?
> None. It is unitless.

Can r be negative?
> Yes. It is negative when one variable tends to go down as the other goes up. When r is positive, one variable tends to go up as the other variable goes up.

Can correlation be calculated if all X values are the same? If all Y values are the same?
> No. If all the X or Y values are the same, it makes no sense to calculate correlation.

Why are there no best-fit lines in Figure 32.1?

Correlation quantifies the relationship but does not fit a line to the data. Chapter 33 explains linear regression, which finds the best-fit line, and reviews the differences between correlation and regression.

If all the X or Y values are transformed to new units, will r change?

No. Multiplying by a factor to change units (e.g., inches to centimeters, milligrams per milliliter to millimolar) will not change the correlation coefficient.

If all the X or Y values are transformed to their logarithms, will r change?

Yes. A logarithmic transformation, or any transformation that changes the relative values of the points, will change the value of r. However, the nonparametric Spearman correlation coefficient (see Chapter 41), which depends only on the order of the values, will not change.

If you interchange X and Y, will r change?

No. X and Y are completely symmetrical in calculating and interpreting the correlation coefficient.

If you double the number of points but r doesn't change, what happens to the CI and P value?

With more data points, the CI will be narrower. With more data points and the same value of r, the P value will be smaller.

Can correlation be used to quantify how closely two alternative assay methods agree?

No. This is a common mistake. If you want to compare two different analysis methods, special methods are needed. Look up the Bland–Altman plot.

Is there a distinction between r^2 and R^2?

No. Upper-and lowercase mean the same thing in this context. However, the correlation coefficient, r, is always written in lower case.

What value of r is considered a "high correlation"? What value is considered a "low correlation"?

There are no standards. A correlation that might seem high to scientists in one field might seem really low to scientists in another.

Can r be expressed as a percentage?

No. Its value ranges from −1 to 1. It is not a fraction, so cannot be expressed as a percentage.

CHAPTER SUMMARY

- The correlation coefficient, r, quantifies the strength of the linear relationship between two continuous variables.
- The value of r is always between −1.0 and 1.0; it has no units.
- A negative value of r means that the trend is negative—that is, Y tends to go down as X goes up.
- A positive value of r means that the trend is positive—that is, X and Y go up together.
- The square of r equals the fraction of the variance that is shared by X and Y. This value, r^2, is sometimes called the coefficient of determination.
- Correlation is used when you measure both X and Y variables, but it is not appropriate if X is a variable you manipulate.

- Note that correlation and linear regression are not the same. In particular, note that the correlation analysis does not fit or plot a line.
- You should always view a graph of the data when interpreting a correlation coefficient.
- Correlation does not imply causation. The fact that two variables are correlated does not mean that changes in one cause changes in the other.

TERMS INTRODUCED IN THIS CHAPTER

- Correlation (p. 319)
- Correlation coefficient (r) (p. 319)
- Coefficient of determination (r^2) (p. 322)
- Covariation (p. 319)

PART G

Fitting Models to Data

CHAPTER 33

Simple Linear Regression

> In the space of 176 years, the Lower Mississippi has shortened itself 242 miles. This is an average of a trifle over one mile and a third per year. Therefore . . . any person can see that seven hundred and forty-two years from now, the lower Mississippi will be only a mile and three-quarter long.
>
> MARK TWAIN, *LIFE ON THE MISSISSIPPI*

One way to think about linear regression is that it fits the "best line" through a graph of data points. Another way to look at it is that linear regression fits a simple model to the data to determine the most likely values of the parameters that define that model (slope and intercept). Chapter 34 introduces the more general concepts of fitting a model to data.

THE GOALS OF LINEAR REGRESSION

Recall the insulin example of Chapter 32. The investigators were curious to understand why insulin sensitivity varies so much among individuals. They measured insulin sensitivity and the lipid content of muscle obtained at biopsy in 13 men. You've already seen that the two variables (insulin sensitivity and the fraction of the fatty acids that are unsaturated with 20–22 carbon atoms [%C20–22]) correlate substantially.

In this example, the authors concluded that differences in the lipid composition affect insulin sensitivity and proposed a simple model—that insulin sensitivity is a linear function of %C20–22. As %C20–22 goes up, so does insulin sensitivity. This relationship can be expressed as an equation:

Insulin sensitivity = Intercept + Slope · %C20 − 22

Let's define the insulin sensitivity to be Y, %C20–22 to be X, the intercept to be b, and the slope to be m. Now the model takes this standard form:

$Y = b + m \cdot X$

The model is not complete, however, because it doesn't account for random variation. The investigators used the standard assumption that random variability from the predictions of the model will follow a Gaussian distribution.

BEST-FIT VALUES	
Slope	37.21 ± 9.296
Y intercept when X = 0.0	−486.5 ± 193.7
X intercept when Y = 0.0	13.08
1/slope	0.02688
95% CI	
Slope	16.75 to 57.67
Y intercept when X = 0.0	−912.9 to −60.17
X intercept when Y = 0.0	3.562 to 15.97
GOODNESS OF FIT	
R^2	0.5929
Sy.x	75.90
IS SLOPE SIGNIFICANTLY NONZERO?	
F	16.02
DFn, DFd	1.000, 11.00
P value	0.0021
Deviation from zero?	Significant
DATA	
Number of X values	13
Maximum number of Y replicates	1
Total number of values	13
Number of missing values	0

Table 33.1. Linear regression results.
These results are from GraphPad Prism; other programs would format the results differently.

We can't possibly know the one true population value for the intercept or the one true population value for the slope. Given the sample of data, our goal is to find those values for the intercept and slope that are most likely to be correct and to quantify the imprecision with CIs. Table 33.1 shows the results of *linear regression*. The next section will review all of these results.

It helps to think of the model graphically (see Figure 33.1). A simple way to view linear regression is that it finds the straight line that comes closest to the points on a graph. That's a bit too simple, however. More precisely, linear regression finds the line that best predicts Y from X. To do so, it only considers the vertical distances of the points from the line and, rather than minimizing those distances, it minimizes the sum of the square of those distances. Chapter 34 explains why.

LINEAR REGRESSION RESULTS

The slope

The best-fit value of the slope is 37.2. This means that when %C20–22 increases by 1.0, the average insulin sensitivity is expected to increase by 37.2 mg/m²/min.

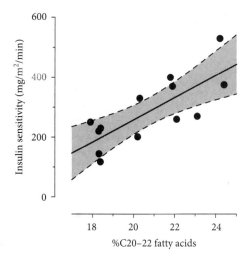

Figure 33.1. The best-fit linear regression line along with its 95% confidence band (shaded).

The CI is an essential (but often omitted) part of the analysis. The 95% CI of the slope ranges from 16.7 to 57.7 mg/m²/min. Although the CI is fairly wide, it does not include zero; in fact, it doesn't even come close to zero. This is strong evidence that the observed relationship between lipid content of the muscles and insulin sensitivity is very unlikely to be a coincidence of random sampling.

The range of the CI is reasonably wide. The CI would be narrower if the sample size were larger.

Some programs report the standard errors of the slope instead of (or in addition to) the CIs. The standard error of the slope is 9.30 mg/m²/min. CIs are easier to interpret than standard errors, but the two are related. If you read a paper that reports the standard error of the slope but not its CI, compute the CI using these steps:

1. Look in Appendix D to find the critical value of the t distribution. The number of df equals the number of data points minus 2. For this example, there were 13 data points, so there are 11 df, and the critical t ratio for a 95% CI is 2.201.
2. Multiply the value from Step 1 by the standard error of the slope reported by the linear regression program. For the example, multiply 2.201 times 9.30. The margin of error of the CI equals 20.47.
3. Add and subtract the value computed in Step 2 from the best-fit value of the slope to obtain the CI. For the example, the interval begins at 37.2 − 20.5 = 16.7. It ends at 37.2 + 20.5 = 57.7.

The intercept

A line is defined by both its *slope* and its Y-*intercept*, which is the value of the insulin sensitivity when the %C20–22 equals zero.

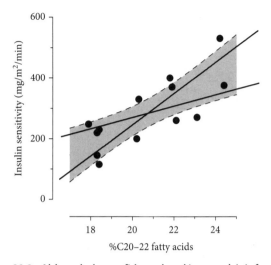

Figure 33.2. Although the confidence band is curved, it is for linear regression and only considers the fits of straight lines.

Two lines that fit within the confidence band. Given the assumptions of the analysis, you can be 95% sure that the true (population) line can be drawn within the shaded area.

For this example, the Y-intercept is not a scientifically relevant value. The range of %C20–22 in this example extends from about 18%z to 24%. Extrapolating back to zero is not helpful. The best-fit value of the intercept is –486.5, with a 95% CI ranging from –912.9 to –60.17. Negative values are not biologically possible, because insulin sensitivity is assessed as the amount of glucose needed to maintain a constant blood level and so must be positive. These results tell us that the linear model cannot be correct when extrapolated way beyond the range of the data.

Graphical results

Figure 33.1 shows the best-fit regression line defined by the values for the slope and the intercept, as determined by linear regression.

The shaded area in Figure 33.1 shows the 95% confidence bands of the regression line, which combine the CIs of the slope and the intercept. The best-fit line determined from this particular sample of subjects (solid black line) is unlikely to really be the best-fit line for the entire (infinite) population. If the assumptions of linear regression are true (which is discussed later in this chapter), you can be 95% sure that the overall best-fit regression line lies somewhere within the shaded confidence bands.

The 95% confidence bands are curved but do not allow for the possibility of a curved (nonlinear) relationship between X and Y. The confidence bands are computed as part of linear regression, so they are based on the same assumptions as linear regression. The curvature simply is a way to enclose possible straight lines, of which Figure 33.2 shows two.

The 95% confidence bands enclose a region that you can be 95% confident includes the true best-fit line (which you can't determine from a finite sample of data). But note that only six of the 13 data points in Figure 33.2 are included within the confidence bands. If the sample was much larger, the best-fit line would be determined more precisely, so the confidence bands would be narrower and a smaller fraction of data points would be included within the confidence bands. Note the similarity to the 95% CI for the mean, which does not include 95% of the values (see Chapter 12).

R^2

The R^2 value (0.5929) means that 59% of all the variance in insulin sensitivity can be accounted for by the linear regression model and the remaining 41% of the variance may be caused by other factors, measurement errors, biological variation, or a nonlinear relationship between insulin sensitivity and %C20–22. Chapter 35 will define R^2 more rigorously. The value of R^2 for linear regression ranges between 0.0 (no linear relationship between X and Y) and 1.0 (the graph of X vs. Y forms a perfect line).

P value

Linear regression programs report a P value. To interpret any P value, the null hypothesis must be stated. With linear regression, the null hypothesis is that there really is no linear relationship between insulin sensitivity and %C20–22. If the null hypothesis were true, the best-fit line in the overall population would be horizontal with a slope of zero. In this example, the 95% CI for slope does not include zero (and does not come close), so the P value must be less than 0.05. In fact, it is 0.0021. The P value answers the question, If that null hypothesis was true, what is the chance that linear regression of data from a random sample of subjects would have a slope as far (or farther) from zero as that which is actually observed?

In this example, the P value is tiny, so we conclude that the null hypothesis is very unlikely to be true and that the observed relationship is unlikely to be caused by a coincidence of random sampling.

With this example, it makes sense to analyze the data both with correlation (see Chapter 32) and with linear regression. The two are related. The null hypothesis for correlation is that there is no correlation between X and Y. The null hypothesis for linear regression is that a horizontal line is correct. Those two null hypotheses are essentially equivalent, so the P values reported by correlation and linear regression are identical.

ASSUMPTIONS: LINEAR REGRESSION

Assumption: The model is correct

Not all relationships are linear (see Figure 33.3). In many experiments, the relationship between X and Y is curved, making simple linear regression inappropriate. Chapter 36 explains nonlinear regression.

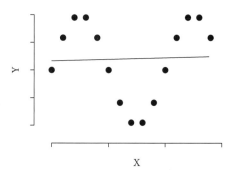

Figure 33.3. Linear regression only looks at linear relationships.
X and Y are definitely related here, just not in a linear fashion. The line
is drawn by linear regression and misses the relationship entirely.

The linear regression equation defines a line that extends infinitely in both directions. For any value of X, no matter how high or how low, the equation can predict a Y value. Of course, it rarely makes sense to believe that a model can extend infinitely. But the model can be salvaged by using the predictions of the model that only fall within a defined range of X values. Thus, we only need to assume that the relationship between X and Y is linear within that range. In the example, we know that the model cannot be accurate over a broad range of X values. At some values of X, the model even predicts that Y will be negative, a biological impossibility. But the linear regression model is useful within the range of X values actually observed in the experiment.

Assumption: The scatter of data around the line is Gaussian
Linear regression analysis assumes that the scatter of data around the model (the true best-fit line) is Gaussian. The CIs and P values cannot be interpreted if the distribution of scatter is far from Gaussian or if some of the values are outliers from another distribution.

Assumption: The variability is the same everywhere
Linear regression assumes that the scatter of points around the best-fit line has the same SD all along the curve. This assumption is violated if the points with high or low X values tend to be farther from the best-fit line. The assumption that the SD is the same everywhere is termed *homoscedasticity*. Linear regression can be calculated without this assumption by differentially weighting the data points, giving more weight to the points with small variability and less weight to the points with lots of variability.

Assumption: Data points (residuals) are independent
Whether one point is above or below the line is a matter of chance and does not influence whether another point is above or below the line.

Assumption: X and Y values are not intertwined

If the value of X is used to calculate Y (or the value of Y is used to calculate X), then linear regression calculations will be misleading. One example is a Scatchard plot, which is used by pharmacologists to summarize binding data. The Y value (drug bound to receptors divided by drug free in solution) is calculated from the X value (free drug), so linear regression is not appropriate. Another example would be a graph of midterm exam scores (X) versus total course grades (Y). Because the midterm exam score is a component of the total course grade, linear regression is not valid for these data.

Assumption: The X values are known precisely

Regression assumes that the true X values are known and that all the variation is in the Y direction. If X is something you measure (rather than control) and the measurement is not precise, the linear regression calculations might be misleading.

COMPARISON OF LINEAR REGRESSION AND CORRELATION

The example data have now been analyzed twice, using correlation (see Chapter 32) and linear regression. The two analyses are similar, yet distinct.

Correlation quantifies the degree to which two variables are related but does not fit a line to the data. The correlation coefficient tells you the extent to which one variable tends to change when the other one does, as well as the direction of that change. The CI of the correlation coefficient can only be interpreted when both X and Y are measured and when both are assumed to follow Gaussian distributions. In the example, the experimenters measured both insulin sensitivity and %C20–22. You cannot interpret the CI of the correlation coefficient if the experimenters manipulated (rather than measured) X.

With correlation, you don't have to think about cause and effect. You simply quantify how well two variables relate to each other. It doesn't matter which variable you call X and which you call Y. If you reversed the definition, all of the results would be identical. With regression, you do need to think about cause and effect. Regression finds the line that best predicts Y from X, and that line is not the same as the line that best predicts X from Y.

It made sense to interpret the linear regression line only because the investigators hypothesized that the lipid content of the membranes would influence insulin sensitivity and so defined %C20–22 to be X and insulin sensitivity to be Y. The results of linear regression (but not correlation) would be different if the definitions of X and Y were swapped (so that changes in insulin sensitivity would somehow change the lipid content of the membranes).

With most data sets, it makes sense to calculate either linear regression or correlation but not both. The example here (lipids and insulin sensitivity) is one for which both correlation and regression make sense. The R^2 is the same whether computed by a correlation or a linear regression program.

LINGO: LINEAR REGRESSION

Model

Regression refers to a method used to fit a model to data. A model is an equation that describes the relationship between variables. Chapter 34 defines the term more fully.

Parameters

The goal of linear regression is to find the values of the slope and intercept that make the line come as close as possible to the data. The slope and the intercept are called *parameters*.

Regression

Why the strange term *regression*? In the 19th century, Sir Francis Galton studied the relationship between parents and children. Children of tall parents tended to be shorter than their parents. Children of short parents tended to be taller than their parents. In each case, the height of the children reverted, or "regressed," toward the mean height of all children. Somehow, the term *regression* has taken on a much more general meaning.

Residuals

The vertical distances of the points from the regression line are called *residuals*. A residual is the discrepancy between the actual Y value and the Y value predicted by the regression model.

Least squares

Linear regression finds the value of the slope and intercept that minimizes the sum of the squares of the vertical distances of the points from the line. This goal gives linear regression the alternative name *linear least squares.*

Linear

The word *linear* has a special meaning to mathematical statisticians. It can be used to describe the mathematical relationship between model parameters and the outcome. Thus, it is possible for the relationship between X and Y to be curved but for the mathematical model to be considered linear.

Simple versus multiple linear regression

This chapter explains simple linear regression. The term *simple* is used because there is only one X variable. Chapter 37 explains multiple linear regression. The term *multiple* is used because there are two or more X variables.

COMMON MISTAKES: LINEAR REGRESSION

Mistake: Concluding there is no relationship between X and Y when R^2 is low

A low R^2 value from linear regression means there is little linear relationship between X and Y. But not all relationships are linear. Linear regression in Figure 33.3 has an R^2 of 0.01, but X and Y are clearly related (just not linearly).

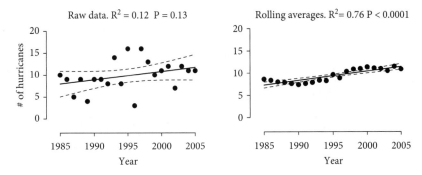

Figure 33.4. Don't smooth.
Smoothing, or computing a rolling average (which is a form of smoothing), appears to reduce the scatter of data and so gives misleading results.

Mistake: Fitting rolling averages or smoothed data

Figure 33.4 shows a synthetic example to demonstrate the problem of fitting smoothed data. The graph plots the number of hurricanes over time. The graph on the left shows actual "data," which jump around a lot. In fact, these are random values (from a Poisson distribution with a mean of 10), and each was chosen without regard to the others. There is no true underlying trend.

The graph on the right of Figure 33.4 shows a *rolling average* (adapted from Briggs, 2008b). This is also called *smoothing* the data. There are many ways to smooth data. For this graph, each value plotted is the average of the number of hurricanes for that year plus average of the number of hurricanes for the prior eight years (thus, a rolling average). The idea of smoothing is to reduce the noise so you can see underlying trends. Indeed, the smoothed graph shows a clear upward trend. The R^2 is much higher, and the P value is very low. But this trend is entirely an artifact of smoothing. Calculating the rolling average makes any random swing to a high or low value look more consistent than it really is, because neighboring values usually also become low or high.

One of the assumptions of linear regression is that each point contributes independent information. When the data are smoothed, the residuals are not independent. Accordingly, the regression line is misleading, and the P value and R^2 are meaningless.

Mistake: Fitting data when X and Y are intertwined

When interpreting the results of linear regression, make sure that the X and Y axes represent separate measurements. If the X and Y values are intertwined, the results will be misleading. Here is an example.

Figure 33.5 shows computer-generated data simulating an experiment in which blood pressure was measured before and after an experimental intervention. The graph on the top left shows the data. Each point represents an individual whose blood pressure was measured before or after an intervention. Each value was sampled from a Gaussian distribution with a mean of 120 and an SD of 10. The data are entirely

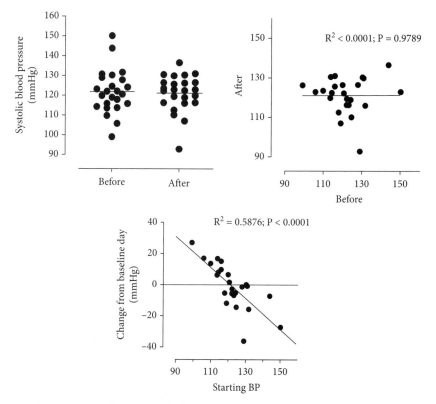

Figure 33.5. Beware of regression to the mean.
(Top left) Actual data (which are simulated). The before and after values are indistinguishable.
(Top right) Graph of one set of measurements versus the others. There is no trend at all.
(Bottom) Graph of the starting (before) blood pressure versus the change (after – before). The
apparent trend is an illusion. All the graph tells you is that if your blood pressure happens to
be very high on one reading, it is likely to be lower on the next. And when it is low on one
reading, it is likely to be higher on the next. Regressions in which X and Y are intertwined like
this are not useful.

random, and the two data sets (top-left graph) look about the same. Therefore, a best-
fit regression line (top-right graph) is horizontal. Although blood pressure levels
varied between measurements, there was no systematic effect of the treatment.
The graph on the bottom shows the same data. But now the vertical axis shows the
change in blood pressure (after–before) plotted against the starting (before) blood
pressure. Note the striking linear relationship. Individuals who initially had low
pressures tended to see an increase; individuals with high pressures tended to see a
decrease. This is entirely an artifact of data analysis and tells you nothing about the
effect of the treatment, only about the stability of the blood pressure levels between
treatments. Chapter 1 gave other examples of regression to the mean.

Graphing a change in a variable versus the initial value of the variable is quite
misleading. Attributing a significant correlation on such a graph to an experimental intervention is termed the *regression fallacy*. Such a plot should not be analyzed

by linear regression, because these data (as presented) violate one of the assumptions of linear regression, that the X and Y values are not intertwined.

Mistake: Not thinking about which variable is X and which is Y

Unlike correlation, linear regression calculations are not symmetrical with respect to X and Y. Switching the labels X and Y will produce a different regression line (unless the data are perfect, with all points lying directly on the line). This makes sense, because the whole point is to find the line that best predicts Y from X, which is not the same as the line that best predicts X from Y.

Consider the extreme case, in which X and Y are not correlated at all. The linear regression line that best predicts Y from X is a horizontal line through the mean Y value. In contrast, the best line to predict X from Y is a vertical line through the mean of all X values, which is 90° different. With most data, the two lines are far closer than that but are still not the same.

Which variable should be X and which should be Y? If one variable was under experimental control, call it X. If both variables are just observed but you suspect a causal relationship between them, call "the cause" X. If you are would like to make predictions of one variable based on the other, call the one you're predicting Y and the variable you're predicting from X.

Mistake: Looking at numerical results of regression without viewing a graph

Figure 33.6 shows four linear regressions designed by Anscombe (1973). The best-fit values of the slopes are identical for all four data sets. The best-fit values of the Y intercept are also identical. Even the R^2 values are identical. But a glance at the graph shows you that the data are very different!

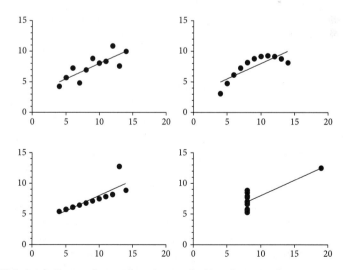

Figure 33.6. Look at a graph as well as at numerical results.

The linear regressions of these four data sets all have identical best-fit values for slope, best-fit values for the Y-intercept, and R^2s. Yet the data are far from identical.

Mistake: Using standard unweighted linear regression when scatter increases as Y increases

It is common for biological data to violate the assumption that variability is the same for all values of Y. Instead, it is common for variability to be proportional to Y. Few linear regression programs can handle this situation, but most nonlinear regression programs can. If you want to fit a linear model to data for which scatter increases with Y, use a nonlinear regression program to fit a straight-line model and choose to weight the data points to account for unequal variation (see Chapter 36).

Mistake: Extrapolating beyond the data

The best-fit values of the slope and intercept define a linear equation that defines Y for any value of X. So, for the example, it is easy enough to plug numbers into the equation and predict what Y (insulin sensitivity) would be if X (lipid composition as %C20–22) were 100%. But the data were only collected with X values between 18% and 24%, and there is no reason to think that the linear relationship would continue much beyond the range of the data. Predictions substantially beyond that range are likely to be very wrong.

Figure 33.7 shows why you must beware of predictions that go beyond the range of the data. The top graph shows data that are fit by a linear regression model very well. The data range from X = 1 to X = 15.

The bottom graphs predict what will happen at later times. The left graph on the bottom shows the prediction of linear regression. The other two graphs show predictions of two other models. Each of these models actually fit the data slightly better than does linear regression, so the R^2 values are slightly higher. The three models make very different predictions at late time points. Which prediction is most likely to be correct? It is impossible to answer that question without more data or at least some information and theory about the scientific context. The predictions of linear regression, when extended far beyond the data, may be very wrong.

Figure 33.8 is a cartoon that makes the same point.

Mistake: Over interpreting a small P value

Figure 33.9 shows simulated data with 500 points fit by linear regression. The P value from linear regression answers the question of whether the regression line fits the data statistically significantly better than does a horizontal line (the null hypothesis). The P value is only 0.0302, so the deviation of the line from a horizontal line is indeed statistically significant. But look at R^2, which is only 0.0094. That means that not even 1% of the overall variation can be explained by the linear regression model. The low P value should convince you that the true slope is unlikely to horizontal. But the discrepancy is tiny. There may possibly be some scientific field in which such a tiny effect is considered important or worthy of follow-up, but in most fields, this kind of effect is completely trivial, even though it is statistically significant. Focusing on the P value can lead you to misunderstand the findings.

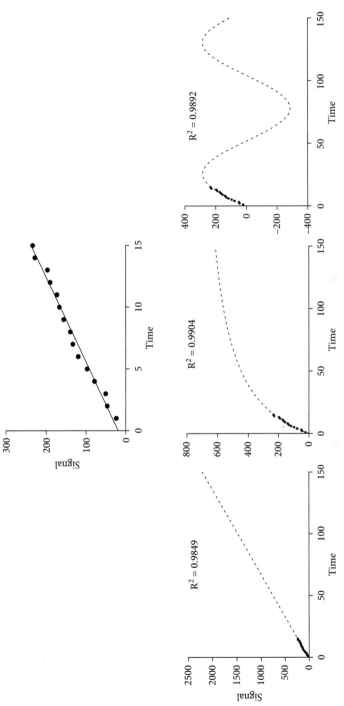

Figure 33.7. Beware of extrapolating models.

(Top) Data that fit a linear regression model very well. What happens at later time points? The three graphs on the bottom show three alternative predictions. (Left) The prediction of linear regression. (Middle and right) Predictions of two other models that actually fit the data slightly better than a straight line does. Predictions of linear regression that extend far beyond the data may be very wrong.

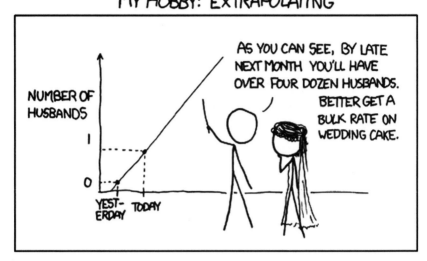

Figure 33.8. Beware of extrapolation.
Source: https://xkcd.com.

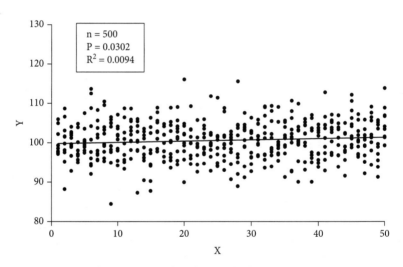

Figure 33.9. Beware of small P values from large sample sizes.
The line was fit by linear regression. It is statistically different from horizontal ($P = 0.03$). But R^2 is only 0.0094, so the regression model explains less than 1% of the variance. The P value is small, so these data would not be expected if the true slope were exactly horizontal. However, the deviation from horizontal is trivial. You'd be misled if you focused on the P value and the conclusion that the slope differs significantly from zero.

The cartoon in Figure 33.10 gives some perspective. If it is easier to find a constellation among the data points than it is to see whether the slope is positive or negative, don't pay too much attention to the small P value.

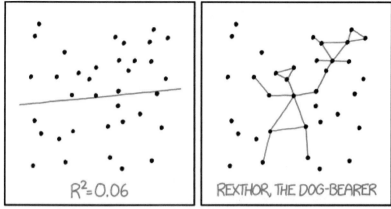

I DON'T TRUST LINEAR REGRESSIONS WHEN IT'S HARDER TO GUESS THE DIRECTION OF THE CORRELATION FROM THE SCATTER PLOT THAN TO FIND NEW CONSTELLATIONS ON IT.

Figure 33.10. A rule of thumb for when to trust linear regression.
Source: https://xkcd.com.

Mistake: Using linear regression when the data are complete with no sampling error

Figure 33.11 shows the number of high school students in the United States who took the advanced placement (AP) exam in statistics each year from 1998 to 2015 (with data from two years missing). The graph shows the total number of students who took the exam. There is no sampling from a population and no random error. The exam is given once a year, so there are no values between the points to interpolate.

The graph shows the linear regression line, but it is not very helpful. The standard error and CI of the slope would be meaningless. Interpolating between years would be meaningless as well, because the exam is only offered once each year. Computing a P value would be meaningless because there is no random sampling. The numbers really do speak for themselves, and the results of linear regression would not add any insights and might even be misleading.

What if you want to extrapolate to predict the number of students who will take the exam the next year (2016)? Using a linear regression line would not be unreasonable. But it might be just as reasonable to simply predict that the increase from 2015 to 2016 (which you are trying to predict) will equal the increase from 2014 to 2015 (which you know). Any prediction requires assuming a model, and there really is no reason to assume that the increase will be linear indefinitely. In fact, it looks as though the increase from year to year is increasing in recent years. You could also use a second- or third-order polynomial model (or some other model) to predict future years. Or maybe only fit to data from the last few years.

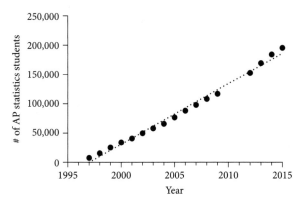

Figure 33.11. No need for linear regression.

The number of high school students who took the advanced placement test in statistics each year from 1998 to 2015. These data are not sampled from a larger population. The graph shows the total number of U.S. students who took the advanced placement statistics exam each year. The dotted line shows the best-fit linear regression line, but it would not be useful to describe the data.

Source: Data from Wikipedia (http://en.wikipedia.org/wiki/AP_Statistics).

You could also look deeper. The number of students taking the exam depends on how many schools offer an AP statistics course, how many students (on average) take that course, and what fraction of the students who take the course end up taking the AP exam. If you wanted to predict future trends, it would make sense to learn how these three different factors have changed over the years.

Mistake: Combining data from two groups into one regression

Regression assumes that the data are independent. If you combine data from two groups, you can violate this assumption badly and be misled. Neither group (closed or open symbols) shows a linear trend (left), but when combined there is a strong linear trend (right). The solid lines are created by linear regression, with the dashed lines showing the 95% confidence bands.

Figure 33.12 shows an example where combining two groups gives a false impression of a strong linear relationship. If you look at the two groups on the left (shown as open or closed symbols) separately, there is no trend. Knowing X does not help you know Y. If you fit separate regression lines to the two data sets separately, both R^2 values are less than 0.01 and both P values are greater than 0.8. But when you fit one regression line to all the data (right side of the figure; all data shown as squares), there appears to be a trend. R^2 equals 0.7 and the P value testing the null hypothesis that the slope is horizontal (zero) is tiny (less than 0.0001). But there really is no trend between X and Y. Rather the figure combines two groups that differ, on average, in both X and Y values.

Figure 33.13 shows the opposite problem. In each group (shown as solid or open symbols), there is a strong linear trend with both R^2 values greater than

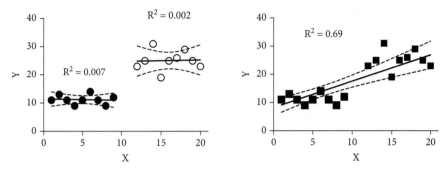

Figure 33.12. Combining two groups into one regression can mislead by creating a strong linear relationship.

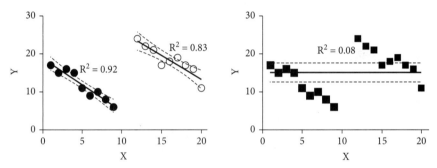

Figure 33.13. Combining two groups into one regression can mislead by hiding a trend. Both groups (closed and open symbols) show a strong linear trend (left), but when combined there is no trend (right).

0.8 (left panel of Figure 33.13). But when you fit one line to all the data, as shown in the graph on the right with all data shown as squares, the regression line is horizontal showing essentially no trend with R^2 equal to 0.08.

Q & A

Do the X and Y values have to have the same units to perform linear regression?
    No, but they can. In the example, X and Y are in different units.

Can linear regression work when all X values are the same? When all Y values are the same?
    No. The whole point of linear regression is to predict Y based on X. If all X values are the same, they won't help predict Y. If all Y values are the same, there is nothing to predict.

Can linear regression be used when the X values are actually categories?

If you are comparing two groups, you can assign the groups to $X = 1$ and $X = 0$ and use linear regression. This is identical to performing an unpaired t test, as Chapter 35 explains. If there are more than two groups, using simple linear regression only makes sense when the groups are ordered and equally spaced and thus can be assigned sensible numbers. If you need to use a categorical variable with more than two possible values, read about indicator variables (also called dummy variables) in a text about multiple linear regression.

Is the standard error of the slope the same as the SEM?

No. The standard error is a way to express the precision of a computed value (parameter). The first standard error encountered in this book happened to be the standard error of the mean (see Chapter 14). The standard error of a slope is quite different. Standard errors can also be computed for almost any other parameter.

What does the variable Ŷ (used in other statistics books) mean?

Mathematical books usually distinguish between the Y values of the data you collected and the Y values predicted by the model. The actual Y values are called Y_i where i indicates the value to which you are referring. For example, Y_3 is the actual Y value of the third subject or data point. The Y values predicted by the model are called Ŷ, pronounced "Y hat." So $\hat{Y}_3$ is the value the model predicts for the third participant or data point. That value is predicted from the X value for the third data point but doesn't take into account the actual value of Y.

Will the regression line be the same if you exchange X and Y?

Linear regression fits a model that best predicts Y from X. If you swap the definitions of X and Y, the regression line will be different unless the data points line up perfectly so that every point is on the line. However, swapping X and Y will not change the value of R^2.

Can R^2 ever be zero? Negative?

R^2 will equal zero if there is no trend whatsoever between X and Y, so the best-fit line is exactly horizontal. R^2 cannot be negative with standard linear regression, but Chapter 36 explains that R^2 can be negative with nonlinear regression.

Do you need more than one Y value for each X value to calculate linear regression? Does it help?

Linear regression does not require more than one Y value for each X value. But there are three advantages to using replicate Y values for each value of X:

- With more data points, you'll determine the slope and intercept with more precision.

- Additional calculations (not detailed here) can test for nonlinearity. The idea is to compare the variation among replicates to the distances of the points from the regression line. If your points are "too far" from the line (given the consistency of the replicates), then you can conclude that a straight line does not really describe the relationship between X and Y.

- You can test the assumption that the variability in Y is the same at all values of X.

If you analyze the same data with linear regression and correlation (see Chapter 12), how do the results compare?

If you square the correlation coefficient (r), the value will equal R^2 from linear regression. The P value testing the null hypothesis that the population correlation coefficient is zero will match the P value testing the null hypothesis that the population slope is zero.

R^2 or r^2?

With linear regression, both forms are used and there is no distinction. With nonlinear and multiple regression, the convention is to always use R^2.

Can R^2 be expressed as a percentage? How about r?

Yes. R^2 is a fraction, so it makes sense to express it as a percentage (but this is rarely done). Note, however that the correlation coefficient r is not a fraction so should not be expressed as a percentage.

Does linear regression depend on the assumption that the X values are sampled from a Gaussian distribution? Y values?

No. The results of linear regression are based on the assumption that the residuals (the vertical distances of the data points from the regression line) are Gaussian but does not assume that either X values or Y values are Gaussian.

CHAPTER SUMMARY

- Linear regression fits a model to data to determine the value of the slope and intercept that makes a line best fit the data.
- Linear regression also computes the 95% CI for the slope and intercept and can plot a confidence band for the regression line.
- Goodness of fit is quantified by R^2.
- Linear regression reports a P value, which tests the null hypothesis that the population slope is horizontal.

TERMS INTRODUCED IN THIS CHAPTER

- Intercept (p. 333)
- Linear regression (p. 332)
- Linear least squares (p. 338)
- Linear (p. 338)
- Parameter (p. 338)
- Regression (p. 338)
- Regression fallacy (p. 340)
- Residuals (p. 338)
- Rolling average (p. 339)
- Simple vs. multiple regression (p. 338)
- Slope (p. 333)
- Smoothed data (p. 339)

CHAPTER 34

Introducing Models

A mathematical model is neither a hypothesis nor a theory.
Unlike scientific hypotheses, a model is not verifiable directly
by an experiment. For all models are both true and false. . . .
The validation of a model is not that it is "true" but that it
generates good testable hypotheses relevant to important
problems.

R. Levins (1966)

Chapter 33 explained how linear regression fits a simple model to
data. This chapter generalizes the concept of fitting a model to data.
The goal in fitting a model is to find the best-fit values of the parameters
that define the model. It is essential to understand these basic ideas of
fitting models before reading about the various kinds of regression in
Chapters 35 through 39.

LINGO: MODELS, PARAMETERS, AND VARIABLES

Model

In general, a *model* is a representation of something else. We study models
because they are less expensive and more accessible than the real thing, as well as
easier to manipulate (perhaps with fewer ethical issues).

A mathematical model is an equation or set of equations that describe, represent, or approximate a physical, chemical, or biological state or process. Using a
model can help you think about chemical and physiological processes or mecha-
nisms so you can design better experiments and comprehend the results.

Creating new and useful models from general principles can be difficult. In
contrast, fitting models to data (a major theme of this book) and simulating data
from models (a minor theme of this book) are much easier tasks.

When you fit a model to your data, you obtain best-fit values (parameter
estimates) that you can interpret in the context of the model.

Random component

A mathematical model must specify both the ideal predictions and how the data
will be randomly scattered around those predictions. The following two equivalent,

crude equations point out that a full model must specify a random component (noise) as well as the ideal component (signal).

Data = Ideal + Random
Response = Signal + Noise

Parameters, statistics, and variables

A model is defined by an equation or a set of equations. The left side of the equals sign defines the outcome, called the *dependent variable* or the *response variable.* The right side of the equation defines the dependent variable, as a function of one or more *independent variables* (also called *explanatory variables*), one or more parameters, and often one or more true constants. The model in Chapter 33 defined insulin sensitivity (the dependent variable) as a function of %C20–22 (the independent variable) and the slope and intercepts (the parameters). It had no true constants.

When fitting a model, you will provide the computer program with a set of values for the dependent and independent variables, and then the program will find estimates of the parameters that make the model do the best possible job of predicting the dependent variable. In some cases, models include true constants, such as 1, 2, 0.693, π, and so on, that just come along for the ride and provide structure to the mathematical model.

Note that each data point has its own values for the independent and dependent variables. The population has one set of true parameters, which are unknown. Fitting the model to sample data generates an estimated value for each parameter in the model, along with a CI for each parameter.

The estimated values of each parameter are given several interchangeable names: parameter estimates, best-fit values of parameters, and sample statistics (which are distinct from population parameters).

Estimate

When regression is used to fit a model to data, it reports values for each parameter in the model. These can be called best-fit values or estimated values. Note that this use of the term *estimate* is very different from the word's conventional use to mean an informed guess. The estimates provided by regression are the result of calculations. The best-fit value is called a *point estimate*. The CI for each parameter is called an *interval estimate*.

Fitting a model to data

Regression fits a model to data. It does this by adjusting the values of the parameters in the model to make the predictions of the model come as close as possible to the actual data, taking into account the mathematical model for random error (a Gaussian distribution in the case of linear regression).

THE SIMPLEST MODEL

To compute an average, add up all the values and then divide by the number of values. That average, also called a mean, is a way to describe a stack of numbers with a single value. But what makes an average special?

When you sample values from a population that follows a Gaussian distribution, each value can be defined by this simple model:

$$Y = \mu + \varepsilon$$

where Y is the dependent variable, which is different for each value (data point); μ (the population mean) is a parameter with a single value, which you are trying to find out (it is traditional to use Greek letters to denote population parameters and regular letters to denote sample statistics); and ε (random error) is different for each data point, randomly drawn from a Gaussian distribution centered at zero. Thus, ε is equally likely to be positive or negative, so each Y is equally likely to be higher or lower than μ. The random variable ε is often referred to as *error*. As used in this statistical context, the term *error* refers to any random variability, whether caused by experimental imprecision or by biological variation.

This kind of model is often written like this:

$$Y_i = \mu + \varepsilon_i$$

The subscript i tells you that each Y and ε has a different value, but the population parameter μ has only a single value.

Note that the right side of this model equation has two parts. The first part computes the *expected* value of Y. In this case, it is always equal to the parameter μ, the population mean. More complicated models would have more complicated calculations involving more than one parameter. The second part of the model takes into account the random error. In this model, the random scatter follows a Gaussian distribution and is centered at zero. This is a pretty standard assumption but is not the only model of scatter.

Now you have a set of values, and you are willing to assume that this model is correct. You don't know the population value of μ, but you want to estimate its value from the sample of data. What value of the parameter μ is most likely to be correct? Mathematicians have proven that if the random scatter follows a Gaussian distribution, then the value of μ that is most likely to be correct is the sample mean or average. To use some mathematical lingo, the sample mean is the *maximum likelihood estimate* of μ.

What I've done so far is to take the simple idea of computing the average and turn it into a complicated process using statistical jargon and Greek letters. This doesn't help you understand the idea of a sample mean, but it warms you up for understanding more complicated models, for which the jargon really is helpful.

THE LINEAR REGRESSION MODEL

Recall the linear regression example of Chapter 33. The investigators wanted to understand why insulin sensitivity varies so much among individuals. They measured insulin sensitivity and the lipid content of muscle obtained at biopsy in 13 men. To find out how much insulin sensitivity increases for every percentage point increase in C20–22, linear regression was used to fit this model to the data:

$$Y = \text{intercept} + \text{slope} \cdot X + \text{scatter}$$

This equation describes a straight line. It defines the dependent variable Y (insulin sensitivity) as a function of the independent variable X (%C20–22), two parameters (the slope and Y-intercept), and scatter (a random factor). It can be rewritten in a more standard mathematical form:

$$Y_i = \beta_0 + \beta_1 \cdot X_i + \varepsilon_i$$

The intercept (β_0) and slope (β_1) are parameters that each have a single true underlying population value. In contrast, the random component of the model takes on a different value for each data point. These random values are assumed to follow a Gaussian distribution with a mean of zero. Any point is just as likely to be above the line as below it but more likely to be close to the line than far from it. Remember that the term *error* refers to any random variability, whether caused by experimental imprecision or biological variation.

Note that this model is very simple and certainly not 100% correct. Although data were only collected with X values ranging from 18 to 24, the model predicts Y values for any value of X. Predictions beyond the range of the data are unlikely to be accurate or useful. This model even predicts Y values when X has values outside the range of zero to 100, although X quantifies the percentage of lipids that is C20–22, so values outside the range of 0% to 100% make no sense at all. Furthermore, the model assumes Gaussian random scatter. As discussed in Chapter 10, this assumption is never 100% true.

It's OK that the model is simple and cannot possibly be 100% accurate. That is the nature of scientific models. If a model is too simple, it won't provide useful results. If a model is too complicated, it won't be possible to collect enough data to estimate all the parameters. Useful models are simple, but not too simple.

WHY LEAST SQUARES?

Linear regression finds the "best" values of the slope and intercept by minimizing the sum of the squares of the vertical distances of the points from the line. Why minimize the sum of the squared distances rather than the sum of the absolute values of the distances?

This question can't really be answered without delving into math, but the answer is related to the assumption that the random scatter follows a Gaussian distribution.

If the goal were to minimize the sum of the absolute distances, the regression would be indifferent to fitting a model for which the distances of two points from the line are each 5.0 versus a model for which the distances are 1.0 and 9.0. The sum of the distances equals 10.0 in both cases. But if the scatter is Gaussian, the first model is far more likely, so the model that places the line equidistant from both points should be preferred. Minimizing the sum of the squares accomplishes this. The sums of the squares from the two models are 50 and 82, so minimizing the sum of the squares makes the first model preferable.

A more rigorous answer is that the regression line determined by the *least-squares method* is identical to the line determined by maximum likelihood calculations. What does that mean? Given any hypothetical set of parameter values, it is possible to compute the chance of observing our particular data. Maximum likelihood methods find the set of parameter values for which the observed data are most probable. Given the assumption that scatter follows a Gaussian distribution (with a uniform SD), it can be proven that the maximum likelihood approach and the least-squares approach generate identical results. More simply (perhaps a bit too simply), minimizing the sum of the squares finds values for the parameters that are most likely to be correct.

OTHER MODELS AND OTHER KINDS OF REGRESSION

Regression includes a large family of techniques beyond linear regression:

- Nonlinear regression (see Chapter 36). As in linear regression, Y is a measured variable and there is a single X variable. However, a graph of X versus Y is curved, and the model has a nonlinear relationship between the parameters and Y.
- Multiple linear regression (see Chapter 37). *Multiple* means there are two or more independent X variables.
- Multiple nonlinear regression.
- Logistic regression (see Chapter 38). Here, Y is a binary variable such as infected/not infected or cancer/no cancer. There may be only one X variable, but logistic regression models often have more than one independent X variables. Even so, the analysis is often called *logistic regression* rather than *multiple logistic regression*.
- Proportional hazards regression. Here, the outcome is survival time. There may be only one X variable, but proportional hazards regression is usually used with several X variables.
- Poisson regression. The outcomes are counts that follow a Poisson distribution (see Chapter 6).

COMMON MISTAKES: MODELS

Mistake: Trying to find a perfect model

Your goal in using a model is not necessarily to describe your system perfectly. A perfect model may have too many parameters and variables to be useful. Rather, your goal is to find as simple a model as possible that comes close to describing your system for your intended purpose. You want a model to be simple enough so it is easy to collect enough data to which you can fit the model, yet complicated enough to do a good job of explaining the data. It is common for new fields of science to begin with simple models and then gradually use more and more complicated models as the systems become better understood.

Mistake: Saying that you will fit the data to the model

Regression does not fit data to a model, a phrase that implies fudging the data to make them comply with the predictions of the model. Rather, the model is fit to the data or the data are fit by the model.

Think of models as shoes in a shoe store and data as the pair of feet that you bring into the store. The shoe salesman fits a set of shoes of varying styles, lengths, and widths (parameter estimates) to your feet to find a good, comfortable fit. The salesman does not bring over one pair of universal shoes along with surgical tools to fit your feet to the shoes.

Mistake: Fitting a model without looking at a graph

It is too easy to be misled if you look only at best-fit parameter values and the R^2. Look at a graph of the data with the fit model superimposed.

Mistake: Being afraid to fit a nonlinear model

Fit the model that makes the most sense for your data. If linear regression does not make sense for your data, don't use it simply because it is easily available. Don't be afraid of nonlinear models (introduced in Chapter 36).

Mistake: Expecting a computer program to find a model for you

If your goal is to interpolate new values, it can make sense to ask the computer to choose a model for you. But in most cases you'll want to fit a model that makes scientific sense so you can interpret the parameter values.

Mistake: Doing regression without having a clear idea of your goal

To think clearly about the results of any analysis, it is essential to keep your goal in mind. Regression can be used for several reasons:

- To estimate the values of the parameters in the model. For linear regression, this would be to find the values for the slope and intercept, and your main task in evaluating the results is to interpret these values in a scientific context.
- To compare models. For linear regression, you might want to compare that fit to a fit of a more complicated (nonlinear) model. Or you might want to fit linear regression to two data sets and compare the two slopes. If this is your goal, your main task in evaluating the results is to see how differently the two models fit, and whether the preferred fit statistically also makes scientific sense.
- To interpolate from a standard curve. The simplest case would be to use a linear regression line as a standard curve. Enter known concentrations of protein as X and optical density (which measures how blue the solution became after adding appropriate chemicals) as Y. Then fit a linear regression line. Finally enter experimentally observed Y values (optical density) and find out what X values (protein concentrations) they correspond to.
- To predict new data. In some cases, the goal is to predict a value (on a continuous scale). In other cases, the goal is to predict an outcome with two possibilities. The best-fit parameters are of no interest, and it rarely makes sense to try to interpret them. Since the goal is to predict, the way to

evaluate such a model is to see how well it predicts future data. When the data sets are huge and the model has a huge number of parameters, this is called *machine learning*.

CHAPTER SUMMARY

- A mathematical model is an equation that describes a physical, chemical, or biological process.
- Models are simplified views of the world. As a scientific field progresses, it is normal for the models to get more complicated.
- Statistical techniques fit a model to the data. It is wrong to say that you fit data to the model.
- When you fit a model to data, you obtain the best-fit values of each parameter in the model along with a CI showing you how precisely you have determined its value.
- If the scatter (error) is Gaussian, fitting the model entails finding the parameter values that minimize the sum of the square of the differences between the data and the predictions of the model. These are called least-squares methods.

TERMS INTRODUCED IN THIS CHAPTER

- Dependent variable (p. 351)
- Error (p. 352)
- Estimate (p. 351)
- Explanatory variable (p. 351)
- Independent variable (p. 351)
- Interval estimate (p. 351)
- Least-squares method (p. 353)
- Machine learning (p. 356)
- Maximum likelihood estimate (p. 352)
- Model (p. 350)
- Parameter (p. 351)
- Point estimate (p. 351)
- Response variable (p. 351)
- Variable (p. 351)

CHAPTER 35

Comparing Models

Beware of geeks bearing formulas.

WARREN BUFFETT

At first glance, comparing the fit of two models seems simple: just choose the model with predictions that come closest to the data. In fact, choosing a model is more complicated than that, and it requires accounting for both the number of parameters and the goodness of fit. The concept of comparing models provides an alternative perspective for understanding much of statistics.

COMPARING MODELS IS A MAJOR PART OF STATISTICS

So far, this book has explained the concepts of P values and hypothesis testing in the context of comparing two groups. Statistics is used in many other situations, and many of these can be thought of as comparing the fits of two models. Chapter 36 will briefly explain how to compare the fits of two different nonlinear models. Chapters 37 and 38 will explain multiple regression models (including logistic and proportional hazards regression models), and a major part of these analyses involves comparing alternative models. One-way ANOVA, explained in Chapter 39, can also be thought of as comparing models.

What's so hard about comparing models? It seems as though the answer is simple: pick the one that comes the closest to the data. In fact, that approach is not useful, because models with more parameters almost always fit the sample data better, even if they don't reflect what is actually going on in the population. A useful comparison must also take into account the number of parameters fit by each model.

The problem of having a model with too few parameters is clear. It won't fit the sample data well.

The problems of having a model with too many parameters is harder to understand. Such a model will indeed fit the sample data well, but the problem is that CIs of the parameters will be wide. If the goal of the model is to obtain parameter values that can be interpreted scientifically, the wide CIs mean that the estimated parameter values are too imprecise to be useful.

LINEAR REGRESSION AS A COMPARISON OF MODELS

Chapter 33 explained how to informally interpret the P value and R^2 from linear regression. Here we use the same example (insulin sensitivity) but switch the focus to comparing two models.

The two models being compared

The linear regression model is shown (again) on the right side of Figure 35.1. The right side of Figure 35.2 shows the variability of points around the linear regression line. Each point in this graph shows the vertical distance of a point from the regression line. Points above the horizontal line at $Y = 0$ represent points that are above the regression line, whereas points below that line represent points that are below the regression line.

The alternative, null hypothesis model is a horizontal line. The left side of Figure 35.1 shows the horizontal line fit through the data. The left side of Figure 35.2 shows how well the null hypothesis model fits the data. Each point in the graph shows the difference between a Y value and the mean of all Y values. Points above the horizontal line at zero represent Y values that are greater than the mean, and points below that horizontal line represent Y values that are less than the mean.

The meaning of R^2

The linear regression model fits the data better. The variability of points around the regression line (right side of Figure 35.1) is less than the variability of points around the null hypothesis horizontal line (left side of Figure 35.1). How much less?

Table 35.1 compares the *sum of squares* (SS). The first row shows the sum of the squared distances of points from the horizontal line that is the null hypothesis. The second row shows the sum of the squared distances of points from the linear regression line.

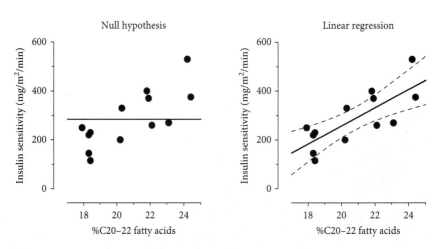

Figure 35.1. Fitting the sample data with two alternative models.

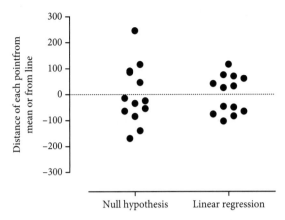

Figure 35.2. **Residuals (distance from line) of each point from both models. The Y-axis has the same units as the Y-axis of Figure 35.1.**

	HYPOTHESIS	SUM OF SQUARES	SCATTER FROM	PERCENTAGE OF VARIATION	
	Null	Horizontal line	155,642	100.0%	
−	Alternative	Regression line	63,361	40.7%	
=	Difference	Improvement	92,281	59.3%	$R^2 = 0.593$

Table 35.1. **Comparing the fit of a horizontal line versus the best-fit linear regression line.** The points are closer to the regression line, so the sum of squares is lower.

The regression line fits the data better than a horizontal line, so the sum of squares is lower. The bottom row in Table 35.1 shows the difference between the fit of the two models. It shows how much better the linear regression model fits than the alternative null hypothesis model.

The fourth column shows the two sums of squares as a percentage of the total. Scatter around the regression line accounts for 40.7% of the variation. Therefore, the linear regression model itself accounts for $100\% - 40.7\% = 59.3\%$ of the variation. This is the definition of R^2, which equals 0.593.

P value

Table 35.2 is rearranged and relabeled to match other statistics books and programs. The focus is no longer on comparing the fit of two models but on dividing the total sum of squares into its components. The first row quantifies how well the linear regression model explains the data. The second row quantifies the scatter of data around the predictions of the model. The third row quantifies the scatter of the data from the predictions (horizontal line) of the null hypothesis.

The important points are that overall variation among data points is quantified with the sum of the squared distances between the point and a prediction

	SOURCE OF VARIATION	SUM OF SQUARES	DF	MS	F RATIO
	Regression	92,281	1	92,281.0	16.0
+	Random	63,361	11	5,760.1	
=	Total	155,642	12		

Table 35.2. Comparing the fit of a horizontal line versus the best-fit linear regression line.
The points are closer to the regression line, so the F ratio is high. Table 35.2 uses the format of ANOVA, which will be explained in Chapter 39.

of a model, and that the sum of squares can be divided into various sources of variation.

The third column of Table 35.2 shows the number of df. The bottom row shows the sum of squares of the distances from the fit of the null hypothesis model. There are 13 data points and only one parameter is fit (the mean), which leaves 12 df. The next row up shows the sum of squares from the linear regression line. Two parameters are fit (slope and intercept), so there are 11 df (13 data points minus two parameters). The top row shows the difference. The linear regression model has one more parameter than the null hypothesis model, so there is only one df in this row. The df, like the sums of squares, can be partitioned so the bottom row is the sum of the values in the rows above.

The fourth column of Table 35.2 divides the sums of squares by the number of df to compute the *mean square* (MS) values, which can also be called *variances*. Note that it is not possible to add the MS in the top two rows to obtain the MS in the bottom row.

Even if the null hypothesis were correct, you'd expect the sum of squares around the regression line to be a bit smaller than the sum of squares around the horizontal null hypothesis line. Dividing by the number of df accounts for this difference. If the null hypothesis were true, the two MS values would be expected to have similar values, so their ratio would be close to 1.0. In fact, for this example, the ratio equals 16.0.

This ratio of the two MS values is called the *F ratio*, named after the pioneering statistician Ronald Fisher. The distribution of F ratios is known when the null hypothesis is true. So for any value of F, and for particular values of the two df values, a P value can be computed. When using a program to find the P value from F, be sure to distinguish the df of the numerator (1 in this example) and the df for the denominator (11 in this example). If you mix up those two df values, you'll get the wrong P value.

The P value, which you already saw in Chapter 33, is 0.0021. From the point of view of probability distributions, this P value answers the following question: If the null hypothesis were true and given an experimental design with one and 11 df, what is the chance that random sampling would result in data with such a strong linear trend that the F ratio would be 16.0 or higher?

UNPAIRED t TEST RECAST AS COMPARING
THE FIT OF TWO MODELS

Chapter 30 explained how to compare the means of two unpaired groups with an unpaired t test. Here we present that same example (bladder relaxation in young and old rats) with a different mindset—comparing the fit of the data to two models.

Unpaired t test as linear regression

To view the unpaired t test as a comparison of the fits of two models, consider it a special case of linear regression.

The example compares two groups, old and young. Let's call the variable that defines age X and assign $X = 0$ to the old group and $X = 1$ to the young group (those values are arbitrary). Figure 35.3 shows the data analyzed by linear regression.

The slope of the regression line is the increase in Y when the X value increases by one unit. The X values denoting old and young groups are one unit apart, so the slope of the best-fit regression line equals the difference between means. The best-fit value of the slope is 23.5%, with a 95% CI ranging from 9.338% to 37.75%. These values match the results reported by the unpaired t test.

The P value from linear regression tests the null hypothesis that the slope is horizontal. This is another way to state the null hypothesis of the unpaired t test (that the two populations share the same mean). Because the two null hypotheses

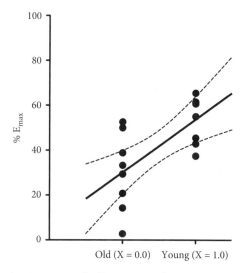

Figure 35.3. Comparing two groups by linear regression.

The data are maximal bladder relaxation in percentages, as explained in Chapter 30. The two groups were assigned arbitrary X values, and the data were fit by linear regression. The best-fit slope of the regression line equals the difference between the two group means, and the CI of the slope equals the CI for the difference between two means. Linear regression reports a two-tailed P value, testing the null hypothesis that the true slope is zero.

are equivalent, the P value determined by linear regression (0.0030) is identical to the P value reported by the t test.

Goodness of fit and R^2

The top row of Table 35.3 and the left side of Figure 35.4 quantify how well the null hypothesis model (a horizontal line) fits the data. Goodness of fit is quantified by the sum of squares of the difference between each value and the grand mean (calculated by ignoring any distinctions between the two groups). Figure 35.4 plots the distance of each value from the grand mean, and Table 35.3 shows the sum of squares.

The second row of Table 35.3 and the right side of Figure 35.4 show how well the alternative hypothesis fits the data. Figure 35.4 plots the distance of each value from its own group mean, and Table 35.3 reports the sum of squares of the difference between each value and the mean of the group from which that value was drawn.

The third row shows the difference in variation. Of all the variation (sum of squares from the null hypothesis grand mean), 54.6% is caused by scatter within

	HYPOTHESIS	SCATTER FROM	SUM OF SQUARES	PERCENTAGE OF VARIATION	
	Null	Grand mean	5,172	100.0	
−	Alternative	Group means	2,824	54.6	
=	Difference	Improvement	2,348	45.4	$R^2 = 0.454$

Table 35.3. The t test example of Chapter 30 recast as a comparison of models.

Values are closer to their group means than the grand mean, so the sum of squares of the alternative hypothesis is lower.

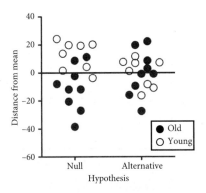

Figure 35.4. How well the t test example data are fit by the two models.

(Left) The fit of the model defined by the null hypothesis that both groups share the same population mean, showing the difference between each value and the grand mean. (Right) How well the data fit the alternative model that the groups have different population means. This part of the graph shows the difference between each value and the mean of its group.

the groups, and 45.5% of the total variation is caused by a difference between the two group means. Therefore, $R^2 = 0.454$.

P value

Determining the P value requires more than partitioning the variance into its components. It also requires accounting for the number of values and the number of parameters fit by each model.

The third column of Table 35.4 shows the number of df. The bottom row shows the fit of the null hypothesis model. There are 17 data points and only one parameter is fit (the grand mean), which leaves 16 df. The next row up quantifies the fit of the alternative model. Two parameters are fit (the mean of each group), so there are 15 df (17 data points minus 2 parameters). The top row shows the difference. The alternative model (two distinct means) has one more parameter than the null hypothesis model (one mean for both groups), so there is only 1 df in this row. The df, like the sum of squares, can be partitioned so the bottom row is the sum of the values in the two rows above. The fourth column divides the sum of squares by the number of df to compute the MS, which could also be called variance. Note that it is not possible to add the MS in the top two rows to obtain the MS in the bottom row.

If the null hypothesis were correct, you'd expect the sum of squares around the individual means to be a bit smaller than the sum of squares around the grand mean. But after dividing by df, the MS values would be expected to be about the same if the null hypothesis were in fact true. Therefore, if the null hypothesis were true, the ratio of the two MS values would be expected to be close to 1.0. If fact, for this example, the ratio (called the F ratio) equals 12.47.

The F distribution under the null hypothesis is known, and so the P value can be computed. The P value is 0.0030. It is the answer to the following question: If the simpler (null hypothesis) model were correct, what is the chance that randomly chosen values would have group means far enough apart to yield an F ratio of 12.47 or higher?

Summary

The t test data can be viewed as comparing how well the data are fit by two models. One model is the null hypothesis—that the two groups of data are sampled from two populations with the same mean. Viewed in terms of linear regression, this model is a

	SOURCE OF VARIATION	SUM OF SQUARES	DF	MS	F RATIO	P VALUE
	Between groups	2,348	1	2,348.0	12.47	0.0030
+	Within groups	2,824	15	188.3		
=	Total	5,172	16			

Table 35.4. The t test example of Chapter 30 recast as a comparison of models.

Values are closer to their group means than the grand mean, so the F ratio is high and the P value is low.

horizontal line with a slope equal to zero. The alternative model is that the population means differ. Viewed as linear regression, the slope is not zero. The goodness of fit of the two models is compared to see whether there is substantial evidence to reject the simpler (null hypothesis) model and accept the more complicated alternative model.

COMMON MISTAKES: COMPARING MODELS

Mistake: Deciding between two models based purely on the P value

When comparing models, look at more than the P values. Also look at how well each model fits the data, at whether the parameters make scientific sense, and at whether all the parameters are defined with reasonable precision.

Mistake: Comparing the fits of models that don't make scientific sense

Only use a statistical approach to compare models that are scientifically sensible. It rarely makes sense to blindly test a huge number of models. If a model isn't scientifically sensible, it probably won't be useful—no matter how high its R^2 may be for a particular data set.

Mistake: Using the F test method to compare unrelated models

The approach described in this chapter can only compare two related models. One model must be a simpler case of the other. The linear regression example does compare two related models, because the null hypothesis model is the same as the other model, with the slope fixed to equal zero. Another way to say this is that the two models are *nested*. Models that are not nested can also be compared, but they require methods that are beyond the scope of this book. To learn about alternative methods, start by reading Burnham and Anderson (2003).

Mistake: Comparing the fits of models to different data sets

When comparing models, both models must be fit to the same data sets. In some cases, two data sets are combined and both models fit both data sets. That's fine. But it is critical that both models "see" all the data. You can't use the methods described here to compare the fit of a model to a data set with the fit of a model to the same data set minus some outliers. You can't compare the fit of a model to a data set with the fit of a model to the same data after being transformed (say to logarithms). You can't compare the fit of a model to data using one weighting scheme with the fit of the model using a different weighting scheme. When comparing the fit of two models, only the model can differ, not the data.

Mistake: Comparing the fits of models whose predictions are indistinguishable in the range of the data

Figure 33.7 showed the predictions of three alternative models (straight line, hyperbola, and sine wave). In the range of the data (X < 15), the three models are indistinguishable. If you had scientific reason to compare the fits of these models, you'd be stuck. Spending your time struggling with statistical software wouldn't help a bit. The data simply cannot distinguish between those three models.

At larger X values (X > 15), the predictions of the models diverge considerably. If you had reason to compare those models, it would be essential to collect data at X values (times) for which the predictions of the model differ.

Q & A

Is it OK to fit lots of models and choose the one that fits best?

There are two problems with this approach. First, in many cases you want to only compare models that make scientific sense. If the model that fits best makes no scientific sense, it won't be helpful in many contexts. You certainly won't be able to interpret the parameters. The second problem is if the best model has too many parameters, you may be overfitting. This means that the model is unlikely to work as well with future data. It is usually better to only compare the fits of two (or a few) models that make scientific sense.

What if I want to compare the fits of two models with the same number of parameters?

The examples presented in this chapter compare the fits of models with different numbers of parameters, so you need to look at the trade-off of better fit versus more complexity in the model. If both your models have the same number of parameters, there is no such trade-off. Pick the one that fits best unless it makes no scientific sense.

CHAPTER SUMMARY

- Much of statistics can be thought of as comparing the fits of two models.
- This chapter looked at examples from Chapter 33 (linear regression) and Chapter 30 (unpaired t test) and recast the examples as comparisons of models.
- Viewing basic statistical tests (t test, regression, ANOVA) as a comparison of two models is a bit unusual, but it makes statistics more sensible for some people.
- Every P value can be viewed as the result of comparing the fit of the null hypothesis with the fit of a more general alternative hypothesis. The P value answers this question: If the null hypothesis were true, what is the chance that the data would fit the alternative model so much better? Interpreting P values is easy once you identify the two models being compared.
- It is only fair to compare the fits of two models to data when the data are exactly the same in each fit. If one of the fits has fewer data points or transformed data, the comparison will be meaningless.

TERMS INTRODUCED IN THIS CHAPTER

- F ratio (p. 360)
- Mean square (MS) (p. 360)
- Nested models (p. 364)
- Sum of squares (SS) (p. 358)

CHAPTER 36

Nonlinear Regression

Models should be as simple as possible, but not more so.

A. EINSTEIN

Nonlinear regression is more general than linear regression. It can fit any model that defines Y as a function of X and one or more parameters. Don't skip this chapter because it seems advanced and complicated. In fact, analyzing data with nonlinear regression is not much harder than analyzing data with linear regression. Be sure to read Chapters 33 (on linear regression) and 35 (on comparing models) before reading this one.

INTRODUCING NONLINEAR REGRESSION

From a scientist's point of view, nonlinear regression is a simple extension of linear regression. While linear regression can only fit one model (a straight line), nonlinear regression can fit any model that defines Y as a function of X. As the name suggests, the relationship between Y and X can be curved.

From a statistician's point of view, nonlinear regression is quite different from linear regression. The methodology is much more complicated, and you really can't understand how it works without using matrix algebra and calculus. Even the term *nonlinear* is more complicated than it seems, as the mathematical definition of *nonlinear* is not simply that the graph of X versus Y is curved. Perhaps for these reasons, nonlinear regression is a topic excluded from almost all introductory statistics books.

Nonlinear regression is commonly used in many fields of science such as pharmacology and physiology. Using nonlinear regression and understanding the results is not much more difficult than using linear regression.

AN EXAMPLE OF NONLINEAR REGRESSION

The data

Chapters 11 and 30 have already discussed data from Frazier, Schneider, and Michel (2006), who measured the degree to which the neurotransmitter norepinephrine relaxes bladder muscle in old and young rats. Strips of bladder muscle were exposed to various concentrations of norepinephrine, and muscle relaxation was measured. The data from each rat were analyzed to determine the maximum

LOG[NOREPINEPHRINE, M]	% RELAXATION
−8.0	2.6
−7.5	10.5
−7.0	15.8
−6.5	21.1
−6.0	36.8
−5.5	57.9
−5.0	73.7
−4.5	89.5
−4.0	94.7
−3.5	100.0
−3.0	100.0

Table 36.1. Bladder muscle relaxation data for one young rat.

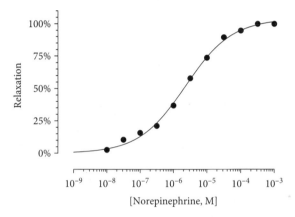

Figure 36.1. Bladder muscle relaxation data for one young rat.
The circles show the data from Table 36.1. The curve was fit by nonlinear regression.

relaxation and the concentration of norepinephrine that relaxes the muscle half that much (EC_{50}).

Table 36.1 and Figure 36.1 show the data from one young rat. Note that the X-axis of Figure 36.1 is logarithmic. Going from left to right, each major tick on the axis represents a concentration of norepinephrine 10-fold higher than the previous major tick.

The model

The first step in fitting a model is choosing a model. In many cases, like this one, a standard model will work fine. Pharmacologists commonly model dose-response (or concentration-effect) relationships using the following equation:

$$Y = \text{Bottom} + \frac{\text{Top} - \text{Bottom}}{1 + 10^{(\text{LogEC}_{50} - X)\ \text{HillSlope}}}$$

To enter this equation into a computer program, use the following syntax. Note that the asterisk (*) denotes multiplication and the carrot (^) denotes exponentiation.

$$Y = Bottom + (Top - Bottom) / (1 + 10 \wedge ((LogEC_{50} - X) * HillSlope))$$

In this equation, X represents the logarithm of the concentration of norepinephrine, and Y is the response the investigators measured, the amount of muscle relaxation. Y is defined as a function of X and four parameters:

- Bottom. This is the value of Y when X is very low, the baseline value of Y when there is no drug added.
- Top. This is the value of Y when X is very high. As the concentration gets higher and higher, the response plateaus at a value called Top.
- $logEC_{50}$. This is the value of X (the logarithm of concentration) that provokes a response halfway between Bottom and Top.
- HillSlope. This quantifies how steep the curve is. The two words ("Hill" and "Slope") are smashed together to form a single parameter in the equation.

After choosing a model, the next step is to decide which of the parameters in the model should be fit to the data and which should be fixed to constant values. In this case, muscle relaxation must be zero with no added norepinephrine. So we'll tell the program not to find a best-fit value of Bottom but rather to set it to a constant value of zero. We don't want to fix the top to 100, because the one goal of the study is to see whether that top plateau differs in old and young rats.

NONLINEAR REGRESSION RESULTS

Interpreting nonlinear regression results is quite straightforward. The whole point of nonlinear regression is to fit a curve to the data, so first look at the curve (Figure 36.1) and then look at the numerical results (Table 36.2).

Best-fit values of parameters
The best-fit values of the parameters should be interpreted in a scientific context.

For the bladder relaxation example, let's focus only on the $logEC_{50}$. The program fits the $logEC_{50}$ and also transforms it back to the EC_{50}, which is reported as 2.3e–006. This notation is commonly used by many programs and means 2.3 times 10 to the power of –6. Because X values are the logarithms of concentration in molar, the EC_{50} is in molar units (M). The best-fit EC_{50} is 0.0000023 M, which can also be written as 2.3 μM. This concentration of norepinephrine relaxes the bladder muscle half as much as the maximum possible relaxation.

CI of the parameters
The program also reports the standard error and the 95% CI of each parameter. The CIs are much easier to interpret and are an essential part of the nonlinear regression results. Given all the assumptions of the analysis (listed later in the chapter), you can be 95% confident that the interval contains the true parameter value.

BEST-FIT VALUES

Bottom	0.0
Top	104
$LogEC_{50}$	–5.64
HillSlope	0.622
EC_{50}	2.30e–006

STANDARD ERRORS

Top	2.06
$LogEC_{50}$	0.0515
HillSlope	0.0358

95% CIs

Top	99.3 to 109
$LogEC_{50}$	–5.76 to –5.52
HillSlope	0.540 to 0.705
EC_{50}	1.75e–006 to 3.02e–006

GOODNESS OF FIT

df	8
R^2	0.997
Absolute sum of squares	43.0

Table 36.2. Results of nonlinear regression as reported by GraphPad Prism.

The 95% CI for the $logEC_{50}$ ranges from –5.76 to –5.52. Pharmacologists are accustomed to thinking in log units of concentration. Most people prefer to see the values in concentration units. The 95% CI of the EC_{50} ranges from 1.75 to 3.02 μM. The CI is fairly narrow and shows us that the EC_{50} has been determined to within a factor of two. That is more than satisfactory for this kind of experiment.

If the CI had been very wide, then you would not have determined the parameter very precisely and would not be able to interpret its value. How wide is too wide? It depends on the context and goals of the experiment.

Some programs don't report the CIs but instead report the standard error of each parameter. Sometimes this is simply labeled the error, and sometimes it is labeled the SD of the parameter. The CI of each parameter is computed from its standard error. When you have plenty of data, the 95% CI extends about two standard errors in each direction.

R^2

R^2 is the fraction of the total variance of Y that is explained by the model. In this example, the curve comes very close to all the data points, so the R^2 is very high, at 0.997.

When $R^2 = 0.0$, the best-fit curve fits the data no better than a horizontal line at the mean of all Y values. When $R^2 = 1.0$, the best-fit curve fits the data perfectly, going through every point. If you happen to fit a really bad model (maybe you made a mistake when choosing a model in a nonlinear regression program or set a nonsensical constraint), R^2 can be negative. That would tell you that the fit of the selected model is even worse than the fit of a horizontal line at the mean of all Y values.

Upper- or lowercase? With linear regression, you'll usually see r^2. With nonlinear and multiple regression, it is always written R^2.

How high should R^2 be? There are no general guidelines. If you are performing a routine set of experiments, you will learn what range of R^2 values to expect and then can troubleshoot if the value is too low. R^2 can be low for many reasons—the presence of outliers, fitting the wrong model, or the presence of lots of experimental or biological variation.

HOW NONLINEAR REGRESSION WORKS

Both linear and nonlinear regression find the values of the parameters that minimize the sum of the squares of the difference between the actual Y value and the Y value predicted by the regression model. Linear regression fits one simple model. Nonlinear regression can fit any model you choose. Linear regression can be thought of as a special case of nonlinear regression.

Although linear and nonlinear regression have the same goal, they work differently. The linear regression method can be completely explained with simple algebra. In contrast, nonlinear regression uses a computationally intensive approach that can only be explained using calculus and matrix algebra.

Nonlinear regression works via an *iterative* or *stepwise approach*. In this method, each parameter is initially provided with an estimated value. Nonlinear regression programs may provide these values automatically, or you may need to enter them yourself. The idea is to generate an initial curve that goes somewhere in the vicinity of the data points. The nonlinear regression method then changes the values of the parameters to move the curve closer to the points. Then it changes the values again. It repeats, or iterates, these steps many times, which is why the method is called *iterative*. When any possible changes to the parameters would make the curve fit more poorly (or fit the same), the method finishes and reports the results.

Because the foundations of nonlinear regression cannot be understood without using matrix algebra and calculus, nonlinear regression has a reputation of being complicated and advanced. But this reputation only applies to the mathematical foundation of the method. Using a nonlinear regression program and interpreting the results is just a tiny bit harder than using linear regression.

ASSUMPTIONS: NONLINEAR REGRESSION

Assumption: The model is correct
When fitting one model, nonlinear regression assumes the model is correct. The next section explains how to compare the fits of two (or more) models.

Assumption: The independent variable has no variability, or at least much less variability than the dependent variable

This assumption is quite reasonable for most experimental work. For example, there will be (or should be) very little error in pipetting different amounts of drugs into wells of a 96-well plate or recording the time at which a blood sample is taken.

Assumption: The scatter of data around the curve is Gaussian

Nonlinear regression analysis assumes that the scatter of data around the curve is Gaussian. The CIs and P values cannot be interpreted if the distribution of the residuals (the difference between the actual and predicted Y value) is far from Gaussian or if some of the values are outliers from another distribution.

Assumption: The variability is the same everywhere

Nonlinear regression assumes that scatter of points around the best-fit line has the same SD all along the curve. This assumption is referred to as *homoscedasticity*. This assumption is violated if, for example, the points with high Y values tend to be farther from the best-fit line than points with small Y values.

Assumption: Data points are independent

Regression assumes that each point is randomly situated above or below the true curve and that no factor influences a batch of points collectively.

Assumption: The value of X is not used to compute Y

If the value of X is used to calculate Y (or the value of Y is used to calculate X), then linear and nonlinear regression calculations will be misleading. One example is a linearized Scatchard plot, used by pharmacologists to summarize binding data. The Y value (drug bound to receptors divided by drug free in solution) is calculated from the X value (free drug), so linear regression is not appropriate. If a Scatchard plot is curved, it is a mistake to fit it with nonlinear regression. Instead, fit an appropriate model to the nonlinearized raw data.

COMPARING TWO MODELS

Chapter 35 explained the general idea of comparing the fit of two models. Applying this idea to nonlinear regression is straightforward.

Figure 36.2 compares the fit of two dose-response models. The solid curve shows the fit previously explained, for which the program determined the best-fit value of the Hill slope (0.622). The dotted curve represents a fit by a simpler model, in which the Hill slope was fixed to a standard value of 1.0. It is obvious that this curve does not fit the data as well.

Figure 36.3 shows the residuals of the data points from each curve. You can see that the residuals from the fixed-slope model are, on average, much larger. You can also see that the residuals from that fit (on the right) are not random. In contrast, the residuals from the variable slope model are smaller (the curve comes closer to the points) and random (no obvious pattern).

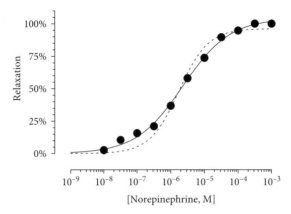

Figure 36.2. Comparing the fit of two models.

The solid curve is the same as that in Figure 36.1. The Hill slope was fit by nonlinear regression and equals 0.622. The data were then fit again, but with the constraint that the Hill slope must equal 1.0. The dotted curve shows this fit, which does not do a good job of fitting the data.

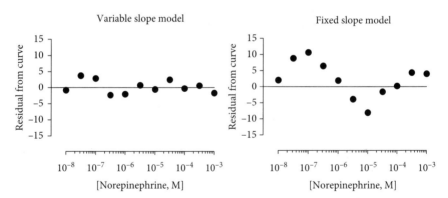

Figure 36.3. Residuals from the two models.

Each circle represents the residual (distance from the curve) of one data point. (Left) The residuals from the fit shown in Figure 36.1, with the Hill slope fit by nonlinear regression (variable slope). (Right) The residuals from the dotted curve of Figure 36.2, in which the Hill slope is fixed to a standard value of 1.0.

Figure 36.4 makes the comparison of the size of the residuals easier to see. Tables 36.3 and 36.4 compare the two fits using a format similar to that used in examples in Chapter 35. The null hypothesis is that the simpler model (fixed slope, one fewer parameter to fit) is correct. In fact, the alternative model fits much better (lower sum of squares) but has one fewer df. The calculations balance the difference in df with the difference in sum of squares.

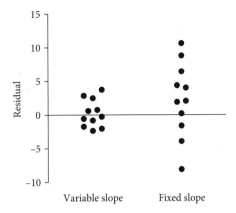

Figure 36.4. Residuals from the two models.

Each circle represents the residual (distance from the curve) of one data point. (Left) The residuals from the solid curve of Figure 36.2, with the Hill slope fit by nonlinear regression. (Right) The residuals from the dotted curve of Figure 36.2, in which the Hill slope is fixed to a standard value of 1.0. This curve does not come as close to the points, so the residuals are larger. Note that the points are displaced horizontally to avoid overlap. Because the residuals from the variable slope model are closer to each other, there is more potential for overlap, so those points have more horizontal displacement.

	HYPOTHESIS	SCATTER FROM	SUM OF SQUARES	DF
	Null	Fixed slope	358.1	9
−	Alternative	Variable slope	43.0	8
=	Difference	Improvement	315.1	1

Table 36.3. The variable slope model fits the data much better, so the sum of squares is much smaller.

SOURCE OF VARIATION	SUM OF SQUARES	DF	MS	F RATIO	P VALUE
Difference	315.1	1	315.1	58.6	< 0.0001
Variable slope model	43.0	8	5.4		
Fixed slope model	358.1	9			

Table 36.4. Computing an F ratio and P value from the fits of the two models.
The F ratio is high, so the P value is tiny.

If the fixed-slope model were correct, it is possible that random scatter of points made the curve fit so badly, but since the P value is less than 0.0001, this would happen less than 0.01% of the time. The low P value is strong evidence that the simple model is inadequate, and that more complicated model (in which the program fits the Hill slope) is preferable.

TIPS FOR UNDERSTANDING MODELS

The first step in nonlinear regression is to choose a model (equation). These tips will help you understand what an equation means. As an example, let's use the Michaelis–Menten equation that describes enzyme activity (Y) as a function of substrate concentration (X):

$$Y = \frac{V_{max} \cdot X}{K_m + X}$$

Tip: Make sure you know the meaning and units of X and Y

Here Y is enzyme activity, which can be expressed in various units, depending on the enzyme. X is the substrate concentration expressed in units of concentration.

Tip: Figure out the units of the parameters

In the example equation, the parameter K_m is added to X. It only makes sense to add things that are expressed in the same units, so K_m must be expressed in the same concentration units as X. This means that the units cancel out in the term $X/(K_m + X)$, so V_{max} is expressed in the same units of enzyme activity as Y.

Tip: Figure out the value of Y at extreme values of X

Because X is a concentration, it cannot be negative. It can, however, be zero. Substitute X = 0 into the equation, and you will see that Y is also zero.

What happens when $\bar{X}$ is very large? As X gets large compared with K_m, the denominator $(X + K_m)$ approaches a value very similar to X. So the ratio $X/(X + K_m)$ approaches 1.0 and Y approaches V_{max}. Therefore, the graph of the model must level off at $Y = V_{max}$ as X gets very large. V_{max} is the maximum enzyme velocity.

Tip: Figure out the value of Y at special values of X

Because K_m is expressed in the same units as X, what is Y when X equals K_m? The ratio $X/(K_m + X)$ equals 0.5, so Y equals half of V_{max}. This means K_m is the concentration of substrate that leads to a velocity equal to half the maximum velocity, V_{max}.

LEARN MORE ABOUT NONLINEAR REGRESSION

Few statistics texts even mention nonlinear regression, and the advanced texts on nonlinear regression tend to be very mathematical. Start with Glantz, Slinker, and Neilands (2016), which includes a great chapter on nonlinear regression. If you like the style of this book, you'll also want to read my book on curve fitting (Motulsky & Christopoulos, 2004) or the GraphPad Curve Fitting Guide (Motulsky, 2016).

COMMON MISTAKES: NONLINEAR REGRESSION

Mistake: Choosing a polynomial model because it seems simpler

Some computer programs offer only one kind of curve fit, *polynomial regression*, which fits the following model to data:

$$Y = \alpha + \beta_1 \cdot X + \beta_2 \cdot X^2 + \beta_3 \cdot X^3 + \beta_4 \cdot X^4 \ldots$$

If the equation ends with the β_2 term, it is called a *second-order* or *quadratic equation*. If you stop after the β_3 term, it is called a *third-order* or *cubic equation*.

Mathematicians and programmers prefer polynomial regression, because the method used to fit polynomial models is much more straightforward than the method used to fit nonlinear models. In fact, although a graph of the polynomial equation is curved, mathematicians consider it a linear model (because Y is linear with respect to each of the parameters).

Few biological or chemical processes follow models described by polynomial equations, so the best-fit parameters can rarely be interpreted in terms of chemistry or biology. Avoid using polynomial regression just because it seems simpler. Instead, use nonlinear regression to fit the model that makes the most scientific sense. This will rarely be a polynomial model.

Mistake: Transforming curved data to a linear form and then using linear regression

Before nonlinear regression became readily available in the early 1980s, scientists often transformed their data to make the graph linear, plotted the transformed linearized data by hand on a sheet of graph paper, drew a straight line by hand with a ruler, calculated the slope and Y-intercept by hand, and then back-calculated the parameter estimates of the original nonlinear model. Later, scientists substituted computerized linear regression for the manual calculations.

Examples of linearizing transforms include Scatchard plots used by pharmacologists, double reciprocal Lineweaver–Burk plots used in enzymology, and the log transform used in kinetic studies. These methods are outdated and should not be used to analyze data. The problem is that the assumptions of linear regression are often badly violated after the data are transformed. Nonlinear regression gives more precise results and is not much harder to perform.

Mistake: Letting a computer program choose the model

Choosing a model should be a scientific decision based on chemistry (or physiology, or genetics, etc.). The choice should not be based solely on the shape of the graph.

Some people don't like thinking about equations, so they want a computer program to automatically fit thousands of equations and choose the model that fits the data best. Because a computer program cannot understand the scientific context of your experiment, a model chosen in this way is unlikely to be scientifically meaningful. The best-fit values of the parameters will probably have no scientific interpretation, so the fit is unlikely to be useful. Don't use a computer program as a way to avoid making scientific decisions.

Mistake: Fitting rolling averages, or smoothed data

As discussed in Chapter 33, smoothing the data (or computing rolling averages) violates the assumption that the residuals are independent and misleads the regression program about the amount of scatter. If you fit smoothed data, R^2 will be falsely high and the CIs will be falsely narrow. The results will be misleading.

Mistake: Using unweighted nonlinear regression when scatter increases as Y increases

It is common for biological data to violate the assumption that variability is the same for all values of Y. Instead, variability is often proportional to Y. Nonlinear regression programs offer weighting, and it is important to choose an appropriate weighting scheme.

Mistake: Using a standard model without thinking about which parameters should be fit and which should be held constant

A crucial step in using nonlinear regression is to decide which parameters should be estimated by nonlinear regression and which should be fixed to constant values based on control data. For example, in the experiment discussed earlier in this chapter, it made sense to constrain the Bottom parameter to a constant value of zero. If you skip this step and fit all the parameters into the model, you may end up with unhelpful results.

Mistake: Using R^2 to assess whether the fit is useful

A high value of R^2 tells you that the curve comes close to the data points. If your goal is to create a standard curve for interpolations, then this fit is useful. But if your goal is to determine best-fit values of parameters, a fit can have a high value of R^2 but still be quite useless. You need to look at whether the best-fit values are scientifically meaningful and have reasonably narrow CIs.

Mistake: Accepting the best-fit values of the parameters, without first looking at the corresponding CIs

Nonlinear regression reports a CI with each parameter. Look at it to see whether the parameter has been determined with reasonable precision. If the CI extends, say, over several orders of magnitude, you won't really care about the best-fit value of the parameter.

Mistake: Comparing two R^2 values to compare two fits

Comparing fits is tricky, and just choosing the fit with the larger R^2 is rarely appropriate. A model with more parameters will almost always fit with a larger R^2. You should not use R^2 to compare the fit of two models. Instead, use the method outlined in Chapter 35, which takes into account the number of parameters being fit to the model.

Q & A

Is nonlinear regression mean the same as *curve fitting*?

> *Curve fitting* is a more general phrase that can include anything from using a flexible ruler, to using cubic spline to nonlinear regression. In contrast, *nonlinear regression* means fitting a specified model to your data.

Is polynomial regression the same as nonlinear regression?

> No. Polynomial models looked curved when you plot X versus Y, but they are actually considered to be linear by mathematicians. Polynomial models can be fit by nonlinear regression programs, but most nonlinear models cannot be fit by polynomial regression programs.

How can I use nonlinear regression if I don't know what model to fit?

> You can't. The idea of nonlinear regression usually is to fit a model that makes scientific sense with the goal of obtaining best-fit values for the parameters that are scientifically meaningful. If you goal is simply to draw an attractive curve, you could try fitting a polynomial model or just try a bunch of standard models.

Can R^2 be negative? Really?

> R^2 is not actually calculated as the square of anything, but rather as the difference between two values, so, yes, it can be negative. This can only happen when the model you chose fits worse than a horizontal line. This could happen if you chose the wrong model by mistake or set an incorrect constraint.

CHAPTER SUMMARY

- Like linear regression, nonlinear regression fits a model to your data to determine the best-fit values of parameters.
- As the name suggests, nonlinear regression fits nonlinear models. For this reason, the details of how the method works are far more complicated than those of linear regression.
- Using a nonlinear regression program is not much harder than using a linear regression program.
- The most important result from nonlinear regression is a graph of the data with the best-fit curve superimposed.
- The most important numerical results are the best-fit values of the parameters and the CIs of each parameter.

TERMS INTRODUCED IN THIS CHAPTER

- Iterative/stepwise approach (p. 370)
- Nonlinear regression (p. 377)
- Polynomial regression (p. 375)
- Second-order/quadratic equation (p. 375)
- Third-order/cubic equation (p. 375)

CHAPTER 37

Multiple Regression

An approximate answer to the right problem is worth a good deal more than an exact answer to an approximate problem.

JOHN TUKEY

In laboratory experiments, you can generally control all the variables. You change one variable, measure another, and then analyze the data with one of the standard statistical tests. But in some kinds of experiments, and in many observational studies, you must analyze how one variable is influenced by several variables. Before reading this chapter, which introduces the idea of multiple regression, first read Chapters 33 (simple linear regression) and 34 (fitting models to data). This chapter explains multiple regression, in which the outcome is a continuous variable. The next chapter explains logistic regression (in which the outcome is binary) and proportional hazards regression (in which the outcome is survival time).

GOALS OF MULTIVARIABLE REGRESSION

Multiple regression extends simple linear regression to allow for multiple independent (X) variables. This is useful in several contexts:

- To assess the impact of one variable after accounting for others. Does a drug work after accounting for age differences between the patients who received the drug and those who received a placebo? Does an environmental exposure increase the risk of a disease after taking into account other differences between people who were and were not exposed to that risk factor?
- To create an equation for making useful predictions. Given the data we know now, what is the chance that this particular man with chest pain is having a myocardial infarction (heart attack)? Given several variables that can be measured easily, what is the predicted cardiac output of this patient?
- To understand scientifically how much changes in each of several variables contribute to explaining an outcome of interest. For example, how do the concentrations of high-density lipoproteins (HDL, good cholesterol), low-density lipoproteins (LDL, bad cholesterol), triglycerides, C-reactive protein, and homocysteine predict the risk of heart disease? One goal might be

to generate an equation that can predict the risk for individual patients (as explained in the previous point). But another goal might be to understand the contributions of each risk factor to aid public health efforts and help prioritize future research projects.

Different regression methods are available for different kinds of data. This chapter explains *multiple linear regression*, in which the outcome variable is continuous. The next chapter explains logistic regression (binary outcome) and proportional hazards regression (survival times).

LINGO

Variables

A regression model predicts one variable Y from one or more other variables X. The Y variable is called the *dependent variable*, the *response variable*, or the *outcome variable*. The X variables are called *independent variables*, *explanatory variables*, or *predictor variables*. In some cases, the X variables may encode variables that the experimenter manipulated or treatments that the experimenter selected or assigned.

Each independent variable can be:

- Continuous (e.g., age, blood pressure, weight).
- Binary. An independent variable might, for example, code for gender by defining zero as male and one as female. These codes, of course, are arbitrary. When there are only two possible values for a variable, it is called a *dummy variable*.
- Categorical, with three or more categories (e.g., four medical school classes or three different countries). Consult more advanced books if you need to do this, because it is not straightforward, and it is easy to get confused. Several dummy variables are needed.

Parameters

The multiple regression model defines the dependent variable as a function of the independent variables and a set of parameters, or *regression coefficients*. Regression methods find the values of each parameter that make the model predictions come as close as possible to the data. This approach is analogous to linear regression, which determines the values of the slope and intercept (the two parameters or regression coefficients of the model) to make the model predict Y from X as closely as possible.

Simple regression versus multiple regression

Simple regression refers to models with a single X variable, as explained in Chapter 33. *Multiple regression*, also called *multivariable regression*, refers to models with two or more X variables.

Univariate versus multivariate regression

Although they are beyond the scope of this book, methods do exist that can simultaneously analyze several outcomes (Y variables) at once. These are called *multivariate methods*, and they include factor analysis, cluster analysis, principal components analysis, and multiple ANOVA (MANOVA). These methods contrast with *univariate methods*, which deal with only a single Y variable.

Note that the terms *multivariate* and *univariate* are sometimes used inconsistently. Sometimes *multivariate* is used to refer to multivariable methods for which there is one outcome and several independent variables (i.e., multiple and logistic regression), and sometimes *univariate* is used to refer to simple regression with only one independent variable.

AN EXAMPLE OF MULTIPLE LINEAR REGRESSION

As you learned in Chapter 33, simple linear regression determines the best linear equation to predict Y from a single variable X. Multiple linear regression extends this approach to find the linear equation that best predicts Y from multiple independent (X) variables.

Study design and questions

Staessen and colleagues (1992) investigated the relationship between lead exposure and kidney function. Heavy exposure to lead can damage kidneys. Kidney function decreases with age, and most people accumulate small amounts of lead as they get older. These investigators wanted to know whether accumulation of lead could explain some of the decrease in kidney function with aging.

The researchers studied 965 men and measured the concentration of lead in blood, as well as creatinine clearance to quantify kidney function. The people with more lead tended to have lower creatinine clearance, but this is not a useful finding. Lead concentration increases with age, and creatinine clearance decreases with age. Differences in age are said to *confound* any investigation between creatinine clearance and lead concentration. To adjust for the confounding effect of age, the investigators used multiple regression and included age as an independent variable. They also included three other independent variables: body mass, logarithm of γ-glutamyl transpeptidase (a measure of liver function), and previous exposure to diuretics (coded as zero or 1).

The X variable about which they cared most was the logarithm of lead concentration. They used the logarithm of concentration rather than the lead concentration itself because they expected the effect of lead to be multiplicative rather than additive—that is, they expected that a doubling of lead concentration (from any starting value) would have an equal effect on creatinine clearance. So why logarithms? The regression model is intrinsically additive. Note that the sum of the two logarithms is the same as the logarithm of the product: $\log(A) + \log(B) = \mathrm{Log}(A \cdot B)$. Therefore, transforming a variable to its logarithm converts a multiplicative effect to an additive one (Appendix E reviews logarithms).

Mathematical model

This book is mostly nonmathematical. I avoid explaining the math of how methods work and only provide such explanation when it is necessary to understand the

	β_0
+	$\beta_1 \times \log(\text{serum lead})$
+	$\beta_2 \times \text{Age}$
+	$\beta_3 \times \text{Body mass index}$
+	$\beta_4 \times \log(\text{GGT})$
+	$\beta_5 \times \text{Diuretics?}$
+	ε (Gaussian random)
=	Creatinine clearance

Table 37.1. The multiple regression model for the example.

β_0 through β_5 are the six parameters fit by the multiple regression program. Each of these parameters has different units (listed in Table 37.2). The product of each parameter times the corresponding variable is expressed in the units of the Y variable (creatinine clearance), which are milliliters per minute (ml/min). The goal of multiple regression is to find the values for the six parameters of the model that make the predicted creatinine clearance values come as close as possible to the actual values.

questions the statistical methods answer. But you really can't understand multiple regression at all without understanding the model that is being fit to the data.

The multiple regression model is shown in Table 37.1. The dependent (Y) variable is creatinine clearance. The model predicts its value from a baseline value plus the effects of five independent (X) variables, each multiplied by a regression coefficient, also called a regression parameter.

The X variables were the logarithm of serum lead, age, body mass, logarithm of γ-glutamyl transpeptidase (a measure of liver function), and previous exposure to diuretics (coded as zero or 1). This last variable is designated as the dummy variable (or indicator variable) because those two particular values were chosen arbitrarily to designate two groups (people who have not taken diuretics and those who have).

The final component of the model, ε, represents random variability (error). Like ordinary linear regression, multiple regression assumes that the random scatter (individual variation unrelated to the independent variables) follows a Gaussian distribution.

The model of Table 37.1 can be written as an equation:

$$Y_i = \beta_0 + \beta_1 \cdot X_{i,1} + \beta_2 \cdot X_{i,2} + \beta_3 \cdot X_{i,3} + \beta_4 \cdot X_{i,4} + \beta_5 \cdot X_{i,5} + \varepsilon_i \qquad (37.1)$$

Notes on this equation:

- Y_i. The subscript i refers to the particular patient. So Y_3 is the predicted creatinine clearance of the third patient.
- $X_{i,j}$. The first subscript (i) refers to the particular participant/patient, and the second subscript enumerates a particular independent variable. For example, $X_{3,5}$ encodes the use of diuretics (the fifth variable in Table 37.2) for the third participant. These values are data that you collect and enter into the program.

VARIABLE	MEANING	UNITS
X_1	log(serum lead)	Logarithms are unitless. Untransformed serum lead concentration was in micrograms per liter.
X_2	Age	Years
X_3	Body mass index	Kilograms per square meter
X_4	log(GGT)	Logarithms are unitless. Untransformed serum GGT level was in units per liter.
X_5	Diuretics?	Unitless. 0 = never took diuretics. 1 = took diuretics.
Y	Creatinine clearance	Milliliters per minute

Table 37.2. Units of the variables used in the multiple regression examples.

- The regression coefficients (β_1 to β_5). These will be fit by the regression program. One of the goals of regression is to find the best-fit value of each regression coefficient, along with its CI. Each regression coefficient represents the average change in Y when you change the corresponding X value by 1.0. For example, β_5 is the average difference in the logarithm of creatinine between those who have diabetes ($X_{i,5} = 1$) and those who don't ($X_{i,5} = 0$).
- The intercept, β_0. This is the predicted average value of Y when all the X values are zero. In this example, the intercept has only a mathematical meaning, but no practical meaning because setting all the X values to zero means that you are looking at someone who is zero years old with zero weight! β_0 is fit by the regression program.
- ε. This is a random variable that is assumed to follow a Gaussian distribution. Predicting Y from all the X variables is not a perfect prediction because there also is a random component, designated by ε.

Goals of regression

Multiple regression fits the model to the data to find the values for the parameters that will make the predictions of the model come as close as possible to the actual data. Like simple linear regression, it does so by finding the values of the parameters (regression coefficients) in the model that minimize the sum of the squares of the discrepancies between the actual and predicted Y values. Like simple linear regression, multiple regression is a least-squares method.

Best-fit values of the parameters

The investigator's goal was to answer the question, After adjusting for effects of the other variables, is there a substantial linear relationship between the logarithm of lead concentration and creatinine clearance?

The best-fit value of β_1 (the parameter for the logarithm of lead concentration) was -9.5 ml/min. This means that, on average, and after accounting for

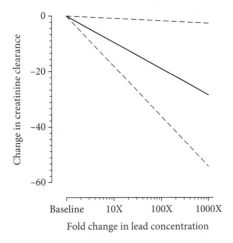

Figure 37.1. The prediction of multiple regression.
One of the variables entered into the multiple regression model was the logarithm of lead concentration. The best-fit value for its coefficient was –9.5. For every one-unit increase in the logarithm of lead concentration, the creatinine clearance is predicted to go down by 9.5 ml/min. When the logarithm of lead concentration increases by one unit, the lead concentration itself increases tenfold. The solid line shows the best-fit value of that slope. The two dashed lines show the range of the 95% CI.

differences in the other variables, an increase in log(lead) of one unit (so that the lead concentration increases tenfold, since they used common base 10 logarithms) is associated with a decrease in creatinine clearance of 9.5 ml/min. The 95% CI ranges from –18.1 to –0.9 ml/min.

Understanding these values requires some context. The study participants' average creatinine clearance was 99 ml/min. So a tenfold increase in lead concentrations is associated with a reduction of creatinine clearance (reduced renal function) of about 10%, with a 95% CI ranging from about 1% to 20%. Figure 37.1 illustrates this model.

The authors also report the values for all the other parameters in the model. For example, the β_5 coefficient for the X variable "previous diuretic therapy" was –8.8 ml/min. That variable is coded as zero if the patient had never taken diuretics and 1 if the patient had taken diuretics. It is easy to interpret the best-fit value. On average, after taking into account differences in the other variables, participants who had taken diuretics previously had a mean creatinine clearance that was 8.8 ml/min lower than that of participants who had not taken diuretics.

Statistical significance

Multiple regression programs can compute a P value for each parameter in the model testing the null hypothesis that the true value of that parameter is zero. Why zero? When a regression coefficient (parameter) equals zero, then the corresponding independent variable has no effect in the model (because the product of the independent variable times the coefficient always equals zero).

The authors of this paper did not include P values and reported only the best-fit values with their 95% CIs. We can figure out which ones are less than 0.05 by looking at the CIs.

The CI of β_1 runs from a negative number to another negative number and does not include zero. Therefore, you can be 95% confident that increasing lead concentration is associated with a drop in creatinine clearance (poorer kidney function). Since the 95% CI does not include the value defining the null hypothesis (zero), the P value must be less than 0.05. The authors don't quantify it more accurately than that, but most regression programs would report the exact P value.

R^2: How well does the model fit the data?

R^2 equals 0.27. This means that only 27% of the variability in creatinine clearance is explained by the model. The remaining 73% of the variability is explained by variables not included in this study, variables included in this study but not in the forms entered in the mode, and random variation.

With simple linear regression, you can see the best-fit line superimposed on the data and visualize goodness of fit. This is not possible with multiple regression. With two independent variables, you could visualize the fit on a three-dimensional graph, but most multiple regression models have more than two independent variables.

Figure 37.2 shows a way to visualize how well a multiple regression model fits the data and to understand the meaning of R^2. Each point represents one participant. The horizontal axis plots each participant's measured creatinine clearance (the Y variable in multiple regression). The vertical axis plots the creatinine clearance value predicted by the model. This prediction is computed from the other variables for that participant and the best-fit parameter values computed by multiple regression, but this calculation does not use the measured value of creatinine clearance. So the graph shows how well the model predicts the actual creatinine clearance. If the prediction were perfect, all the points would align on a 45-degree line with the predicted creatinine clearance matching the actual creatinine clearance. You can see that the predictions for the example data are far from perfect. The predicted and actual values are correlated, with R^2 equal to 0.27. By definition, this is identical to the overall R^2 computed by multiple regression.

R^2 versus adjusted R^2

R^2 is commonly used as a measure of goodness of fit in multiple linear regression, but it can be misleading. Even if the independent variables are completely unable to predict the dependent variable, R^2 will be greater than zero. The expected value of R^2 increases as more independent variables are added to the model. This limits the usefulness of R^2 as a way to quantify goodness of fit, especially with small sample sizes.

In addition to reporting R^2, which quantifies how well the model fits the data being analyzed, most programs also report an *adjusted R^2,* which estimates how well the model is expected to fit new data. This measure accounts for the number of independent variables and is always smaller than R^2. How much smaller depends on the relative numbers of participants and variables. This study

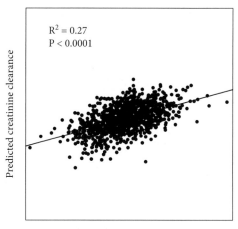

Actual creatinine clearance

Figure 37.2. The meaning of R^2 in multiple regression.

The authors did not post the raw data, so this graph does not accurately represent the data in the example. Instead, I simulated some data that look very much like what the actual data would have looked like. Each of the 965 points represents one man in the study. The horizontal axis shows the actual creatinine clearance for each participant. The vertical axis shows the creatinine clearance computed by the multiple regression model from that participant's lead level, age, body mass, log γ-glutamyl transpeptidase, and previous exposure to diuretics. The prediction is somewhat useful, because generally people who actually have higher creatinine clearance levels are predicted to have higher levels. However, there is a huge amount of scatter. If the model were perfect, each predicted value would be the same as the actual value, all the points would line up on a 45-degree line, and R^2 would equal 1.00. Here the predictions are less accurate, and R^2 is only 0.27.

has far more participants (965) than independent variables (5), so the adjusted R^2 is only a tiny bit smaller than the unadjusted R^2, and the two are equal to two decimal places (0.27).

ASSUMPTIONS

Assumption: Sampling from a population

This is a familiar assumption of all statistical analyses. The goal in all forms of multiple regression is to analyze a sample of data to make more general conclusions (or predictions) about the population from which the data were sampled.

Assumption: Linear effects only

The multiple regression model assumes that increasing an X variable by one unit increases (or decreases) the value of Y (multiple regression) by the same amount, regardless of the values of the X variables.

In the multiple regression (lead exposure) example, the model predicts that (on average) creatinine clearance will decrease by a certain amount when the logarithm of lead concentration increases by 1.0 (which means the lead concentration increases tenfold, since the common or base 10 logarithm of 10 is 1). The assumption of linear effects means that increasing the logarithm of lead concentration by 2.0 will have twice the impact on creatinine clearance as increasing it by 1.0 and that the decrease in creatinine clearance by a certain concentration of lead does not depend on the values of the other variables.

Assumption: No interaction beyond what is specified in the model

Two of the independent variables in the multiple linear regression model are the logarithm of lead concentration and age. What if lead has a bigger impact on kidney function in older people than it does in younger people? If this were the case, the relationship between the effects of lead concentration and the effects of age would be called an *interaction*. Standard multiple regression models assume there is no interaction. It is possible to include interaction terms in the model, and this process is explained later in the chapter.

Assumption: Independent observations

The assumption that data for each participant provide independent information about the relationships among the variables should be familiar by now. This assumption could be violated, for example, if some of the participants were identical twins (or even siblings).

Assumption: The random component of the model is Gaussian

For any set of X values, multiple linear regression assumes that the random component of the model follows a Gaussian distribution, at least approximately. In other words, it assumes that the residuals (the differences between the actual Y values and Y values predicted by the model) are sampled from a Gaussian population. Furthermore, it assumes that the SD of that scatter is always the same and is unrelated to any of the variables.

AUTOMATIC VARIABLE SELECTION

How variable selection works

The authors of our example stated that they collected data for more variables for each participant and that the fit of the model was not improved when the model also accounted for smoking habits, mean blood pressure, serum ferritin level (a measure of iron storage), residence in urban versus rural areas, or urinary cadmium levels. Consequently, they omitted these variables from the model whose fit they reported. In other words, they computed a P value for each independent variable in the model, removed variables for which P values were greater than 0.05, and then reran the model without those variables.

What is automatic variable selection?

Many multiple regression programs can choose variables automatically. One approach (called *all-subsets regression*) is to fit the data to every possible model (recall that each model includes some X variables and may exclude others) and then find the one that is the best. With many variables and large data sets, the computer time required for this approach is prohibitive. To conserve computer time when working with huge data sets, other algorithms use a stepwise approach. One approach (called *forward-stepwise selection* or a *step-up procedure*) is to start with a very simple model and add new X variables one at a time, always adding the X variable that most improves the model's ability to predict Y. Another approach (*backward-stepwise selection* or a *step-down procedure*) is to start with the full model (including all X variables) and then sequentially eliminate those X variables that contribute the least to the model.

The problems of automatic variable selection

The appeal of automatic variable selection is clear. You just put all the data into the program, and it makes all the decisions for you. The problem is multiple comparisons. How many models does a multiple regression program compare when given data with k independent variables and instructed to use the all-subsets method to compare the fit of every possible model? Each variable can be included or excluded from the final model, so the program will compare 2^k models. For example, if the investigator starts with 20 variables, then automatic variable selection compares 2^{20} models (more than a million), even before considering interactions.

When you read a paper presenting results of multiple regression, you may not even know the number of variables with which the investigator started. Peter Flom (2007) explains why this ignorance makes it impossible to interpret the results of multiple regression with stepwise variable selection:

> If you toss a coin ten times and get ten heads, then you are pretty sure that something weird is going on. You can quantify exactly how unlikely such an event is, given that the probability of heads on any one toss is 0.5. If you have 10 people each toss a coin ten times, and one of them gets 10 heads, you are less suspicious, but you can still quantify the likelihood. But if you have a bunch of friends (you don't count them) toss coins some number of times (they don't tell you how many) and someone gets 10 heads in a row, you don't even know how suspicious to be. That's stepwise.

The consequences of automatic variable selection are pervasive and serious (Harrell 2015; Flom 2007):

- The final model fits too well. R^2 is too high.
- The best-fit parameter values are too far from zero. This makes sense. Since variables with low absolute values have been eliminated, the remaining variables tend to have absolute values that are higher than they should be.
- The CIs are too narrow, so you think you know the parameter values with more precision than is warranted.
- When you test whether the parameters are statistically significant, the P values are too low, so cannot be interpreted.

Simulated example of variable selection

Chapter 23 already referred to a simulated study by Freedman (1983) that demonstrated the problem with this approach (his paper was also reprinted as an appendix in a text by Good and Hardin, 2006). He simulated a study with 100 participants and recorded data from 50 independent variables for each. All values were simulated, so it is clear that the outcome is not associated with any of the simulated independent variables. As expected, the overall P value from multiple regression was high, as were most of the individual P values (one for each independent variable).

He then chose the 15 independent variables that had the lowest P values (less than 0.25) and reran the multiple regression program using only those variables. The resulting overall P value from multiple regression was tiny (0.0005). The contributions of 6 of the 15 independent variables were statistically significant ($P < 0.05$).

If you didn't know these were all simulated data with no associations, the results might seem impressive. The tiny P values beg you to reject the null hypotheses and conclude that the independent variables can predict the dependent variable.

The problem is essentially one of multiple comparisons, an issue already discussed in Chapter 23. With lots of variables, it is way too easy to be fooled. You can be impressed with high R^2 values and low P values, even though there are no real underlying relationships in the population.

Should you mistrust all analyses with variable selection?

It may make sense to use statistical methods to decide whether to include or exclude one or a few carefully selected independent variables. But it does not make sense to let a statistics program test dozens or hundreds (or thousands) of possible models in hope that the computer can work magic.

When reading the results of research that depends on multiple regression, read the paper carefully to see how many variables were collected but not used in the final model. If a lot of variables were collected but not ultimately used, be skeptical of the results. In some cases, there is no problem, as the extra variables were collected for other purposes and the variables used in the model were selected based on a plan created according to the goal of the study. But watch out if the investigators first included many variables in the model but report results with variables that were selected because they worked the best.

In some cases, the goal of the study is exploration. The investigators are not testing a hypothesis but rather are looking for a hypothesis to test. Variable selection can be part of the exploration. But any model that emerges from an exploratory study must be considered a hypothesis to be tested with new data. Deciding how to construct models is a difficult problem. Says Gelman (2012), "This is a big open problem in statistics: how to make one's model general enough to let the data speak, but structured enough to hear what the data have to say."

SAMPLE SIZE FOR MULTIPLE REGRESSION

How many participants are needed to perform a useful multiple regression analysis? It depends on the goals of the study and your assumptions about the distributions of the variables. Chapter 26 explained the general principles of sample size determination, but applying these principles to a multiple regression analysis is not straightforward.

One approach to determining sample size is to follow a rule of thumb. The only firm rule is that you need more cases than variables, a lot more. Beyond that, rules of thumb specify that the number of participants (n) should be somewhere between 10 and 20 (or even 40) times the number of variables. You must count the number of variables you have at the start of the analysis. Even if you plan to use automatic variable selection to reduce the number of variables in the final model, you can't use that lower number as the basis to calculate a smaller sample size.

The rules of thumb are inconsistent because they don't require you to state the goal of the study. More sophisticated approaches to calculating sample size for multiple regression require you to specify your goal (Kelley & Maxwell, 2008). Is your goal to test the null hypothesis that the true R^2 is zero? Is it to test the null hypothesis that one particular regression coefficient equals zero? Is it to determine particular regression coefficients with a margin of error smaller than a specified quantity? Is it to predict future points within a specified margin of error? Once you have articulated your goal, along with corresponding α and power as explained in Chapter 26, you can then use specialized software to compute the required sample size.

MORE ADVANCED ISSUES WITH MULTIPLE REGRESSION

Multicollinearity

When two X variables are highly correlated, they both convey essentially the same information. For example, one of the variables included in the multiple regression example of Chapter 37 is body mass index (BMI), which is computed from an individual's weight and height. If the investigators had entered weight and height separately into the model, they probably would have encountered *collinearity*, because people who are taller also tend to be heavier.

The problem is that neither variable adds much to the fit of the model after the other one is included. If you removed either height or weight from the model, the fit probably would not change much. But if you removed both height and weight from the model, the fit would be much worse.

The multiple regression calculations assess the *additional* contribution of each variable after accounting for all the other independent variables. When variables are collinear, each variable makes little individual contribution. Therefore, the CIs for the corresponding regression coefficients are wider and the P values for those parameters are larger.

One way to reduce collinearity (or *multicollinearity*, when three or more variables are entangled) is to avoid entering related or correlated independent variables into your model. An alternative is the approach used in the example study. In this case, the researchers combined height and weight in a biologically sensible way into a single variable (BMI).

Interactions among independent variables

Two of the independent variables in the multiple linear regression model are the logarithm of lead concentration and age. What if the effects of lead concentration matter more with older people? This kind of relationship, as mentioned earlier, is called an *interaction*.

NAME	EXAMPLE
Longitudinal or repeated measures	Multiple observations of the same participant at different times.
Crossover	Each participant first gets one treatment, then another.
Bilateral	Measuring from both knees in study of arthritis or both ears in a study of tinnitus, and entering the two measurements into the regression separately.
Cluster	Study pools results from three cities. Patients from the same city are more similar to each other than they are to patients from another city.
Hierarchical	A clinical study of a surgical procedure uses patients from three different medical centers. Within each center, several different surgeons may do the procedure. For each patient, results might be collected at several time points.

Table 37.3. Study designs that violate the assumption of independence.

To include interaction between age and the logarithm of serum lead concentration, add a new term to the model equation with a new parameter multiplied by the product of age (X_2) times the logarithm of lead (X_1):

$$Y = \beta_0 + \beta_1 \cdot X_1 + \beta_2 \cdot X_2 + \beta_3 \cdot X_3 + \beta_4 \cdot X_4 + \beta_5 \cdot X_5 + \beta_{1,2} \cdot X_1 \cdot X_2 + \varepsilon$$

If the CI for the new parameter ($\beta_{1,2}$) does not include zero, then you will conclude that there is a significant interaction between age and log(lead). This means that the effects of lead change with age. Equivalently, the effect of age depends on the lead concentrations.

Correlated observations

One of the assumptions of multiple regression is that each observation is independent. In other words, the deviation from the prediction of the model is entirely random. Table 37.3 (adapted from Katz, 2006) is a partial list of study designs that violate this assumption and so require specialized analysis methods.

COMMON MISTAKES: MULTIPLE REGRESSION

Mistake: Comparing too many model variations

In addition to choosing which variables to include or exclude from the model (explained earlier in the chapter), multiple regression analyses also let investigators choose among many variations of the same model:

- With and without interaction terms
- With and without transforming some independent variables
- With and without transforming the dependent variable
- Including and excluding outliers or influential points
- Pooling all the data or analyzing some subsets separately
- Rerunning the model after defining a different variable to be the dependent (outcome) variable

When you fit the data many different ways and then report only the model that fits the data best, you are likely to come up with conclusions that are not valid. This is essentially the same problem as choosing which variables to include in the model, previously discussed.

Mistake: Preselecting variables to include in the model

To decide which variables to enter into a multiple regression program, some investigators first look at the correlation between each possible independent variable and the outcome variable and then only enter the variables that are strongly correlated with the outcome. Selecting variables this way is not so different than letting a program select variables. The results will be affected the same way, with an R^2 that is too high, CIs that are too narrow and P values that are too low.

Mistake: Too many independent variables

The goal of regression, as in all of statistics, is to analyze data from a sample and make valid inferences about the overall population. That goal cannot always be met using multiple regression techniques. It is too easy to reach conclusions that apply to the fit of the sample data but are not really true in the population. When the study is repeated, the conclusions will not be reproducible.

This problem is called *overfitting* (Babyak, 2004). It happens when you ask more questions than the data can answer—when you have too many independent variables in the model compared to the number of participants. The problem with overfitting is that the modeling process ends up fitting aspects of the data caused by random scatter or experimental quirks. Such a model won't be reproduced when tested on a new set of data.

How many independent variables is too many? For multiple regression, a rule of thumb is to have at least 10 to 20 participants per independent variable. Fitting a model with five independent variables thus requires at least 50 to 100 participants.

Mistake: Including redundant or highly correlated independent variables

The problem of multicollinearity was discussed earlier in the chapter. This problem is most severe when two variables in a model are redundant or highly correlated. Say your study includes both men and women so you have one independent variable "Woman" that equals 1 for females and zero for males, and another variable "Man" that equals zero for females and 1 for males. You've introduced collinearity because the two variables encode the same information. Only one variable is needed. An example of correlated variables would be including both weight and height in a model, as people who are taller also tend to weigh more. One way around this issue would be to compute the BMI from height and weight and only include that single variable in the mode, rather than including both height and weight.

Mistake: Dichotomizing without reason

A common mistake is to convert a continuous variable to a binary one without good reason. For example, if you want to include blood pressure in the model, you have two choices: include blood pressure itself in the model or dichotomize blood pressure to a binary variable that encodes whether the patient has hypertension (high blood pressure) or not. One problem with the latter approach is that it requires deciding on a somewhat arbitrary definition of whether someone has hypertension. Another problem is that it treats patients with mild and severe hypertension as the same, and so information is lost.

Mistake: Assuming that regression proves causation

In the example from the beginning of this chapter, the investigators chose to place creatinine clearance (a measure of kidney function) on the left side of the model (Y) and the concentration of lead on the right (X) and therefore concluded that lead affects kidney function. It is conceivable that the relationship goes the opposite direction, and that damaged kidneys somehow accumulate more lead. This is a fundamental problem of observational studies. The best way to overcome any doubts about cause and effect is to do an experiment. While it wouldn't be ethical to expose people to lead to see what would happen to their renal function, such experiments can be done with animals.

Also beware of confounding variables. An absurd example mentioned by Katz (2006) makes this concept clear. A multiple regression model to find risk factors associated with lung cancer might identify carrying matches as a significant risk factor. This would prove that people who carry matches in their pocket are more likely to get lung cancer than people who don't. But, of course, carrying matches doesn't cause cancer. Rather, people who carry matches are also likely to smoke, and that does cause cancer.

Mistake: Not thinking about the model

It is too easy to run a multiple regression program without thinking about the model. Does the assumption of linearity make sense? Does it make sense to assume that the variables don't interact? You probably know a lot about the scientific context of the work. Don't forget all that when you pick a standard multiple regression model.

Q & A

Do you always have to decide which variable is the outcome (dependent variable) and which variables are the predictors (independent variables) at the time of data collection?

> No. In some cases, the independent and dependent variables may not be distinct at the time of data collection. The decision is sometimes made only at the time of data analysis. But beware of these analyses. The more ways you analyze the data, the more likely you are to be fooled by overfitting and multiple comparisons.

Does it make sense to compare the value of one best-fit parameter with another?

> No. The units of each parameter are different, so they can't be directly compared. If you want to compare, read about standardized parameters in a more advanced book. Standardizing rescales the parameters so they become unitless and can then be compared. A variable with a larger standardized parameter has a more important impact on the dependent variable.

Do all the independent variables have to be expressed in the same units?

No. Usually they are in different units.

What possible values can R^2 and adjusted R^2 have?

Like in linear and nonlinear regression, R^2 is the fraction of variation explained by the model, so it must have a value between 0.0 and 1.0. The adjusted R^2, despite its name, is not the square of anything. It is the R^2 minus an adjustment factor based on the number of variables and sample size. In rare circumstances (poor fit with small sample size), this adjustment can be larger than R^2, so the adjusted R^2 can be negative.

In what units are the parameters expressed?

Each parameter is expressed in the Y units divided by the X units of the variable associated with that parameter.

How is regression related to other statistical methods?

Chapter 35 pointed out that an unpaired t test can be recast as linear regression. Similarly, ANOVA can be recast as multiple regression. It is also possible to use logistic or proportional hazards regression to compare two groups, replacing the methods explained in Chapters 27 through 29 (the results won't be exactly the same, but they should be close). Essentially all of statistics can be recast as a form of fitting some kind of model using an appropriate kind of regression.

I fit a model and the P value associated with of the variables is small enough so that variable is considered to have a statistically significant effect. Now I include one more variable in the model. Is it possible that that variable will no longer have a statistically significant effect?

Yes. Each P value considers the effect of one variable, accounting for all the others. Adding a new variable will change the results for all the variables.

What if a variable has a nonsignificant effect in the first model. Now I add another variable to the model. Is it possible that the first variable will now have a statistically significant effect?

Yes.

How does ANCOVA fit in?

This book does not discuss ANCOVA. It is a model that is equivalent to multiple linear regression when at least one independent variable is categorical and at least one is continuous.

How are the CIs calculated?

Some programs report the standard error of each parameter instead of its CI. Computing the CI of the best-fit value of a model parameter works just like computing the CI of the mean if you know the SEM (see Chapter 12). Compute the margin of error by multiplying the reported standard error by a value obtained from the t distribution (see Appendix D). This value depends only on the level of confidence desired (95% is standard) and the number of df (equal to the number of participants in the study minus the number of parameters fit by the model). For 95% confidence and with plenty of df (common in multiple regression), the multiplier approximates 2.0. Add and subtract this margin of error from the best-fit value to obtain the CI.

Where can I learn more about multiple regression and related analyses?

The texts by Katz (2006) and Campbell (2006) are concise, clear, practical, and nonmathematical. The books by Glantz, Slinker, and Neilands (2016) and Vittinghoff and colleagues (2007) have more depth and more math, but they remain clear and practical.

CHAPTER SUMMARY

- Multiple variable regression is used when the outcome you measure is affected by several other variables.
- This approach is used when you want to assess the impact of one variable after correcting for the influences of others to predict outcomes from several variables or to try to tease apart complicated relationships among variables.
- Multiple linear regression is used when the outcome (Y) variable is continuous. Chapter 38 explains methods used when the outcome is binary.
- Beware of the term *multivariate*, which is used inconsistently.
- Automatic variable selection is appealing but the results can be misleading. It is a form of multiple comparisons.

TERMS INTRODUCED IN THIS CHAPTER

- Adjusted R^2 (p. 384)
- All-subsets regression (p. 386)
- Automatic variable selection (p. 387)
- Backwards-stepwise selection (or step-down procedure) (p. 387)
- Collinearity (p. 389)
- Confounding variable (p. 380)
- Dichotomize (p. 391)
- Dummy variable (p. 379)
- Forward-stepwise selection (or step-up procedure) (p. 387)
- Interaction (p. 386)
- Multicollinearity (p. 389)
- Multiple linear regression (p. 379)
- Multivariable/multiple regression (p. 379)
- Multivariate methods (p. 380)
- Overfitting (p. 391)
- Regression coefficients (p. 379)
- Simple linear regression (p. 379)
- Univariate methods (p. 380)

CHAPTER 38

Logistic and Proportional Hazards Regression

The plural of anecdote is not data.

ROGER BRINNER

This chapter briefly explains regression methods used when the dependent variable is not continuous. Logistic regression is used when the dependent variable is binary (i.e., has two possible values). Proportional hazards regression is used when the dependent variable is survival time. Read Chapter 37 (multiple regression) before reading this one.

LOGISTIC REGRESSION

When is logistic regression used?

As was discussed in Chapter 37, methods that fit regression models with two or more independent variables are called *multiple regression methods*. Multiple regression is really a family of methods, with the specific type of regression used depending on what kind of outcome is being measured (see Table 38.1). *Logistic regression* is used when the outcome (dependent variable) has two possible values. Because it is almost always used with two or more independent variables, it should be called *multiple logistic regression*. However, the word *multiple* is often omitted but assumed.

Study design of example study

Bakhshi and colleagues (2008) studied factors that influence the prevalence of obesity in Iranian women. They collected data from 14,176 women. For each woman, they recorded height and weight and computed BMI. When the BMI exceeded a threshold, that person was defined to be obese. In this way, every woman was defined to be obese or not obese. Since there are two categories, these data are appropriately analyzed by logistic regression.

The researchers also collected data on the variables listed in Table 38.2: urban versus rural, age, years of education, married or not, and economic class. The only thing confusing about Table 38.2 is the encoding of economic class. They defined four economic classes (lower, lower middle, upper middle, higher) and encoded them using three binary (also called *dummy*) variables. The first variable encodes

TYPE OF REGRESSION	WHERE?	TYPE OF DEPENDENT (Y) VARIABLE	EXAMPLE OF Y VARIABLES
Linear	Chapters 33 and 37	Continuous (interval or ratio)	• Enzyme activity • Renal function (creatinine clearance) • Weight
Logistic	Chapter 38	Binary or dichotomous	• Death during surgery • Graduation
Proportional hazards	Chapter 38	Elapsed time to a one-time event	• Recurrence of cancer • Months until death • Days until patient is weaned from ventilator • Quarters in school before graduation

Table 38.1. Different kinds of regression for different kinds of outcomes.

VARIABLE	MEANING	UNITS
X_1	Place of residence	Rural = 0; urban = 1
X_2	Age	Years
X_3	Years of education	Years
X_4	Smoking	No = 0; yes = 1
X_5	Married	No = 0; yes = 1
X_6	Lower-middle economic index	No = 0; yes = 1
X_7	Upper-middle economic index	No = 0; yes = 1
X_8	Higher economic index	No = 0; yes = 1

Table 38.2. Variables used in the logistic regression example.

whether a person belongs to lower middle class (1 = yes; 0 = no). The second variable encodes membership in the upper middle class, and the third variable encodes membership in the higher economic class. Lower-class status is encoded by setting all three of those variables to zero. This is one of several ways to encode categorical variables with more than two possible outcomes.

The goal of logistic regression

The goal of logistic regression for this study is to create an equation that can be used to estimate the probability that someone will be obese as a function of the independent variables.

Review: Odds versus probabilities

Logistic regression actually works with odds rather than probability. Chapter 2 explained how the two are related. The key point is that odds and probability are two alternative ways to express precisely the same concept. The odds of an event occurring equals the probability that that event will occur divided by the probability that it will not occur. Every probability can be expressed as odds. Every odds can be expressed as a probability.

The model: Multiplying odds ratios

The logistic regression model computes the odds that someone is obese (in this example) from baseline odds and from odds ratios computed for each independent variable. Conceptually, the model looks like this:

$$\text{Odds} = (\text{Baseline Odds}) \cdot OR_1 \cdot OR_2 \cdot OR_3 \cdot OR_4 \cdot OR_5 \cdot OR_6 \cdot OR_7 \cdot OR_8$$

$$(38.1)$$

The baseline odds answers this question, If every single independent X variable equaled zero, what are the odds of obesity? For this example, this baseline odds value is not really useful on its own, because one of the variables is age. So the baseline odds, interpreted literally, would be the odds of obesity for someone who is zero years old. Sometimes investigators define the X variable differently to make the baseline odds more meaningful. For this example, they could have encoded that variable as Age – 20, so X = 0 would encode people who are 20 years old, and X = 1 would encode people who are 21 years old, and so on.

Each of the other odds ratios correspond to one independent variable. The first independent variable is for place of residence, coded as zero for rural and 1 for urban. The corresponding odds ratio answers the question, how much does knowing whether a particular person is urban or rural change the odds of obesity? If $X_1 = 0$, the person in question lives in a rural area and the baseline odds do not change. If $X_1 = 1$, the person in question lives in an urban area and the odds ratio estimated by logistic regression will tell us if the urban residents have the same odds of being obese as the rural residents (OR = 1), if the odds are greater (OR > 1), or if the odds are smaller (OR < 1). The goal of logistic regression is to estimate a value for that odds ratio and all the others.

The next independent variable (X_2) is age, which is a continuous variable. The corresponding odds ratio (OR_2) answers this question: for each additional year of age, by how much do the odds of obesity increase or decrease? If the OR associated with age equals 1.0, then age is not related to obesity, since multiplying the baseline odds by 1.0 does not change its value. If the OR > 1, then the odds of obesity increase by a set percentage for each additional year of age. For instance, an OR of 1.03 would mean that the odds of obesity increase by 3% as a person grows older by one year. Conversely, an OR < 1 indicates a decline in the odds of obesity associated with each additional year of age.

How logistic regression works

The goal of logistic regression is to find values for all the coefficients in an equation that estimates the odds that someone will have the outcome being tracked as a function of all the independent variables. However logistic regression equations do not use Equation 38.1. Instead, the equation is transformed so it is almost identical to Equation 37.1, used for multiple regression.

$$Y_i = \beta_0 + \beta_1 \cdot X_{i,1} + \beta_2 \cdot X_{i,2} + \beta_3 \cdot X_{i,3} + \beta_4 \cdot X_{i,4} + \beta_5 \cdot X_{i,5}$$
$$+ \beta_6 \cdot X_{i,6} + \beta_7 \cdot X_{i,7} + \beta_8 \cdot X_{i,8} \tag{38.2}$$

The equation is easy to understand:

- Y_i is the natural logarithm of the odds for a particular participant. Logistic regression is based on logarithms, because the logarithm of the odds can take on any value, either negative or positive, whereas odds must be zero or positive, and the probability itself must be between zero and 1.
- $X_{i,j}$. The first subscript (i) refers to the particular participant/patient, and the second subscript enumerates a particular independent variable. For example, $X_{3,5}$ encodes whether the third patient is married (the fifth variable in Table 38.2). These values are data that you collect and enter into the program.
- β_0 is the natural logarithm of the baseline odds.
- β_1 is the natural logarithm of the odds ratio for the first independent variable.
- β_2 is the natural logarithm of the odds ratio for the second independent variable.
- β_3 to β_8 are the natural logarithms of the odds ratios for the corresponding independent variables.

Equation 37.1 for multiple linear regression ended with a term ε that added Gaussian scatter to the result. The Y value in Equation 38.2 is the natural logarithm of odds, which can be transformed to a probability. Since it implicitly embodies uncertainty, there is no need to explicitly add a random term to the model.

Because there is no Gaussian distribution, the method of least squares is not used to determine the values of the odds ratios. Instead, logistic regression finds the values of the odds ratios using what is called a *maximum likelihood method* (briefly explained in Chapter 34).

Interpreting the results: Odds ratios and CIs

The key results of logistic regression are shown in Table 38.3. For each independent variable, logistic regression reports an odds ratio, along with a 95% CI.

SYMBOL	VARIABLE	ODDS RATIO	95% CI
OR_1	Urban?	2.13	1.915 to 2.369
OR_2	Age	1.02	1.017 to 1.026
OR_3	Years of education	0.98	0.968 to 0.993
OR_4	Smoking?	0.65	0.468 to 0.916
OR_5	Married?	1.48	1.312 to 1.668
OR_6	Lower middle class?	1.37	1.206 to 1.554
OR_7	Upper middle class?	1.29	1.136 to 1.468
OR_8	Higher class?	1.25	1.094 to 1.425

Table 38.3. Results of logistic regression.

The first independent variable has two possible values (rural = 0; urban = 1) and so is a binary, or dummy, variable. The corresponding odds ratio is 1.0 for rural women and 2.13 for urban women. This means that a woman who lives in a city has a bit more than twice the odds of being obese as someone who lives in a rural area but shares other attributes (age, education, married or not, social class). The 95% CI ranges from 1.9 to 2.4.

The next independent variable is age. The odds ratio for this variable is 1.02. Every year, the odds of obesity goes up about 2%. Don't forget that odds ratios multiply. The odds ratio for a 40-year-old compared to a 25-year-old is 1.02^{15}, or 1.35. The exponent is 15 because $40 - 25 = 15$. That means a 40-year-old person has about 35% greater odds of being obese than a 25-year-old.

The next independent variable is education. The odds ratio is 0.98. Since that is less than 1.0, having more education (in this population) makes one less slightly likely to be obese. For each additional year of formal education, the odds of being obese decline by 2% (since we multiply by the OR of 0.98).

Interpreting the results: P values for each independent variable

The investigators in this study did not report any P values, because the CIs were sufficient to clearly present their findings. However, it is easy to learn something about the P values by looking at the CIs of the odds ratios. Remember that if a null hypothesis were true, the corresponding odds ratio would equal 1.0. None of the eight CIs in Table 38.3 include 1.0, so all of the corresponding P values must be less than 0.05. Each P value would answer the question, Assuming the model is correct and the population odds ratio (for this particular X variable) equals 1.0, what is the chance that random sampling would lead to an odds ratio as far (or farther) from 1.0 as that observed in this study?

Interpreting the results: Relationship between CIs and P values

The relationship between CIs and P values should be familiar by now. If the 95% CI of the odds ratio does not contain 1.0 (the value that denotes no effect), then the P value will be less than 0.05. If the 95% CI of the odds ratio does contain 1.0, then the P value must be greater than 0.05.

PROPORTIONAL HAZARDS REGRESSION

Proportional hazards regression is similar to logistic regression but is used when the outcome is elapsed time to an event and is often used for analyses of survival times. To understand the brief following discussion, first read the discussions about analyzing survival data in Chapters 4 and 29.

An example

Rosman and colleagues (1993) investigated whether diazepam (also known as Valium) would prevent febrile seizures in children. They recruited about 400 children who had had at least one febrile seizure. Their parents were instructed to give medication to the children whenever they had a fever. Half were given diazepam and half were given a placebo (because there was no standard therapy).

The researchers asked whether treatment with diazepam would delay the time of first seizure. Their first analysis was simply to compare the two survival curves with a log-rank test (see Chapter 29). The P value was 0.064, so the difference between the two survival curves was not considered to be statistically significant. They did not report the hazard ratio for this analysis.

Next, the researchers did a more sophisticated analysis that adjusted for differences in age, number of previous febrile seizures, the interval between the last seizure and entry into the study, and developmental problems. The proportional hazards regression, which accounts for all such differences among kids in the study, found that the relative risk was 0.61, with a 95% CI extending from 0.39 to 0.94. This means that at any time during the two-year study period, kids treated with diazepam had only 61% of the risk of having a febrile seizure compared with those treated with a placebo. This reduction was statistically significant, with a P value of 0.027. If diazepam were truly ineffective, there would have been only a 2.7% chance of seeing such a low relative risk in a study of this size. This example shows that the results of proportional hazards regression are easy to interpret, although the details of the analysis are complicated.

The proportional hazards model

Note some confusing lingo. The method used to analyze these data is given the name *survival analysis*, although the event being tracked is time to a febrile seizure, not death. Most descriptions of proportional hazards regression assume that the event is death, but the same principles apply to studying time to any event—in this example, time to first seizure.

The slope of the survival curve, called the *hazard*, is the rate of death (or whatever event is being tracked) in a short time interval. For example, if 20% of patients with a certain kind of cancer are expected to die in the first year after diagnosis, then the average hazard during that time is 20% per year. When comparing two groups, investigators often assume that the *ratio* of hazard functions is constant over time. For example, the hazard among treated patients might be one-half the hazard among control patients. The hazards (death rates) may change over the course of the study, but at any particular time, the treated patients' risk of dying is one-half the risk of the control patients. Another way to say this is that the two hazard functions are proportional to one another—hence the name *proportional hazards regression*. If the assumption is true, then the difference between two survival curves can be quantified by a single number, a relative risk. If the ratio is 0.5, then the relative risk of dying in one group is half the risk of dying in the other group.

Proportional hazards regression, which is also called *Cox regression* after the person who developed the method, uses regression methods to fit the relative risk associated with each independent variable, along with a CI and a P value testing the null hypothesis that the population relative risk is 1.0.

ASSUMPTIONS: LOGISTIC REGRESSION

Assumption: Sampling from a population

This is a familiar assumption of all statistical analyses. The goal in all forms of multiple regression is to analyze a sample of data to make more general conclusions (or predictions) about the population from which the data were sampled.

Assumption: Linear effects only

The multiple regression model assumes that increasing an X variable by one unit increases (or decreases) the value of Y (the logarithm of odds) by the same amount, regardless of the values of the X variables.

Assumption: No interaction beyond what is specified in the model

It is possible to include interaction terms in the model, and this process was briefly explained in Chapter 37.

Assumption: Independent observations

The assumption that data for each participant provide independent information about the relationships among the variables should be familiar by now. This assumption could be violated, for example, if some of the participants were identical twins (or even siblings).

Assumption: The random component of the model is binomial

For any set of X values, multiple linear regression assumes that the random component of the model follows a binomial distribution, at least approximately. All participants with the same set of X values share the same probabilities of having each of the two possible outcomes.

COMMON MISTAKES: LOGISTIC REGRESSION

Most common mistakes: Same as listed in Chapter 37

The list of common mistakes for logistic and proportional hazards regression are the same as those for multiple linear regression (see Chapter 37):

- Comparing too many model variations
- Preselecting variables to include in the model
- Too many independent variables
- Dichotomizing independent variables without reason
- Assuming that regression proves causation
- Not thinking about the model

Mistake: Converting from logarithms incorrectly

By tradition, the logistic regression model is written using natural logarithms. If you want to convert from log(odds ratio) to odds ratio, you therefore need to use the exponential function. In Excel, this is the exp() function. Most calculators are labeled the same way. Don't take 10 to the power of the log(odds ratios). That would work if the logarithms were common (base 10) logs, but logistic regression always uses natural logs.

Mistake: Mixing up odds ratios for a particular participant and odds ratios for an independent variable

The Y variable in the logistic regression equation (Equation 38.2) is the natural logarithm of the odds that a participant with a defined set of independent (X) variables will have the outcome being tabulated. Each parameter on the right side of the equation is the odds ratio for that parameter. If you increase that X value by 1.0, it is the amount by which Y (log odds for that participant) increases. Don't confuse the odds ratio for a participant with the odds ratio for a particular independent variable.

Q & A

As its name suggests, proportional hazards regression assumes proportional hazards. Is this assumption always reasonable?
> No. For example, you would not expect the hazard functions of medical and surgical therapy for cancer to be proportional. You might expect surgical therapy to have the higher hazard at early times (because of deaths during the operation or soon thereafter) and medical therapy to have the higher hazard at later times. In such situations, proportional hazards regression should be avoided or used only over restricted time intervals for which the assumption is reasonable.

In what units are the parameters expressed?
> Each parameter is expressed as an odds ratio.

Do all the independent variables have to be expressed in the same units?
> No. Usually they are in different units.

How does one decide how many subjects one needs for a study to be analyzed by logistic regression?
> Chapter 37 explained the principles of figuring out the necessary sample size for data to be analyzed by multiple regression. The principles are the same for data to be analyzed by logistic or proportional hazards regression, with one big difference: the calculated value is not the number of participants, but the number of participants with the rarer of the two possible outcomes being monitored.
>
> For example, imagine that you have five independent variables and decide to use the rule of thumb for multiple regression that you need 15 times more participants per treatment group than independent variables. For multiple regression, you'd need 75 participants per group. But for logistic regression, you would need to go a step further. Say you are planning a study for which the outcome is whether the patient had a stroke (in a defined period of time). If you expect about 10% of the participants to have strokes, you will need 750 participants per group so you can expect 75 strokes per group. This would also be the case if you expected 90% of the patients to have strokes, as the calculation must be based on the rarer of the two outcomes.

Does R² quantify goodness-of-fit in logistic and proportional hazards regression?

No. There are several other ways to quantify goodness of fit. Read a more advanced book for details.

Does the discussion on variable selection in multiple regression in Chapter 37 also apply to logistic and proportional hazards regression?

Yes. The problems of automatic variable selection apply to all kinds of regression with multiple independent variables.

How about the discussions in Chapter 37 about collinearity, interactions, and correlated observations?

Those discussions apply to all regressions with multiple independent variables.

CHAPTER SUMMARY

- Logistic regression is used when the outcome is binary (has two possible values).
- Proportional hazards regression is used when the outcome is survival time.
- These methods are used for the same purposes for which multiple regression is used: to assess the impact of one variable after correcting for influences of others, to predict outcomes from several variables, or to try to tease apart complicated relationships among variables.
- The assumptions and traps of logistic and proportional hazards regression are about the same as those for multiple regression.

TERMS INTRODUCED IN THIS CHAPTER

- Logistic regression (or multiple logistic regression) (p. 395)
- Proportional hazards regression (or Cox regression) (p. 399)

PART H

The Rest of Statistics

CHAPTER 39

Analysis of Variance

Let my dataset change your mindset.

HANS ROSLIN

One-way ANOVA compares the means of three or more groups, as-suming that all values are sampled from Gaussian populations. It is called one-way ANOVA, because the groups are categorized by one scheme (say treatment). Two-way ANOVA is used when the data are categorized by two schemes (say treatment and gender). This chapter explains the idea of ANOVA, the specifics of one-way ANOVA, and an example of two-way ANOVA. Chapter 40 will explain the follow-up multiple comparisons tests.

COMPARING THE MEANS OF THREE OR MORE GROUPS

Example data

Hetland and coworkers (1993) were interested in hormonal changes in women runners. Among their investigations, they measured the level of luteinizing hor-mone (LH) in nonrunners, recreational runners, and elite runners. Because the distribution of hormone levels appeared to be lognormal (see Chapter 11), the investigators transformed all the values to the logarithm of concentration and per-formed all analyses on the transformed data. This was a smart decision, because it made the values actually analyzed (the logarithms) come close to approximating a Gaussian distribution.

They didn't publish the raw data, but the summarized data are shown in Table 39.1 and Figure 39.1. The left side of Figure 39.1 shows the means and SDs. Assuming the data (as shown, already converted to logarithms) are sampled from a Gaussian distribution, the SD error bars should include approximately two-thirds of the data points. You can see that there is a lot of variation within each group and a huge amount of overlap between groups. Certainly, you could not use the LH level to predict whether someone was a runner.

The right side of Figure 39.1 shows the means with SEM error bars. Because the sample sizes are fairly large, you can view these SEM error bars as approxi-mately 68% CIs (see Chapter 14). Looked at this way, it seems as though the mean of the nonrunners might be distinct from the mean of the two groups of runners.

GROUP	LOG(LH)	SD	SEM	N
Nonrunners	0.52	0.25	0.027	88
Recreational runners	0.38	0.32	0.034	89
Elite runners	0.40	0.26	0.049	28

Table 39.1. LH levels in three groups of women.

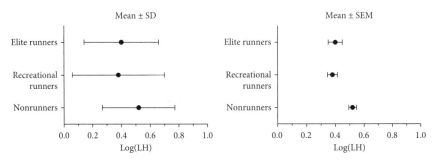

Figure 39.1. Data from Table 39.1 shown as a graph of the mean with SD (left) or SEM (right).

What's wrong with computing several t tests?

Your first thought when analyzing these data might be to use unpaired t tests (see Chapter 30). You'd need one t test to compare nonrunners with recreational runners, another t test to compare nonrunners with elite runners, and yet another t test to compare recreational runners with elite runners. The main problem with this approach is multiple comparisons (see Chapters 22 and 23). As you include more groups in the study, you increase the chance of obtaining one or more P values less than 0.05 just by chance. If the null hypothesis were true (all three populations have the same mean), there would be a 5% chance that each particular t test would yield a P value less than 0.05. But with three comparisons, the chance that any one (or more) of the values would be less than 0.05 would be far higher than 5%.

Interpreting a P value from one-way ANOVA

The data were not analyzed with t tests, but rather with *one-way ANOVA*, which compares all the groups at once. The one-way ANOVA reports that the P value is 0.0039.

To interpret any P value, the first step is to articulate the null hypothesis. Here, the null hypothesis is that the mean concentration of LH is the same in all three populations. The P value answers this question: If the null hypothesis were true, what is the chance of randomly picking samples with mean values as different (or more different) than those observed in this study?

If the three sets of data were really sampled from identical populations, there is only a 0.4% chance that the three means would be as far apart as actually observed or even further apart.

ASSUMPTIONS: ONE-WAY ANOVA

One-way ANOVA is based on the same assumptions as the unpaired t test:

- The samples are randomly selected from, or at least representative of, the larger populations.
- The observations within each sample are obtained independently. The relationships between all the observations in a group should be the same. You don't have independent measurements if some of the LH measurements are from the same person measured on several occasions or if some of the subjects are twins (or sisters).
- The data are sampled from populations that approximate a Gaussian distribution. In this case, the assumption applies to the logarithms of LH, so the assumption is that LH values are sampled from a lognormal distribution (see Chapter 11).
- The SDs of all populations are identical. This assumption is less important with large samples and when the sample sizes are equal. In this example, the data were all transformed to logarithms before the analysis was done, so the assumption refers to the log-transformed values.

HOW IT WORKS: ONE-WAY ANOVA

You can interpret ANOVA without knowing how it works, but you are less likely to use ANOVA inappropriately if you have some idea of how it works. The following descriptions give you a general idea of the process. The first two sections explain alternative ways to look at the method. These approaches are equivalent and produce the same results. I find the first approach easier to understand, but the second is more traditional.

Determine sum of squares by fitting alternative models

Comparing three or more means with one-way ANOVA can be viewed as comparing the fit of the data to two different models using the ideas presented in Chapter 35. The two models are as follows:

- Null hypothesis: all populations share the same mean, and nothing but random sampling causes any differences between sample means.
- Alternative hypothesis: all populations do not share the same means. At least one population has a mean different than the rest.

The top row of Table 39.2 quantifies how well the null hypothesis fits the data. Goodness of fit is quantified by the sum of squares of the difference between each value and the grand mean (which ignores any distinctions among the three groups).

The second row of Table 39.2 shows how well the alternative hypothesis fits the data. Table 39.2 reports the sum of squares of the difference between each value and the mean of the sample from which that value came.

The third row of Table 39.2 shows the difference. Of all the variation (sum of squares from the null hypothesis grand mean), 94.7% is the result of variation

HYPOTHESIS	SCATTER FROM	SUM OF SQUARES	PERCENTAGE OF VARIATION	
Null	Grand mean	17.38	100.0	
− Alternative	Group means	16.45	94.7	
= Difference		0.93	5.3	$R^2 = 0.053$

Table 39.2. One-way ANOVA as a comparison of models.

	SOURCE OF VARIATION	SUM OF SQUARES	DF	MS	F RATIO	P VALUE
	Between groups	0.93	2	0.46	5.69	0.0039
+	Within groups (error, residual)	16.45	202	0.081		
=	Total	17.38	204			

Table 39.3. ANOVA table showing the F ratio.

within the groups, leaving 5.3% of the total variation as the result of differences between the group means.

To determine the P value requires more than dividing the variance into its components. As you'll see, it also requires accounting for the number of values and the number of groups.

Alternative approach: Partitioning the sum of squares

A more common way to think about ANOVA is that it portions the variability into different components.

The first step in ANOVA is to compute the total variability among all the values (ignoring which value came from which group). This is done by summing the squares of the difference of each value from the grand mean. This is shown in the bottom row of Table 39.3. The total sum of squares is 17.38.

Some of the variation comes from differences *among* the group means because the group means are not all the same. Sum the squares of each group mean from the grand mean and weight by sample size to get the sum of squares resulting from treatment. This is shown in the top row of Table 39.3. The sum of squares among group means is 0.93.

The rest of the variation comes from variably *within* each group, quantified by summing the squares of the differences of each value from its group mean. These numbers are shown in the second row of Table 39.3. The sum of squares within groups is 16.45. This is also called the *residual sum of squares* or the *error sum of squares*.

It isn't obvious (although it can be proven with simple algebra), but the sum of squares resulting from treatment and the sum of squares within the groups always add up to the total sum of squares.

Determining P from F

The third column of Table 39.3 shows the number of df. The bottom row, labeled "Total," is for the null hypothesis model. There are 205 values and only one parameter was estimated (the grand mean), which leaves 204 df. The next row up shows the sum of squares from the group means. Three parameters were fit (the mean of each group), so there are 202 df (205 data points minus three parameters). The top row shows the difference. The alternative model (three distinct means) has two more parameters than the null hypothesis model (one grand mean), so there are two df in this row. The df, like the sums of squares, can be partitioned so that the bottom row is the sum of values in the rows above.

The fourth column divides the sum of squares by the number of df to compute the mean square (MS), which could also be called the variance. Note that it is not possible to add the MSs of the top two rows to obtain a meaningful MS for the bottom row. Because the MS for the null hypothesis is not used in further calculations, it is left blank.

To compute a P value, you must take into account the number of values and the number of groups. This is done in the last column of Table 39.3.

If the null hypothesis were correct, each MS value would estimate the variance among values, so the two MS values would be similar. The ratio of those two MS values is called the *F ratio*, after Ronald Fisher, the pioneering statistician who invented ANOVA and much of statistics.

If the null hypothesis were true, F would be likely to have a value close to 1. If the null hypothesis were not true, F would probably have a value greater than 1. The probability distribution of F under the null hypothesis is known for various df and can be used to calculate a P value. The P value answers this question: If the null hypothesis were true, what is the chance that randomly selected data (given the total sample size and number of groups) would lead to an F ratio this large or larger? For this example, $F = 5.690$, with 2 df in the numerator, 202 df in the denominator, and $P = 0.0039$.

R^2

Look at the first column of Table 39.2, which partitions the total sum of squares into its two component parts. Divide the sum of squares resulting from differences between groups (0.93) by the total sum of squares (17.38) to determine the fraction of the sum of squares resulting from differences between groups, 0.053. This is called eta squared, η^2, which is interpreted in the same way as R^2. Only 5.3% of the total variability in this example is the result of differences between group means. The remaining 94.7% of the variability is the result of differences within the groups.

The low P value means that the differences among group means would be very unlikely if in fact all the population means were equal. The low R^2 means that the differences among group means are only a tiny fraction of the overall variability. You could also reach this conclusion by looking at the left side of Figure 39.1, where the SD error bars overlap so much.

REPEATED-MEASURES ONE WAY ANOVA

The difference between ordinary and *repeated-measures ANOVA* is the same as the difference between the unpaired and paired t tests. Use repeated-measures ANOVA to analyze data collected in three kinds of experiments:

- Measurements are made repeatedly for each subject, perhaps before, during, and after an intervention, and these repeated measurements make up the groups being compared by ANOVA.
- Subjects are recruited as matched sets (often called blocks), which are matched for variables such as age, postal code, or diagnosis. Each subject in the set receives a different intervention (or placebo).
- A laboratory experiment is run several times, each time with several treatments (or a control and several treatments) handled in parallel.

More generally, you should use a repeated-measures test whenever the value of one subject in the first group is expected to be closer to a particular value in the other group than to a random subject in the other group.

When the experimental design incorporates matching, use the repeated-measures test, because it is usually more powerful than ordinary ANOVA. Of course, the matching must be done as part of the protocol before the results are collected. The decision about matching is a question of experimental design and should be made long before the data are analyzed.

Although the calculations for the repeated-measures ANOVA are different from those for ordinary ANOVA, the interpretation of the P value is the same. Similar kinds of multiple comparisons tests (see Chapter 39) are performed.

AN EXAMPLE OF TWO-WAY ANOVA

Data

Figure 39.2 shows results from an experiment with four groups of animals. Half received inactive treatment (as a control), and half received an active treatment. For each of those groups, half were treated for a short duration, and half for a longer duration. Note that rather than plotting a mean and error bar, the figure shows all the data. This is a better way to show the data than showing means and standard deviations (or SEMs).

The results are easy to see without any calculation. With a short duration, the treatment made little, if any, difference. With the longer duration of active treatment, the response went up by about a factor of three.

How many factors?

The prior example was *one-way ANOVA* because the subjects were categorized in one way, by amount of exercise. In this new example, the data are divided in two ways, so the data would be analyzed using *two-way ANOVA*, also called *two-factor ANOVA*. That is because each data point is either from an animal given either an inactive or active treatment (one factor) given for either a short or long duration (the second factor).

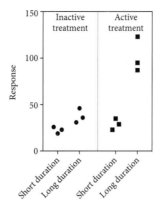

Figure 39.2. Sample two-way ANOVA data.

If male and female animals were both included in the study, you'd need three-way ANOVA. Now each value would be categorized in three ways: treatment, duration, and gender.

Three null hypotheses and three P values

Two-way ANOVA simultaneously tests three null hypotheses and so computes three P values.

Interaction between the two factors (treatment and duration)

The first null hypothesis is that there is no *interaction* between the two factors.

In our example, the null hypothesis is that the active treatment causes the same change in response if given for short or long duration. Equivalently, the null hypothesis is that the difference in effects between short and long duration is the same in animals given active or inactive treatment.

The interaction P value for this example is 0.0017. Because it is so small, you will reject the null hypothesis of no interaction. This small P value tells us that the effect of treatment depends on duration. Accordingly, there is no point interpreting the other two P values described in the following discussion.

First factor (treatment)

This null hypothesis is that the population means are identical for animals given placebo and active treatment (pooling the animals treated for short or long duration). Since the interaction is so strong, it really doesn't make sense to ask about the average effect of treatment, combining both durations. The effect depends on duration.

Second factor (age)

This null hypothesis is that the population means are identical animals given a treatment (active or not) for short versus long durations. But since the interaction is so strong, it really doesn't make sense to ask about the average effect of treating for a longer duration, averaging placebo and active treatment. The active treatment has a much bigger effect than the inactive control.

HOW TWO-WAY ANOVA WORKS

Table 39.4 shows the basic idea upon which two-way ANOVA is based. The bottom row of the sum-of-squares column quantifies the total variation among the 12 values, paying no attention to which row or column they are in. It is the sum-of-squares of the squared difference between each value and the grand mean of all values. The rest of the table breaks this variability into four components: Interaction (defined previously in this chapter), differences between rows (on the average), differences between columns (on the average), and variation among replicates in the same row and column. The last column, MS, is computed as the sum-of-squares divided by the relevant number of df. Each MS is divided by the residual MS to compute an F ratio, and this is used to obtain a P value. If you want to understand two-way ANOVA in detail, you'll need to read a more detailed explanation in a more mathematical text.

REPEATED MEASURES TWO-WAY ANOVA

The previous example had no repeated measures. Each value represents a datum from a different animal (12 of them). As with one-way ANOVA, special methods have been developed to deal with repeated measures in two-way ANOVA. In the example, there are three ways the data could be repeated measures:

- The same animals were first given inactive treatment and the response measured. Then they were given the active treatment, and the response measured again. Since each animal is measured twice, the design would require only six animals.
- The same animals are first measured for a response after a short duration and then later after a long duration. Since each animal is measured twice, the design would require only six animals.
- Repeated measures in both factors. Each animal was first given the inactive treatment and measured for a short and then a long duration. Then the same animal was given the active treatment and measured for a short and then a long duration. Since each animal is measured four times, the design would require only three animals.

	VARIATION SOURCE	SUM OF SQUARES	DF	MEAN SQUARE
	Interaction	12896	1	12896
+	Between rows	5764	1	5764
+	Between columns	3710	1	3710
+	Among replicates (residual)	928	8	116
=	Total	12896	11	

Table 39.4. Sources of variation in two-way ANOVA.

If you analyze data with repeated measures two-way ANOVA, make very sure the program knows which factor is repeated, or if both are. If you set up the program incorrectly, the results will be meaningless.

COMMON MISTAKES: ANOVA

Using ANOVA when you want to compare the SD or variances between groups

Don't be misled by the word *variance* in the name of the method, analysis of variance. In this case, *variance* refers to the statistical method being used, not the hypothesis being tested. ANOVA asks about differences among group means. ANOVA, as usually performed, does not test whether the variances of the groups are different. Just the opposite. ANOVA assumes that the variances are all the same and uses that assumption as part of its comparison among means.

Using ANOVA when one of the factors is quantitative, such as when comparing time courses or dose response curves

When one of the factors has more than two levels, ANOVA pays no attention to their order. If one of the factors is time or dose, ANOVA ignores the fact that times and doses are numerical. It even ignores the fact that they have a natural order. The whole point of an experiment may have been to look at a trend or at a dose–response relationship, but the ANOVA calculations completely ignore the order of the time points or doses. If you randomly scramble the time points or doses, ANOVA would report identical results. ANOVA treats different time points, or different doses, exactly the same way it would treat different drugs, different genotypes, or different countries. ANOVA may still make sense when data were collected at only a few time points or doses, but often regression is a better analysis with such data.

Forgetting that time is a factor

Say you measure a response after three treatments, and for each treatment make the measurement at five different times. You have two factors, treatment and time, so need to use two-way ANOVA (or some other method). A common mistake is to try to analyze these data with one-way ANOVA, considering only treatment to be a factor and ignoring the fact that time is also one.

Q & A

Can one-way ANOVA be done with two groups?
> One-way ANOVA is usually only done for three or more groups, but it could be done for only two groups. Many programs won't allow this, but it certainly is mathematically possible. Although the approach seems very different, one-way ANOVA for two groups is mathematically equivalent to an unpaired t test and the two methods will compute identical P values.

Are the results valid if sample size differs between groups?

Yes. ANOVA does not require that all samples have the same number of values. Two of the assumptions of ANOVA—that the data come from Gaussian populations and that these populations have equal SDs—matter much more when sample size varies a lot between groups. If you have very different sample sizes, a small P value from ANOVA may be caused by non-Gaussian data (or unequal variances) rather than differences among means.

Is the F ratio always positive?

Yes. Because the ANOVA calculations deal with sums of squares, the F ratio is always positive.

Is the F ratio always greater than 1.0?

No. It can be a fraction less than 1.0.

If the P value is small, can we be sure that all the means are distinct?

No. A P value can be small when all groups are distinct or if one is distinct and the rest are indistinguishable. Chapter 40 will show you how to compare individual pairs of means.

Is the P value from one-way ANOVA one- or two-tailed?

Neither. With ANOVA, the concept of one- and two-tailed P values does not really apply. Because the means of the groups can be in many different orders, the P value has many tails.

If I need to use a program to compute a P value from F, does the order in which the two df values are entered matter?

Yes. The calculation of a P value from F requires knowing the number of df for the numerator of the F ratio and the number of df for the denominator. If you mistakenly swap the two df values, the P value will be incorrect.

Do the one-way ANOVA calculations require raw data?

No. One-way ANOVA (but not repeated-measures ANOVA) can be computed without raw data, so long as you know the mean, sample size, and SD (or SEM) of each group.

Can all kinds of ANOVA also be done using regression techniques?

Yes. ANOVA compares the fit of several models to the data, and this can be done with regression techniques as well. The answers will be fundamentally identical, but they will look very different.

What is the difference between fixed-effects and random-effects ANOVA?

Most programs perform Type I ANOVA, also known as fixed-effects ANOVA. This method tests for differences among the means of the particular groups from which you have collected data. Type II ANOVA, also called random-effects ANOVA, is less commonly used. It assumes that you have randomly selected groups from an infinite (or at least large) number of possible groups and that you want to reach conclusions about differences among *all* the groups, not just the groups from which you collected data.

CHAPTER SUMMARY

- One-way ANOVA compares the means of three or more groups.
- The P value tests the null hypothesis that all the population means are identical.

- ANOVA assumes that all data are randomly sampled from populations that follow a Gaussian distribution and have equal SDs.
- The difference between ordinary and repeated-measures ANOVA is the same as the difference between the unpaired and paired t tests. Repeated-measures ANOVA is used when measurements are made repeatedly for each subject or when subjects are recruited as matched sets.
- Two-way ANOVA, also called two-factor ANOVA, determines how a response is affected by two factors. For example, you might measure a response to three different drugs in both men and women.

TERMS INTRODUCED IN THIS CHAPTER

- F ratio (p. 411)
- One-way analysis of variance (ANOVA) (p. 408)
- Repeated-measures ANOVA (p. 412)
- Residual sum of squares (or error sum of squares) (p. 410)
- Two-way ANOVA (or two-factor ANOVA) (p. 412)

CHAPTER 40

Multiple Comparison Tests after ANOVA

When did "skeptic" become a dirty word in science?

MICHAEL CRICHTON (2003)

Ohypothesis that all groups were sampled from populations with
identical means. This chapter explains how multiple comparison tests let
you dig deeper to see which pairs of groups are statistically distinguish-
able. Before reading this chapter, read Chapters 22 and 23 on multiple
comparisons and Chapter 39 on one-way ANOVA.

MULTIPLE COMPARISON TESTS FOR THE EXAMPLE DATA

Goal

Chapter 39 analyzed a sample data set comparing LH levels (actually, their
logarithms) in three groups of women. One-way ANOVA reported a very small
P value. If the null hypothesis that all groups were sampled from populations with
equal means were true, it would be quite rare for random sampling to result in so
much variability among the sample means.

Multiple comparison tests dig deeper to find out which group means differ
from which other group means, taking into account multiple comparisons
(see Chapter 22). To avoid getting fooled by bogus statistically significant conclusions, the significance level is redefined to apply to an entire family of comparisons, rather than to each individual comparison (see Table 40.1). This is one
of the most confusing parts of statistics. Using this new definition of statistical
significance will reduce the chance of obtaining false reports of statistical significance (i.e., fewer Type I errors) but at the cost of reducing the power to detect real
differences (i.e., more Type II errors).

Multiple comparison CIs

For this example, the goal is to compare every mean with every other mean. One
appropriate test is called *Tukey's test* (or, more generally, the *Tukey–Kramer test*,
which allows for unequal sample size). The results include both CIs and conclusions about statistical significance.

Figure 40.1 plots the 95% CI for the difference between each mean and every other mean. The original data were expressed as the logarithm of LH concentration, so these units are also used for the confidence intervals. These are tabulated in Table 40.2.

These are multiple comparisons CIs, so the 95% confidence level applies to the entire family of comparisons, rather than to each individual interval. Given the assumptions of the analysis (listed in Chapter 39), there is a 95% chance that all three of these CIs include the true population value, leaving only a 5% chance that any one or more of the intervals does not include the population value. Because the 95% confidence level applies to the entire set of intervals, it is impossible to correctly interpret any individual interval without seeing the entire set.

SITUATION	MEANING OF A 5% SIGNIFICANCE LEVEL	ERROR RATE
One comparison	If the null hypotheses were true, there is a 5% chance that random sampling would lead to the incorrect conclusion that there is a real difference.	Per comparison
A family of comparisons	If all null hypotheses were true, there is a 5% chance that random sampling would lead to one or more incorrect conclusions that there is a real difference between the mean of two groups.	Per experiment or familywise

Table 40.1. Statistical significance is redefined for multiple comparisons.

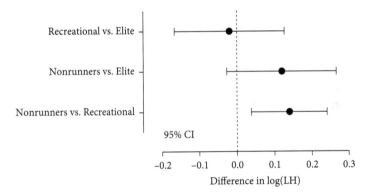

Figure 40.1. 95% CIs for Tukey's multiple comparisons test.
The 95% confidence level applies to the entire family of comparisons rather than to an individual comparison.

TUKEY'S MULTIPLE COMPARISONS TEST	DIFFERENCE BETWEEN MEANS	95% CI OF DIFFERENCE
Recreational runners vs. elite runners	−0.0200	−0.17 to 0.13
Nonrunners vs. elite runners	0.1200	−0.027 to 0.27
Nonrunners vs. recreational runners	0.1400	0.038 to 0.24

Table 40.2. Multiple comparisons CIs.

CIs as ratios

For this example, the data were entered into the ANOVA program as the logarithm of the concentration of LH (shown in Chapter 39). Therefore, Figure 40.1 and Table 40.2 show differences between two logarithms. Many find it easier to think about these results without logarithms, and it is easy to convert the data to a more intuitive format.

The trick is to note that the difference between the logarithms of two values is mathematically identical to the logarithm of the ratio of those two values (logarithms and antilogarithms are reviewed in Appendix E). Transform each of the differences (and each confidence limit) to its antilogarithm, and the resulting values can be interpreted as the ratio of two LH levels. Written as equations:

$$\log(A) - \log(B) = \log\left(\frac{A}{B}\right)$$

$$10^{(\log(A) - \log(B))} = \frac{A}{B}$$

Each row in Table 40.3 shows the ratio of the mean LH level in one group divided by the mean in another group, along with the 95% CI of that ratio. These are plotted in Figure 40.2.

TUKEY'S MULTIPLE COMPARISONS TEST	RATIO	95% CI OF RATIO
Recreational runners vs. elite runners	0.96	0.68 to 1.35
Nonrunners vs. elite runners	1.32	0.94 to 1.86
Nonrunners vs. recreational runners	1.38	1.09 to 1.74

Table 40.3. Multiple comparisons CIs, expressed as ratios rather than differences.

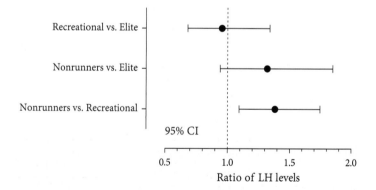

Figure 40.2. 95% CIs for Tukey's multiple comparisons test, expressed as ratios of LH levels rather than as differences between mean log(LH) concentrations.
The values correspond to Table 40.3.

		STATISTICALLY SIGNIFICANT?		
	DIFFERENCE	$\alpha = 0.05$	$\alpha = 0.01$	$\alpha = 0.001$
Recreational–elite	–0.02000	No	No	No
Nonrunners–elite	0.1200	No	No	No
Nonrunners–recreational	0.1400	Yes	Yes	No

Table 40.4. Statistical significance of multiple comparisons.

Statistical significance

If a 95% CI for the difference between two means includes zero (the value specified in the null hypothesis), then the difference is not statistically significant ($P > 0.05$). Two of the three 95% CIs shown in Table 40.2 and Figure 40.1 include zero, and so these comparisons are not statistically significant at the 5% significance level. The CI for the remaining comparison (nonrunners vs. recreational runners) does not include zero, so that difference is statistically significant.

The values in Table 40.2 are the differences between two logarithms, which are mathematically identical to the logarithms of the ratios. Transforming to antilogarithms creates the table of ratios and CIs of ratios, shown in Table 40.3.

The null hypothesis of identical populations corresponds to a ratio of 1.0. Only one comparison (nonrunners vs. recreational runners) does not include 1.0, so that comparison is statistically significant.

The 5% significance level is a familywise significance level, meaning that it applies to the entire set, or family, of comparisons (previously defined in Chapter 22). If the overall null hypothesis (values from all groups were sampled from populations with identical means) is true, there is a 5% chance that one or more of the comparisons will be statistically significant and a 95% chance that none of the comparisons will be statistically significant.

Table 40.4 shows the conclusions about statistical significance at three different significance levels.

The significance levels (α) apply to the entire family of three comparisons, but -the yes/no conclusions apply to each comparison individually.

THE LOGIC OF MULTIPLE COMPARISONS TESTS

Multiple comparisons tests account for multiple comparisons

If all null hypotheses are true and you make several comparisons without any special corrections, then about 5% of all those comparisons will generate a statistically significant result. The chance of making one or more Type I errors (declaring statistical significance when the null hypotheses really are all true) will be greater than 5%. This problem of multiple comparisons was discussed in Chapter 22.

The confidence level applies to the entire family of CIs. Given certain assumptions, there is a 95% chance that *all* of the CIs include the true population value, leaving only a 5% chance that one or more of the intervals do not include the population value.

Similarly, the 5% significance level applies to the entire family of comparisons. If the null hypothesis is true (all data were sampled from populations with identical means), there is a 95% chance that none of the comparisons will be declared statistically significant, leaving a 5% chance that one or more comparisons will (erroneously) be declared statistically significant.

It is easy to get confused when you think about significance levels that apply to a family of comparisons. It really doesn't make sense to think about whether one particular comparison is statistically significant (accounting for the rest). Instead, use a different mindset. The method divides the family of comparisons into two piles: those that are deemed to be statistically significant and those that are not statistically significant. Think about the set of differences deemed to be statistically significant, not about individual comparisons.

With more groups, CIs between means become wider
The example includes three groups and three possible pairwise comparisons of means. If there were five groups, then there would be 10 possible comparisons between sample means (AB, AC, AD, AE, BC, BD, BE, CD, CE, DE). Table 40.5 and Figure 40.3 show how the number of possible comparisons increases with the number of groups.

If you include more groups in your experiment, the CIs will be wider and a difference must be larger before it will be deemed statistically significant. In other words, When you include more treatment groups in your experimental design, you lose statistical power to detect differences between selected group means.

NO. OF GROUPS	NO. OF PAIRWISE COMPARISONS
3	3
4	6
5	10
6	15
7	21
8	28
9	36
10	45
11	55
12	66
13	78
14	91
15	105
16	120
17	136
18	153
19	171
20	190
k	$k(k-1)/2$

Table 40.5. Number of possible pairwise comparisons between group means as a function of the number of groups.

The number of possible pairwise comparisons is very large when you have many groups.

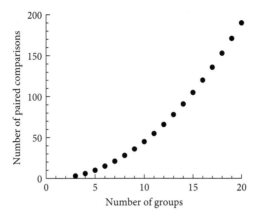

Figure 40.3. Number of possible pairwise comparisons between group means as a function of the number of groups.

Multiple comparisons tests use data from all groups, even when only comparing two groups

The result of most multiple comparison tests depends on all the data in all the groups, not just the two groups being compared. To see why, read about the use of the pooled SD in the following "How It Works" section.

Some multiple comparisons tests account for intertwined comparisons

Once you have compared the mean of Group A with the mean of Group B and the mean of Group A with the mean of Group C, you will already know something about the comparison of Group B with Group C. Consider three groups (with equal n) for which the mean of A is greater than the mean of B, the mean of B is greater than the mean of C, and both comparisons are statistically significant. Without any calculations, you know that the mean of A must be significantly greater than the mean of C. Most multiple comparisons methods account for this interlinking of comparisons.

Multiple comparisons tests can declare the difference between pairs of means to be statistically significant, but they usually do not compute exact P values

When a t test compares two means, it reports an exact P value. You can then decide to compare that P value to a preset significance level to decide whether that the difference is statistically significant.

Multiple comparisons tests following ANOVA are different. Statistical significance is determined by accounting for multiple comparisons. The significance level doesn't apply to any one particular comparison but rather to the entire family of comparisons. Accordingly, exact P values are usually not computed for each comparison. It is common, however, for statistics programs to report statistical significance at several different familywise significance levels (as is done in Table 40.4).

TUKEY'S MULTIPLE COMPARISONS TEST	MULTIPLICITY-ADJUSTED P VALUE
Recreational runners vs. elite runners	0.9440
Nonrunners vs. elite runners	0.1307
Nonrunners vs. recreational runners	0.0037

Table 40.6. Multiplicity-adjusted P values computed with GraphPad Prism, version 6.

Some programs compute multiplicity-adjusted P values for each comparison

Some programs (including GraphPad Prism) can also report multiplicity-adjusted P values. Each multiplicity-adjusted P value is for a particular comparison, but it takes into account all the comparisons (Wright, 1992). It is the overall significance level (α) at which that particular comparison would be right at the border of statistical significance (accounting for multiple comparisons). For the sample data, the multiplicity-adjusted P values computed as part of Tukey's multiple comparisons are shown in Table 40.6.

Note that each adjusted P value is computed from the entire data set so thus takes into account the entire family of comparisons. Consequently, the adjusted P value for one particular comparison would have a different value if there were a different number of comparisons or if the data in the other comparisons were changed. The adjusted P value cannot be compared to an individual P value computed by a t test.

OTHER MULTIPLE COMPARISONS TESTS

Statisticians have developed numerous multiple comparisons methods to accompany one-way ANOVA. The previous example used Tukey's test, which compares the mean of each group with the mean of every other group. This section explains some (but certainly not all) alternative tests.

Dunnett's test: Compare each mean with the mean of a control group

Dunnett's test compares the mean of each group to the mean of a control group without comparing the other groups among themselves. For example, it would be used in an experiment that tests the effects of six different drugs with a goal of defining which drugs have any effect but not of comparing the drugs with each other.

Because Dunnett's test makes fewer comparisons than Tukey's method, it generates narrower CIs and has more power to detect differences. It is very useful.

The decision to use Dunnett's test (along with a definition of which group is the control group) should be part of the experimental design. It isn't fair to first do Tukey's test to make all comparisons and then switch to Dunnett's test to get more power.

Bonferroni's test: Compare preselected pairs of means

Bonferroni's approach was explained in Chapter 22. Although it can be used to compare all pairs of means, it should not be used in this way because Tukey's test has more power. Similarly, it should not be used to compare each group against a control, because Dunnett's test has more power for that purpose. Bonferroni's multiple comparisons test should be used when the experiment design requires comparing only selected pairs of means. By making a limited set of comparisons, you get narrower CIs and more statistical power to detect differences.

It is essential that you select those pairs as part of the experimental design, before collecting the data. It is not fair to first look at the data and then decide which pairs you want to compare. By looking at the data first, you have implicitly compared all groups.

Holm's test: Powerful, but no CIs

The Holm multiple comparisons test is a powerful and versatile multiple comparisons test. It can be used to compare all pairs of means, compare each group mean to a control mean, or compare preselected pairs of means. It is not restricted to being used as a follow-up to ANOVA but instead can be used in any multiple comparisons context. The test is quite logical and easy to understand.

The Holm multiple comparison method starts with a set of P values and then determines which of these P values are low enough for the corresponding comparison to be declared statistically significant. It doesn't adjust the P values themselves but simply determines the conclusion for each one. The threshold for determining significance depends on the rank of the P value. The threshold for the smallest P value is much smaller than the threshold for the largest P value.

The test is powerful, understandable, and versatile. What's not to like? The Holm test can only be used to decide which comparisons are statistically significant and which are not. Unlike the tests developed by Tukey, Bonferroni, and Dunnett, Holm's test cannot also compute a set of CIs (Westfall et al., 1999, p. 40).

The Holm–Sidak test is a modified version of Holm's test that is slightly more powerful.

Scheffe's test: More general comparisons

Scheffe's multiple comparisons test can make more elaborate comparisons than the other tests. For example, you might want to compare the mean of all treated groups with the mean of a control group. Or you might want to compare the mean of Groups A and B with the mean of Groups C, D, and E. These comparisons are sometimes called contrasts.

Scheffe's method can test any number of this kind of comparison. This increased versatility comes with a price, however. To allow for the huge number of possible comparisons, the CIs generated by Scheffe's method are wider than those generated by other methods. Accordingly, the test has less statistical power to detect differences than do the other multiple comparisons tests. The precise comparisons do not need to be defined as part of the experimental design. It is OK to test comparisons that didn't occur to you until you saw the data.

Test for a trend: Is group mean correlated with group order?

The different groups compared by ANOVA often have a natural order—for example, ages, time intervals, or doses. However, one-way ANOVA calculations completely ignore this order. ANOVA analyzes data from different doses or different ages in exactly the same way that it analyzes data from different species or different drugs. If you randomly shuffled the doses or ages, the ANOVA results wouldn't change.

The *test for trend* computes the correlation coefficient between the outcome and group order, along with a P value testing the null hypothesis that there is no trend. If this P value is small, there is a statistically significant trend (correlation) between group order and the outcome.

To learn more, see Altman (1990). Fancier tests for trend can look for nonlinear trends. Look up *polynomial contrasts* in an advanced ANOVA text for more details.

HOW IT WORKS: MULTIPLE COMPARISONS TESTS

Chapter 12 briefly explained how a CI of a mean is computed. The margin of error of the CI of a mean is the product of two values. The first is the SEM, which is computed from the SD and sample size. The second is a critical value from the t distribution, which depends on the desired confidence level (95%) and the number of df $(n - 1)$.

Chapter 30 extended that idea to the CI for the difference between two means. The standard error of the difference between means is computed from both SDs and both sample sizes.

When computing most multiple comparisons tests, the standard error for the difference between two means is not computed from the SDs of those two groups but rather from the pooled SD of *all* the groups. ANOVA software rarely reports this pooled SD, but reports instead the pooled variance, which is labeled "mean square within groups" or "mean square residual." The square root of this value is the pooled SD. The standard error of the difference between two means is computed from that value (which is the same for all comparisons) and the sample size of the two groups being compared (which might not be the same for all comparisons).

The margin of error of the CI is computed by multiplying the standard error of the difference by a critical value that depends on the choice of test, the number of df, the degree of confidence desired, and (importantly) the number of comparisons.

WHEN ARE MULTIPLE COMPARISONS TESTS NOT NEEDED?

Never correct for multiple comparisons?

Correcting for multiple comparisons reduces the risk of a Type I error, but at the cost of increasing the risk of a Type II error (see Table 40.7). Rothman (1990) argues that this trade-off is not worth it and recommends that researchers not correct for multiple comparisons. Instead, it is essential to report all the comparisons

Risk of Type I error (labeling a chance difference as statistically significant)	Lower
Risk of Type II error (missing a real difference)	Higher
Width of CIs	Wider

Table 40.7. Consequences of correcting for multiple comparisons compared to individual comparisons.

you made so that the people reading the research can informally correct for multiple comparisons. This recommendation is sensible but not mainstream.

Don't correct when some comparisons are more important than others?

Correcting for multiple comparisons can lead to ambiguous situations. Imagine that you are running a research program to investigate the mechanism of action of a drug. You want to know whether it blocks a certain receptor. You set up the proper assay and run some test experiments to make sure everything works. Now imagine two alternative scenarios.

Scenario 1: You run an experiment testing only your one drug against a control. The drug works as you predicted, and the results are statistically significant. Your research moves forward.

Scenario 2: You run an experiment testing not only your drug but also two additional drugs. There is no reason to expect these drugs to block the receptor you are studying, but it would be very interesting if they did. Setting up the assay and running various controls is a lot of work, but testing three drugs is not a whole lot harder than testing one drug, so you give it a try. Every once in a while this kind of exploratory experiment can move research forward. Not this time. The two extra drugs don't work. Not a big disappointment and not too much time wasted.

The data for the main drug are exactly the same in both scenarios. In the second scenario, two other drugs are tested at the same time. To account for multiple comparisons, the analysis would use Dunnett's method, so the 5% significance level would apply to the family of three comparisons rather than to each individual comparison. With this analysis, the result for the main drug would not be statistically significant. Because more comparisons were made, the method would become more cautious about concluding statistical significance.

In this example, the correction for multiple corrections doesn't really make sense. The experiment was done to test one main hypothesis, and the other two hypotheses were add-ons. When one or a few comparisons are clearly defined in advance as being critical, some statisticians advocate not using any correction for multiple comparisons. This approach is called *planned comparisons*, but that term is ambiguous, because many multiple comparisons are planned in advance.

The statistical principles of this approach are fairly straightforward and do not generate much controversy. However, that doesn't make it easy to decide what to do in a particular situation. Reasonable statisticians disagree.

COMMON MISTAKES: MULTIPLE COMPARISONS

Mistake: Following ANOVA, using the unpaired t test to compare two groups at a time rather than using multiple comparisons tests

If you perform multiple t tests this way, it is too easy to reach misleading conclusions. Use multiple comparisons designed for this situation.

Mistake: If your experiment has three or more treatment groups, compare the largest mean with the smallest mean using a single unpaired t test

You won't know which group has the largest mean and which has the smallest mean until after you have collected the data. So although you are formally just doing one t test to compare two groups, you are really comparing all the groups. Use one-way ANOVA, followed by multiple comparisons tests.

Mistake: Focusing only on declarations of statistical significance and ignoring CIs

Most multiple comparisons tests can report CIs corrected for multiple comparisons. In many situations, these are more informative (and easier to understand) than declarations of significance.

Mistake: Using multiple comparisons tests to determine the earliest time point (or smallest concentration) at which an effect is significant

If you have collected data over time, or at multiple concentrations of a drug, analyze the data by fitting a line or curve with regression. Some people do one-way ANOVA followed by multiple comparisons to determine the earliest time point (or lowest concentration) at which the difference from control (zero time or zero concentration) is statistically significant. This is rarely a helpful calculation. If you changed sample size, you'd get a different result.

Q & A

What is the difference between a multiple comparisons test and a post hoc test?
> The term *multiple comparisons test* applies whenever several comparisons are performed at once with a correction for multiple comparisons. The term *post hoc test* refers to situations in which you can decide which comparisons you want to make after looking at the data. Often, however, the term is used informally as a synonym for multiple comparisons test. *Posttest* is an informal, but ambiguous, term. It can refer to either all multiple comparisons tests or only to post hoc tests.

If one-way ANOVA reports a P value less than 0.05, are multiple comparisons tests sure to find a significant difference between group means?
> Not necessarily. The low P value from the ANOVA tells you that the null hypothesis that all data were sampled from one population with one mean is unlikely to be true. However, the difference might be a subtle one. It might be that the mean of Groups A and B is significantly different than the mean of Groups C, D, and E.

Scheffe's posttest can find such differences (called contrasts), and if the overall ANOVA is statistically significant, Scheffe's test is sure to find a significant contrast.

The other multiple comparisons tests compare group means. Finding that the overall ANOVA reports a statistically significant result does not guarantee that any of these multiple comparisons will find a statistically significant difference.

If one-way ANOVA reports a P value greater than 0.05, is it possible for a multiple comparisons test to find a statistically significant difference between some group means?

Yes. Surprisingly, this is possible.

Are the results of multiple comparisons tests valid if the overall P value for the ANOVA is greater than 0.05?

It depends on which multiple comparisons test you use. Tukey's, Dunnett's, and Bonferroni's tests mentioned in this chapter are valid even if the overall ANOVA yields a conclusion that there are no statistically significant differences among the group means.

Does it make sense to only focus on multiple comparisons results and ignore the overall ANOVA results?

It depends on the scientific goals. ANOVA tests the overall null hypothesis that all the data come from groups that have identical means. If that is your experimental question—do the data provide convincing evidence that the means are not all identical—then ANOVA is exactly what you want. If the experimental questions are more focused and can be answered by multiple comparisons tests, you can safely ignore the overall ANOVA results and jump right to the results of multiple comparisons.

Note that the multiple comparisons calculations all use the mean square result from the ANOVA table. Consequently, even if you don't care about the value of F or the P value, the multiple comparisons tests still require that the ANOVA table be computed.

Can I assess statistical significance by observing whether two error bars overlap?

If two standard error bars overlap, you can be sure that a multiple comparisons test comparing those two groups will find no statistical significance. However, if two standard error bars do not overlap, you can't tell whether a multiple comparisons test will or will not find a statistically significant difference. If you plot SD, rather than SEM, error bars, the fact that they do (or don't) overlap will not let you reach any conclusion about statistical significance.

Do multiple comparisons tests take into account the order of the groups?

No. With the exception of the test for trend mentioned in this chapter, multiple comparisons tests do not consider the order in which the groups were entered into the program.

Do all CIs between means have the same length?

If all groups have the same number of values, then all the CIs for the difference between means will have identical lengths. If the sample sizes are unequal, then the standard error of the difference between means depends on sample size. The CI for the difference between two means will be wider when sample size is small and narrower when sample size is larger.

Three groups of data (a, b, c) are analyzed with one-way ANOVA followed by Tukey multiple comparisons to compare all pairs of means. Now another group (d) is added, and the ANOVA is run again. Will the comparisons for a–b, a–c, and b–c change?

Probably. When comparing the difference between two means, that difference is compared to a pooled SD computed from all the data, which will change when you add another treatment group. Also, the increase in number of comparisons will lower the threshold for a P value to be deemed statistically significance and widen confidence intervals.

Why wasn't the Newman–Keuls test used for the example?

Like Tukey's test, the Newman–Keuls test (also called the Student–Newman–Keuls test) compares each group mean with every other group mean. Some prefer it because it has more power. I prefer Tukey's test, because the Newman–Keuls test does not really control the error rate as it should (Seaman, Levin, & Serlin, 1991) and cannot compute CIs.

Chapter 22 explains the concept of controlling the FDR. Is this concept used in multiple comparisons after ANOVA?

This is not a standard approach to handling multiple comparisons after ANOVA, but some think it should be.

CHAPTER SUMMARY

- Multiple comparisons tests follow ANOVA to find out which group means differ from which other means.
- To prevent getting fooled by bogus statistically significant conclusions, the significance level is usually defined to apply to an entire family of comparisons rather than to each individual comparison.
- Most multiple comparisons tests can report both CIs and statements about statistical significance. Some can also report multiplicity-adjusted P values.
- Because multiple comparisons tests correct for multiple comparisons, the results obtained by comparing two groups depends on the data in the other groups and the number of other groups in the analysis.
- To choose a multiple comparison test, you must articulate the goals of the study. Do you want to compare each group mean to every other group mean? Each group mean to a control group mean? Only compare a small set of pairs of group means? Different multiple comparison tests are used for different sets of comparisons.

TERMS INTRODUCED IN THIS CHAPTER

- Contrast (p. 425)
- Dunnett's test (p. 424)
- Holm's test (p. 425)
- Multiplicity-adjusted P value (p. 424)
- Planned comparisons (p. 427)
- Post hoc test (p. 428)
- Post test (p. 428)
- Scheffe's test (p. 425)
- Test for trend (p. 426)
- Tukey's test (or Tukey–Kramer test) (p. 418)

CHAPTER 41

Nonparametric Methods

Statistics are like a bikini. What they reveal is suggestive, but
what they conceal is vital.

AARON LEVENSTEIN

Many of the methods discussed in this book are based on the assumption that the values are sampled from a Gaussian distribution. Another family of methods makes no such assumption about the
population distribution. These are called *nonparametric* methods. The
nonparametric methods used most commonly work by ignoring the
actual data values and instead analyzing only their ranks. Computer-
intensive resampling and bootstrapping methods also do not assume a
specified distribution, so they are also nonparametric.

NONPARAMETRIC TESTS BASED ON RANKS

The idea of nonparametric tests
The unpaired t test and ANOVA are based on the assumption that the data are
sampled from populations with a Gaussian distribution. Similarly, the paired t test
and repeated measures ANOVA assume that the set of differences (between paired
or matched values) are sampled from a Gaussian population. Because these tests
are based on an assumption about the distribution of values in the population that
can be defined by parameters, they are called *parametric tests*.

Nonparametric tests make few assumptions about the distribution of the
populations. The most popular forms of nonparametric tests are based on a really
simple idea. Rank the values from low to high and analyze only those ranks, ignoring the actual values. This approach ensures that the test isn't affected much by
outliers (see Chapter 25) and doesn't assume any particular distribution.

Comparing two unpaired groups: Mann–Whitney test
The *Mann–Whitney test* is a nonparametric test used to compare two unpaired
groups to compute a P value. It works by following these steps:

1. Rank all the values without paying attention to the group from which the
   value is drawn. In the example from Chapter 30 (comparing the bladder
   muscle of old and young rats), two values tie for ranks of 4 and 5, so both
   are assigned a rank of 4.5. Table 41.1 and Figure 41.1 show the ranks.

OLD	YOUNG
3.0	10.0
1.0	13.0
11.0	14.0
6.0	6.0
4.5	15.0
8.0	17.0
4.5	9.0
12.0	7.0
2.0	

Table 41.1. Ranks of the data shown in Table 30.1.

The smallest value has a rank of 1, and the largest value has a rank of
17. Note that two values tie for ranks of 4 and 5, so both are assigned a
rank of 4.5. These ranks are plotted in Figure 41.1 and are displayed in
the same order as the actual values are displayed in Table 30.1.

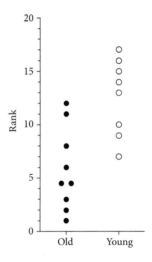

Figure 41.1. Plot of the data (ranks) of Table 41.1.

2. Sum the ranks in each group. In the example data, the sum of the ranks
    of the old rats is 52 and the sum of the ranks of the young rats is 101.
    The values for the younger rats tend to be larger and thus tend to have
    higher ranks.
3. Calculate the mean rank of each group.
4. Compute a P value for the null hypothesis that the distribution of ranks
    is totally random. If there are ties, different programs might calculate a
    different P value.

Under the null hypothesis, it would be equally likely for either of the two
groups to have the larger mean ranks and more likely to find the two mean ranks
close together than far apart. Based on this null hypothesis, the P value is computed
by answering the question, If the distribution of ranks between two groups were

distributed randomly, what is the probability that the difference between the mean ranks would be this large or even larger? The answer is 0.0035. It certainly is not impossible that random sampling of values from two identical populations would lead to sums of ranks this far apart, but it would be very unlikely. Accordingly, we conclude that the difference between the young and old rats is statistically significant.

Although the Mann–Whitney test makes no assumptions about the distribution of values, it is still based on some assumptions. Like the unpaired t test, the Mann–Whitney test assumes that the samples are randomly sampled from (or representative of) a larger population and that each value is obtained independently. But unlike the t test, the Mann–Whitney test does not assume anything about the distribution of values in the populations from which the data are sampled.

The Mann–Whitney test is equivalent to a test developed by Wilcoxon, so the same test is sometimes called the *Wilcoxon rank-sum test*. Don't mix this test up with the nonparametric test for paired data discussed in the next section.

Comparing two paired groups: Wilcoxon matched-pairs signed-rank test

The *Wilcoxon matched-pairs signed-rank test* (often referred to simply as *Wilcoxon's test*) compares two paired groups. It tests the null hypothesis that there is no difference in the populations and so the differences between matched pairs will be randomly positive or negative.

Like the paired t test (see Chapter 31), Wilcoxon's test can be used in several situations:

- When a variable is measured for each subject before and after an intervention.
- When subjects are recruited as pairs matched for variables such as age, postal code, or diagnosis. One of each pair receives one intervention, and the other receives an alternative treatment.
- When twins or siblings are recruited as pairs. Each receives a different treatment.
- When each run of a laboratory experiment has a control and treated preparation handled in parallel.
- When a part of the body on one side is treated with a control treatment and the corresponding part of the body on the other side is treated with the experimental treatment (e.g., different treatments for right and left eyes).

It works via the following steps:

1. Calculate the difference between each matched pair, keeping track of the sign. A decrease is negative and an increase is positive.
2. Rank the absolute value of the differences, temporarily ignoring the sign.
3. Add up the ranks of all the positive differences and the ranks of all the negative differences. For the example data concerning Darwin's plants (see Chapter 31), the sums of the ranks are 96 and 24.
4. Compute the difference between those two sums (here, 72).

5. Computing a P value answers the question, If the null hypothesis were true, what would be the chance of randomly choosing samples such that the sums of the absolute values of the positive and negative ranks would differ by 72 or more? The answer is 0.0413.

Like the paired t test, Wilcoxon's test assumes that the pairs are randomly selected from (or at least representative of) a larger population and that each pair is selected independently from the others. Unlike the paired t test, Wilcoxon's test does not assume a Gaussian distribution of differences.

Nonparametric correlation

One nonparametric method for quantifying correlation is called *Spearman's rank correlation*. Spearman's rank correlation is based on the same assumptions as ordinary (Pearson) correlations, outlined in Chapter 32, with two exceptions. Rank correlation does not assume Gaussian distributions and does not assume a linear relationship between the variables. However, Spearman's (nonparametric) rank correlation does assume that any underlying relationship between X and Y is monotonic (i.e., either always increasing or always decreasing).

Spearman's correlation separately ranks the X and Y values and then computes the correlation between the two sets of ranks. For the insulin sensitivity example of Chapter 32, the nonparametric correlation coefficient, called r_s, is 0.74. The P value, which tests the null hypothesis that there is no rank correlation in the overall population, is 0.0036.

An easy way to think about how the two kinds of correlation are distinct is to recognize that Pearson correlation quantifies the *linear* relationship between X and Y, while Spearman quantifies the *monotonic* relationship between X and Y.

Nonparametric ANOVA

The nonparametric test analogous to one-way ANOVA is called the *Kruskal–Wallis test*. The nonparametric test analogous to repeated-measures one-way ANOVA is called *Friedman's test*. These tests first rank the data from low to high and then analyze the distribution of the ranks among groups.

THE ADVANTAGES AND DISADVANTAGES
OF NONPARAMETRIC TESTS

The advantage of nonparametric tests is clear. They don't require the assumption of sampling from a Gaussian population and so can be used when the validity of that assumption is dubious. And when that assumption is false, nonparametric tests have more power than parametric tests to detect differences.

So why not always use nonparametric tests? There are three reasons.

Nonparametric tests are less powerful when the data are Gaussian

Because nonparametric tests only consider the ranks and not the actual data, they essentially throw away some information, so they are less powerful. If there truly is a difference between Gaussian populations, the P value is likely to be higher with a nonparametric test. How much higher? It depends on sample size.

With large samples, the nonparametric tests are nearly as powerful as the parametric tests when the data are really sampled from Gaussian populations. This is assessed by a value called the *asymptotic relative efficiency*. For example, with large samples drawn from Gaussian populations, the asymptotic relative efficiency of the Mann–Whitney test is 95%. This means that the power of a Mann–Whitney test is equal to the power of a t test with 95% as many data points. The other nonparametric tests have similarly high asymptotic relative efficiencies.

With small samples from Gaussian populations, nonparametric tests have much less power than parametric tests. When the populations are not Gaussian, however, nonparametric tests can be (depending on the distribution) much more powerful than parametric tests (Sawilowsky, 2005).

With tiny samples, nonparametric tests always report P values greater than 0.05 and so have zero power. To ever compute a two-tailed P value less than 0.05:

- The Mann–Whitney test requires eight or more values (in total, both groups).
- Wilcoxon's matched pairs test requires six or more data pairs.
- Spearman's correlation requires five or more XY pairs.

Nonparametric results are usually not reported with CIs

This book has emphasized the importance of CIs. Most programs, however, report only P values and not CIs for nonparametric tests.

Some nonparametric tests can be extended to compute CIs, but this requires additional assumptions. For example, the Mann–Whitney test can be extended to provide a CI for the difference between medians (Graphad Prism does this). However, this requires assuming that the distributions of the two populations have the same shape, with only the distributions being shifted (and thus having different medians). This assumption is not needed to interpret the P value.

Nonparametric tests are not readily extended to regression models

Chapter 35 showed that t tests can be recast as regression. A t test compares two means, and so can simple linear regression. Multiple regression can compare two means after adjusting for differences in other variables. Nonparametric tests cannot be readily extended in this way. There are nonparametric ways to fit curves (look up splines or kernel methods), but these do not estimate values of parameters.

CHOOSING BETWEEN PARAMETRIC AND NONPARAMETRIC TESTS: DOES IT MATTER?

Does it matter whether you choose a parametric or nonparametric test? The answer depends on sample size. There are four cases you must consider (Table 41.2).

DISTRIBUTION	TEST	SMALL SAMPLES	LARGE SAMPLES
Gaussian population	Nonparametric	Misleading. Nonparametric tests have little power with small samples.	Little problem. With large samples, nonparametric tests are nearly as powerful as parametric tests.
Non-Gaussian population	Parametric	Misleading. With small samples, parametric tests are not very robust to violations of the Gaussian assumption.	Little problem. With large samples, parametric tests are robust to violations of the Gaussian assumption.
Not sure	Normality tests	Not very helpful with small samples.	Helpful.

Table 41.2. The problem with small samples.

Using a nonparametric test with a small data set sampled from a Gaussian population

Nonparametric tests lack statistical power with small samples. The P values will tend to be high.

Using a nonparametric test with a large data set sampled from a Gaussian population

Nonparametric tests work well with large samples from Gaussian populations. The P values will tend to be a bit too large, but the discrepancy is small. In other words, nonparametric tests are only slightly less powerful than parametric tests with large samples.

Using a parametric test with a small data set sampled from a non-Gaussian population

The central limit theorem (discussed in Chapter 10) doesn't apply to small samples, so the P value may be inaccurate.

Using a parametric test with a large data set sampled from a non-Gaussian population

The central limit theorem ensures that parametric tests work well with large samples even if the data are sampled from non-Gaussian populations. In other words, parametric tests are robust to mild deviations from Gaussian distributions, so long as the samples are large. But there are two snags:

- It is impossible to say how large is large enough, because it depends on the nature of the particular non-Gaussian distribution. Unless the population distribution is really weird, you are probably safe choosing a parametric test when there are at least two-dozen data points in each group.
- If the population is far from Gaussian, you may not care about the mean or differences between means. Even if the P value provides an accurate answer to a question about the difference between means, that question may be scientifically irrelevant.

Summary

Large data sets present no problems. It is usually easy to tell whether the data are likely to have been sampled from a Gaussian population (and normality tests can help), but it doesn't matter much, because in this case, nonparametric tests are so powerful and parametric tests are so robust.

Small data sets present a dilemma. It is often difficult to tell whether the data come from a Gaussian population, but it matters a lot. In this case, nonparametric tests are not powerful and parametric tests are not robust.

SAMPLE SIZE FOR NONPARAMETRIC TESTS

Depending on the nature of the distribution, nonparametric tests might require either more or fewer subjects than the corresponding test that assumes a Gaussian distribution. A general rule of thumb is this: if you plan to use a nonparametric test, compute the sample size required for a parametric test (see Chapter 26) and add 15% (Lehman, 2007).

NONPARAMETRIC TESTS THAT ANALYZE VALUES (NOT RANKS)

There is another approach to analyzing data that makes no assumptions about the shape of the population distribution and also avoids the need to analyze ranks. These tests go by the names *permutation tests*, *randomization tests*, and *bootstrapping*.

The idea of all these tests is to analyze the actual data, not ranks, but not to assume any particular population distribution. All of these tests rely on intensive calculations that require computers and so are called *computer-intensive methods*. Because they don't make any assumptions about population distributions, these tests are nonparametric (even though that term is sometimes used only to refer to tests that analyze ranks).

To compute a P value, the parametric methods start with an assumption (usually Gaussian distribution) about the population and then use math to figure out the distribution of all possible samples from that population.

The computer-intensive nonparametric methods have a completely different outlook. They make no mathematical assumption about the population beyond the fact that the sample data were selected from it. They work by brute force rather than elegant math.

Randomization or permutation tests work by shuffling. Computers run many simulated "experiments." Each one uses the sample data but changes the labels for the groups. These labels are randomly shuffled among the values. If you are comparing a control and a treated group, each shuffling changes the selection of which values are labeled "control" and which are labeled "treated." Analyzing this set of pseudosamples can lead to valid insights into your data.

With small data sets, these methods can systematically inspect every possible way to shuffle the labels among values. With large data sets, the number of possible rearrangements becomes astronomical, and so these software programs examine a large number (often 1,000 to 10,000) of randomly chosen permutations, which is enough to get valid results.

Another approach called *bootstrapping* or *resampling* has already been mentioned briefly in Chapter 13. The idea of bootstrapping is that all you know for sure about the population is that the sample was obtained from it. Statistical inference requires thinking about what would happen if you picked many samples from that population. Without making any parametric assumptions, this outcome can be approximated by resampling from that single sample. The next two paragraphs give you a taste for how this approach works.

This resampling is done with replacement. Imagine you write each value on a card and place all of the cards in a hat. Mix well, pick a card, and record its value. Then put it back in the hat, mix the cards well again, and pick the next card. Note that some values (cards) may get selected more than once and other values may not be selected at all. Of course, the resampling process is actually done with computerized random-number generators.

Resampling produces many pseudosamples, each of which is the same size as the original sample. The samples are different because some values are repeated and others are omitted. Comparing the actual data with the distribution of bootstrap samples can lead to statistical conclusions expressed as P values and CIs.

This may seem like magic. It sure doesn't seem that useful statistical conclusions will emerge from analysis of sets of pseudosamples created by resampling, or bootstrapping (the two terms are essentially interchangeable). That accounts for the strange name, *bootstrapping*. At first glance, analyzing data this way seems about as (un)helpful as trying to get out of a hole by pulling up on the straps of your boots.

Nevertheless, the theorems have been proven, the simulations have been run, and plenty of real-world experiences have validated these approaches. They work! Many think computer-intensive methods are the future of statistics because they are so versatile. They can be easily adapted for use in new situations, and they don't require making or believing assumptions about the distribution of the population (beyond the assumption that the sample is representative). To learn more, start with a short text by Wilcox (2001) and then a longer book by Manly (2006).

COMMON MISTAKES: NONPARAMETRIC TESTS

Mistake: Using a normality test to automate the decision of when to use a nonparametric test

The decision of when to use a nonparametric test is not straightforward, and reasonable people can disagree about when to use nonparametric tests. Many think that the choice of using a nonparametric test can be automated: First, perform a normality test (see Chapter 24). If the data pass, use a parametric test. If the data fail the normality test, then use a nonparametric test.

This approach is not recommended for the following reasons:

- When analyzing data from a series of experiments, all data should be analyzed the same way (unless there is some reason to think they aren't comparable). In this case, results from a single normality test should not be used to decide whether to use a nonparametric test.
- Data sometimes fail a normality test because the values were sampled from a lognormal distribution (see Chapter 11). In this case, transforming the data to logarithms will create a Gaussian distribution. In other cases, transforming data to reciprocals or using other transformations can often convert a non-Gaussian distribution to a Gaussian distribution.
- Data can fail a normality test because of the presence of an outlier (see Chapter 25). In some cases, it might make sense to analyze the data without the outlier using a conventional parametric test rather than a nonparametric test.
- The decision of whether to use a parametric or nonparametric test is most important with small data sets (because the power of nonparametric tests is so low). But with small data sets, normality tests have little power, so an automatic approach would give you false confidence.

The decision of when to use a parametric test and when to use a nonparametric test really is a difficult one, requiring thinking and perspective. As a result, this decision should not be automated.

Mistake: Referring to "nonparametric" data

The term *nonparametric* refers to a set of statistical tests. It cannot be used to describe data sets.

Mistake: Running both parametric and nonparametric tests and picking the result with the smallest P value

P values can only be interpreted when the test is chosen as part of the experimental design. If you run two (or more) tests and report the results of the one that gives the lowest P value, you cannot interpret that P value at face value.

Q & A

Can data be nonparametric?

No. *Nonparametric* is an adjective that can only be applied to statistical tests, not to data.

If I am sure my data are not Gaussian, should I use a nonparametric test?

Not necessarily. It may be possible to transform the data in a way that makes the population Gaussian. Most commonly, a log transform of data from a lognormal distribution (see Chapter 11) will create a Gaussian population.

Are the chi-square test and Fisher's exact test nonparametric?

The term *nonparametric* is sometimes used for these tests, but there is no distinction between parametric and nonparametric tests when analyzing dichotomous data. Most people would not call these tests *nonparametric*, but some do.

Are the computer-intensive methods (bootstrapping, resampling, permutations) considered to be nonparametric tests?

Those computer-intensive tests don't depend on any assumptions about the distribution of the population, so in that technical sense, they certainly are nonparametric, and that is why I include them in this chapter. But the term *nonparametric* is used inconsistently. Some textbooks of nonparametric tests include computer-intensive methods, and some don't.

Does the Mann–Whitney test compare medians?

Only if you assume that the two populations have identically shaped distributions. The distributions don't have to be Gaussian or even specified, but you do have to assume that the shapes are identical. Given that assumption, the only possible way for two populations to differ is by having different medians (i.e., the distributions are the same shape but shifted). If you don't make that assumption, then it is not correct to say that a Mann–Whitney test compares medians.

Can nonparametric tests be used when some values are "off scale"?

If some values are too high or too low to be quantified, parametric tests cannot be used, because those values are not known. If they simply are ignored, the test results will be biased because the largest (or smallest) values were thrown out. In contrast, a nonparametric test can work well when a few values are too high (or too low) to measure. Assign values too low to measure an arbitrary very low value and assign values too high to measure an arbitrary very high value. Because the nonparametric test only knows about the relative ranks of the values, it won't matter that you didn't enter those extreme values precisely.

When the decision isn't clear, should I choose a parametric test or a nonparametric test?

When in doubt, some people choose a parametric test because they aren't sure the Gaussian assumption is violated. Others choose a nonparametric test because they aren't sure the Gaussian assumption is met. Ideally, the decision should be based partly on experience with the same kind of data in other experiments. Reasonable people disagree. It really is a hard decision.

Which set of tests gives lower P values, parametric or nonparametric?

It depends on the data.

If you transform all data to logs and then rerun a Mann–Whitney test, will the results change?

No. The Mann–Whitney test only looks at the rank of each value and ignores the values themselves. Transforming values to logarithms will not change their ranking, so will not change the results of a Mann–Whitney test. The only exception would be if any values were negative or zero. Since the logarithm of negative numbers and zero is not defined, these values would be essentially removed from the data set during a log transform. In that case, the results of a Mann–Whitney test would be different after the transform, because it would analyze only a subset of the values.

CHAPTER SUMMARY

- ANOVA, t tests, and many statistical tests assume that you have sampled data from populations that follow a Gaussian bell-shaped distribution.

- Biological data never follow a Gaussian distribution precisely, because a Gaussian distribution extends infinitely in both directions and so includes both infinitely low negative numbers and infinitely high positive numbers! But many kinds of biological data follow a bell-shaped distribution that is approximately Gaussian. Because ANOVA, t tests, and other statistical tests work well even if the distribution is only approximately Gaussian (especially with large samples), these tests are used routinely in many fields of science.
- An alternative approach does not assume that data follow a Gaussian distribution. In this approach, values are ranked from low to high, and the analyses are based on the distribution of ranks. These tests, called nonparametric tests, are appealing because they make fewer assumptions about the distribution of the data.
- Computer-intensive methods (resampling, bootstrapping, permutation tests) also do not rely on any assumption about the distribution of the population and so are also nonparametric.

TERMS INTRODUCED IN THIS CHAPTER

- Asymptotic relative efficiency (p. 435)
- Computer-intensive methods (p. 437)
- Friedman's test (p. 434)
- Kruskall–Wallis test (p. 434)
- Mann–Whitney test (or Wilcoxon rank-sum test) (p. 431)
- Nonparametric tests (p. 431)
- Parametric tests (p. 431)
- Permutation tests (p. 437)
- Randomization tests (p. 437)
- Spearman's rank correlation (p. 434)
- Wilcoxon matched-pairs signed-rank test (or Wilcoxon's test) (p. 433)

CHAPTER 42

Sensitivity, Specificity, and Receiver Operating Characteristic Curves

We're very good at recognizing patterns in randomness but we never recognize randomness in patterns.

DANIËL LAKENS

This chapter explains how to quantify false positive and false negative results from laboratory tests. Although this topic is not found in all basic statistics texts, deciding whether a clinical laboratory result is normal or abnormal relies on logic very similar to that used in deciding whether a finding is statistically significant or not. Learning the concepts of sensitivity and specificity presented here is a great way to review the ideas of statistical hypothesis testing and Bayesian logic explained in Chapter 18.

DEFINITIONS OF SENSITIVITY AND SPECIFICITY

This chapter discusses the accuracy of clinical tests that can report two possible results: normal or abnormal. There are two ways such a test can be wrong.

- It can report a *false negative* result—reporting that the result is normal when the patient has the disease for which he or she is being tested. In Table 42.1, false negative results are tabulated in cell C.
- It can report a *false positive* result—reporting that the result is abnormal when the patient really does not have the disease for which he or she is being tested. In Table 42.1, false positive results are tabulated in Cell B.

The accuracy of a diagnostic test is quantified by its sensitivity and specificity.

- The *sensitivity* is the fraction of all those with the disease who get a positive test result. In Table 42.1, sensitivity equals A/(A + C). Sensitivity measures how well the test identifies those with the disease, that is, how sensitive it is. If a test has high sensitivity, it will pick up nearly everyone with the disease.
- The *specificity* is the fraction of those without the disease who get a negative test result. In Table 42.1, specificity equals D/(B + D). Specificity

	DISEASE PRESENT	DISEASE ABSENT	TOTAL
Abnormal (positive) test result	A	B (false positives)	A + B
Normal (negative) test result	C (false negative)	D	C + D
Total	A + C	B + D	A + B + C + D

Table 42.1. The results of many hypothetical lab tests, each analyzed to reach a decision to call the results normal or abnormal.

The top row tabulates results for patients without the disease, and the second row tabulates results for patients with the disease. You can't actually create this kind of table from a group of patients unless you run a "gold standard" test that is 100% accurate.

measures how well the test identifies those who don't have the disease, that is, how specific it is. If a test has very high specificity, it won't mistakenly give a positive result to many people without the disease.

THE PREDICTIVE VALUE OF A TEST

Definitions of positive and negative predictive values

Neither the specificity nor the sensitivity answers the most important questions: If the test is positive (abnormal test result, suggesting the presence of disease), what is the chance that the patient really has the disease? And if the test is negative (normal test result), what is the chance that the patient really doesn't have the disease? The answers to those questions are quantified by the *positive predictive value* and the *negative predictive value*. Based on Table 42.1,

$$\text{Positive Predictive Value} = \frac{\text{True positives}}{\text{All positive results}} = \frac{A}{A+B}$$

$$\text{Negative Predictive Value} = \frac{\text{True negatives}}{\text{All negative results}} = \frac{D}{C+D}$$

The sensitivity and specificity are properties of the test. In contrast, the positive predictive value and negative predictive value are determined by the characteristics of the test and the prevalence of the disease in the population being studied. The lower the prevalence of the disease is, the lower the ratio of true positives to false positives is. This is best understood by example.

Background to porphyria example

Acute intermittent porphyria is an autosomal dominant disease that is difficult to diagnose clinically. It can be diagnosed by detecting reduced levels of the enzyme porphobilinogen deaminase. However, the levels of the enzyme vary in both the normal population and patients with porphyria, so the test does not lead to an exact diagnosis.

Using the definition that levels less than 98 units are abnormal, 82% of patients with porphyria have an abnormal test result. This means that the sensitivity of the test is 82%. Additionally, 3.7% of normal people have an abnormal test result. This means that the specificity of the test is 100% – 3.7%, or 96.3%.

What is the probability that a patient with fewer than 98 units of enzyme activity has porphyria? In other words, what is the positive predictive value of the test? The answer depends on who the patient is. We'll work through two examples.

Predictive values when a test is used for random screening

We'll compute the predictive values in two contexts. In this context, the test was done to screen for the disease, so the people being tested had no particular risk for the disease. Porphyria is a rare disease, with a prevalence of about one in 10,000. Because the people being tested were not selected because of family history or clinical suspicion, you would expect about 1 in 10,000, or 0.01%, to have the disease.

The test gave a positive result for a patient. Knowing the test result, what is the probability that this patient has the disease? To find out, we must enter numbers into Table 42.1 to create Table 42.2.

1. Assume a population of 1,000,000. All we care about are ratios of values, so the total population size is arbitrary. $A + B + C + D = 1,000,000$.
2. Because the prevalence of the disease is 1/10,000, the total number in the disease present column is $0.0001 \times 1,000,000$, or 100. $A + C = 100$.
3. Subtract the 100 diseased people from the total of 1,000,000, leaving 999,900 people without the disease. $B + D = 999,900$.
4. Calculate the number of people with disease present who also test positive. This equals the total number of people with the disease (100) times the sensitivity (0.82). $A = 82$.
5. Calculate the number of people without the disease who test negative. This equals the total number of people without the disease (999,900) times the specificity (0.963). $D = 962,904$.
6. Calculate the number of people with the disease who test negative: $C = 100 - 82 = 18$.
7. Calculate the number of people without the disease who test positive: $B = 36,996$.
8. Calculate the two row totals. $A + B = 37,078$. $C + D = 962,922$.

If you screen 1 million people, you will expect to find an abnormal test result in 37,078 people. Only 82 of these cases will have the disease. Therefore, when someone has a positive test, the chance is only $82/37,078 = 0.22\%$ that he or she

	DISEASE PRESENT	DISEASE ABSENT	TOTAL
Abnormal test result	82	36,996	37,078
Normal test result	18	962,904	962,922
Total	100	999,900	1,000,000

Table 42.2. Expected results of screening 1 million people with the porphyria test from a population with a prevalence of 0.01%.

Most of the abnormal test results are false positives.

has the disease. That is the positive predictive value of the test. Because only about 1 in 500 of the people with positive test results have the disease, the other 499 of 500 positive tests are false positives.

Of the 962,922 negative tests results, only 18 are false negatives. The predictive value of a negative test is 99.998%.

Predictive value when testing siblings

The disease is autosomal dominant, so there is a 50% chance that each sibling has the gene. If you test many siblings of people who have the disease, you will expect about half to have the disease. Table 42.3 shows the predicted results when you test 1,000 siblings of patients. Positive results are expected in 429 (500 × 82% + 500 × 3.7%) of the people tested. Of these individuals, 410 will actually have the disease and 19 will be false positives. The predictive value of a positive test, therefore, is 410/429, which is about 96%. Only about 4% of the positive tests are false positives.

Another example HIV testing

The DoubleCheckGold™ HIV 1&2 test rapidly detects antibodies to human immunodeficiency virus (abbreviated HIV), the cause of AIDS, in human serum or plasma (Alere, 2014). Its sensitivity is 99.9%, which means that 99.9% of people who are infected with HIV will test positive. Its specificity is 99.6%, which means that 99.6% of people without the infection will test negative, which leaves 100.0% − 99.6% = 0.4% of them who will test positive even though they don't have the infection.

What happens if we test 1 million people in a population where the prevalence of HIV infection is 10%? The number of individuals with HIV will be 100,000 (10% of 1 million), and 99,900 of those will test positive (99.9% of 100,000). The number of people without HIV will be 900,000. Of those, 0.4% will test positive, which is 3,600 people. In total, there will be 99,900 + 3,600 = 103,500 positive tests. Of these, 3,600/103,500 = 0.035 = 3.5% will be false positives.

What if the prevalence of HIV is only 0.1%? The number of individuals with HIV will be 1,000 (0.1% of 1 million), and 999 of those will test positive (99.9% of 1,000). The number of people without HIV will be 999,000. Of those, 0.4% will test positive, which is 3,996 people. In total, there will be 999 + 3,996 = 4,995 positive tests. Of these, 3,996/4,995 = 0.80 = 80% will be false positives (Table 42.4).

	DISEASE PRESENT	DISEASE ABSENT	TOTAL
Abnormal test result	410	19	429
Normal test result	90	481	571
Total	500	500	1,000

Table 42.3. Expected results of testing 1,000 siblings of someone with porphyria.
In this group, the prevalence will be 50% and few of the abnormal test results will be false positives.

These examples demonstrate that the fraction of the positive tests that are false positives depends on the prevalence of the disease in the population you are testing.

Analogy to statistical tests

The analogies between interpreting laboratory tests and statistical hypothesis testing are shown in Tables 42.5 and 42.6.

Interpreting a positive or negative test requires knowing who is tested (what the prevalence is). Similarly, as explained in Chapters 18 and 19, the interpretation of statistical significance depends on the scientific context (the prior probability).

	DISEASE PRESENT	DISEASE ABSENT	TOTAL
Abnormal test result	999	3,996	4,995
Normal test result	1	995,004	995,005
Total	1,000	999,000	1,000,000

Table 42.4. **Expected results of testing a million people with an HIV test in a population where the prevalence of HIV is 0.1%.**
Even though the test seems to be quite accurate (sensitivity = 99.9%; specificity = 99.6%), 80% of the abnormal test results will be false positives (3,996/4,995 = 80%).

LAB TEST	DISEASE PRESENT	DISEASE ABSENT
Decision: Abnormal result		False positive
Decision: Normal result	False negative	

STATISTICAL HYPOTHESIS TEST	NULL HYPOTHESIS IS FALSE	NULL HYPOTHESIS IS TRUE
Decision: Reject null hypothesis (significant)		Type I error
Decision: Do not reject null hypothesis (not significant)	Type II error	

Table 42.5. **False positive and false negative lab results are similar to Type I and Type II errors in statistical hypothesis testing.**

	IF . . .	WHAT IS THE CHANCE THAT . . .
Lab test: Sensitivity	The patient has disease . . .	. . . a lab result will be positive?
Statistical hypothesis test: Power	There is a difference between populations . .	. . . a result will be statistically significant?
Lab test: Specificity	The patient does not have the disease . . .	. . . a lab result will be negative?
Statistical hypothesis test: 1-alpha	There is no difference between populations . . .	. . . a result will not be statistically significant?

Table 42.6. **Relationship between sensitivity and power and between specificity and alpha.**

RECEIVER OPERATING CHARACTERISTIC (ROC) CURVES

The trade-off

When evaluating a diagnostic test, it is often difficult to decide where to set the threshold that separates a clinical diagnosis of normal from one of abnormal.

On one hand, if the threshold is set high (assuming that the test value increases with disease severity), some individuals with low test values or mild forms of the disease will be missed. The sensitivity will be low, but the specificity will be high. Few of the positive tests will be false positives, but many of the negative tests will be false negatives.

On the other hand, if the threshold is set low, most individuals with the disease will be detected, but the test will also mistakenly diagnose many normal individuals as abnormal. The sensitivity will be high, but the specificity will be low. Few of the negative tests will be false negatives, but many of the positive tests will be false positives.

The threshold can be set to have a higher sensitivity or a higher specificity, but not both (until a better diagnostic test is developed).

What an ROC curve plots

A *receiver operating characteristic* (ROC) *curve* visualizes the trade-off between high sensitivity and high specificity (Figure 42.1). Why the odd name? ROC curves were developed during World War II, within the context of determining whether a blip on a radar screen represented a ship or an extraneous noise. The radar-receiver operators used this method to set the threshold for military action.

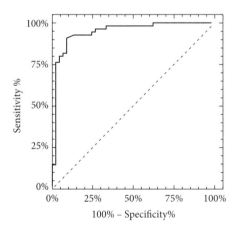

Figure 42.1. An ROC curve.

The solid line shows the trade-off between sensitivity and specificity. Each point along the curve represents a different possible threshold value defining an abnormal test result. The dotted diagonal line shows results from a test that has no ability to discriminate between patients and controls. For the dotted line, the fraction of patients diagnosed as having the disease (sensitivity) equals the fraction of controls given that diagnosis (1 – specificity).

The bottom left of the ROC curve is one silly extreme when the test never, ever returns a diagnosis that the person has the disease. At this extreme, every patient is incorrectly diagnosed as healthy (sensitivity = 0%) and every control is correctly diagnosed as healthy (specificity = 100%). At the other extreme, which is in the upper-right corner, the test always returns a diagnosis that the person tested has the disease. Every patient is correctly diagnosed (sensitivity = 100%) and every control is incorrectly diagnosed (specificity = 0%).

Where is the best cut-off?

Each point on an ROC curve shows the sensitivity and specificity for one possible threshold value for deciding when a test result is abnormal. Which is the best threshold to use? It depends on the consequences or costs of false negative and false positive diagnoses. If the two are equally bad, then the best threshold is the one that corresponds to the point on the ROC curve that is closest to the upper-left corner of the graph. But in most cases, the consequences of false positives and false negatives are not comparable, so it will be difficult to decide what cut-off makes the most sense. That decision must be made by someone who understands the disease and the test. It is not a decision to be made by a computer program.

BAYES REVISITED

Interpreting clinical laboratory tests requires combining what you know about the clinical context and what you learn from the lab test. This is simply Bayesian logic at work. Bayesian logic has already been discussed in Chapter 18.

Probability versus odds

As you may recall from Chapter 2, chance can be expressed either as a probability or as odds.

- The *probability* that an event will occur is the fraction of times you expect to see that event over the course of many trials.
- The *odds* are the probability that the event will occur divided by the probability that the event will not occur.

A probability is a fraction and always ranges from zero to one. Odds range from zero to infinity. Any probability can be expressed as odds. Any odds can be expressed as a probability. To convert between odds and probability (for situations in which there are only two possible outcomes), use the following equations:

$$\text{Odds} = \frac{\text{probability}}{1 - \text{probability}}$$

$$\text{Probability} = \frac{\text{odds}}{1 + \text{odds}}$$

If the probability is 0.50, or 50%, then the odds are 50:50, or 1:1. If you repeat the experiment often, you will expect to observe the event (on average) in one of every two trials (probability = 1/2). That means you'll observe the event once for every time it fails to happen (odds = 1:1).

If the probability is 1/3, the odds equal $1/3/(1 - 1/3) = 1:2 = 0.5$. On average, you'll observe the event once in every three trials (probability = 1/3). That means you'll observe the event once for every two times it fails to happen (odds = 1:2).

The likelihood ratio

The *likelihood ratio* is the probability of obtaining a positive test result in a patient with the disease divided by the probability of obtaining a positive test result in a patient without the disease. The probability of obtaining a positive test result in a patient with the disease is the sensitivity. The probability of obtaining a positive test result in someone without the disease is (1 – specificity). So the likelihood ratio equals sensitivity divided by (1 – specificity).

Using this equation, we can rework the examples. The test used in the intermittent porphyria example has a sensitivity of 82% and a specificity of 96.3%. Thus, the likelihood ratio is $0.82/(1.0 - 0.963) = 22.2$. A person with the condition is 22.2 times more likely to get a positive test result than a person without the condition.

Bayes as an equation

Bayes's equation for clinical diagnosis can be written in two forms:

$$\text{Posttest Odds} = \text{Pretest odds} \cdot \frac{\text{sensitivity}}{1 - \text{specificity}}$$

$$\text{Posttest Odds} = \text{Pretest odds} \cdot \text{Likelihood ratio}$$

The posttest odds are the odds that a patient has the disease, taking into account both the test results and your prior knowledge about the patient. The pretest odds are the odds that the patient has the disease as determined from information you know before running the test.

Table 42.7 reworks the two examples with intermittent porphyria, which has a likelihood ratio of 22.2, using the second equation shown previously in this section.

Who was tested?	PRETEST		POSTTEST	
	Probability	Odds	Odds	Probability
Random screen	0.0001	0.0001	0.0022	0.0022
Sibling	0.50	1.0000	22.2	0.957

Table 42.7. The porphyria calculations computed using Bayes's equation.

The pretest probability is first converted to odds. Then the posttest odds are computed by multiplying the pretest odds by the likelihood ratio, which equals 22.2. Finally, the odds are converted to probabilities.

COMMON MISTAKES

Mistake: Automating the decision about which point on an ROC curve to use as a cut-off.

The ROC curve plots the trade-offs between sensitivity and specificity. Which combination is the best to define a critical value of a lab test? It depends on the consequences of making a false positive or a false negative diagnosis. That decision needs to be made in a clinical (or in some situations, scientific) context and should not be automated.

Mistake: Thinking that a single value can quantify the accuracy of a test.

There are many ways to quantify accuracy.

Q & A

How is the accuracy of a test defined?
"Accuracy" is a fairly ambiguous term. This chapter defines the accuracy of a test in four different ways.

Why is this chapter in a statistics text?
One reason is that terms like specificity and sensitivity get used in scientific papers, and there is a fuzzy border between epidemiology and biostatistics. The other reason is that interpreting statistical significance works very much like interpreting lab results, and one needs to take into account prior probability (prevalence for lab tests) to properly interpret the results. Review Chapters 18 and 19 and see if they make more sense now that you understand how to interpret clinical lab results.

CHAPTER SUMMARY

- When you report a test result as positive or negative, you can be wrong in two ways: A positive result can be a false positive, and a negative result can be a false negative.
- The sensitivity of a test is the fraction of all those with a disease who correctly get a positive result.
- The specificity is the fraction of those without the disease who correctly get a negative test result.
- The positive predictive value of a test answers the question, If the test is positive (abnormal test result, suggesting the presence of disease), what is the chance that the patient really has the disease? The answer depends in part on the prevalence of the disease in the population you are testing.
- The negative predictive value of a test answers the question, If the test is negative (normal test result, suggesting the absence of disease), what is the chance that the patient really does not have the disease? The answer depends in part on the prevalence of the disease in the population you are testing.

- A receiver operating characteristic (ROC) curve visualizes the trade-off between high sensitivity and high specificity depending on where you designate the cut-off between normal and abnormal test results.
- The likelihood ratio is the probability of obtaining a positive test result in a patient with the disease divided by the probability of obtaining a positive test result in a patient without the disease.

TERMS INTRODUCED IN THIS CHAPTER

- False negative (p. 442)
- False positive (p. 442)
- Likelihood ratio (p. 449)
- Negative predictive value (p. 443)
- Positive predictive value (p. 443)
- Receiver operating characteristic (ROC) curve (p. 447)
- Sensitivity (p. 442)
- Specificity (p. 442)

CHAPTER 43

Meta-analysis

> For all meta-analytic tests (including those for bias): If it looks
> bad, it's bad. If it looks good, it's not necessarily good.
>
> DANIËL LAKENS

When many studies have been performed on a particular topic, it is natural to want to combine the results. This chapter explains the methods and pitfalls of meta-analysis.

INTRODUCING META-ANALYSIS

What is meta-analysis?

Meta-analysis is used to combine evidence from multiple studies, usually clinical trials testing the effectiveness of therapies or tests. The pooled evidence from multiple studies can give much more precise and reliable answers than any single study can.

A large part of most meta-analyses is a descriptive review that points out the strengths and weaknesses of each study. In addition, a meta-analysis pools the data to report an overall effect size (perhaps a relative risk or odds ratio), an overall CI, and an overall P value. Meta-analysis is often used to summarize the results of a set of clinical trials.

Why meta-analysis is difficult?

Investigators planning a meta-analysis face many problems, summarized in Table 43.1. Meta-analysis is not simple.

The promise of meta-analysis

If a huge, well-designed clinical trial has been conducted, there is little need for a meta-analysis. The results from the large study will be compelling. Meta-analyses are needed when the individual trials have conflicting, ambiguous, or unconvincing conclusions. The hope is that combining the trials in a meta-analysis will lead to a persuasive conclusion. Many researchers are convinced that meta-analyses are hugely important and helpful (Cumming, 2011; Goldacre, 2013). Others doubt that meta-analysis can consistently overcome the huge challenges listed in Table 43.1 and produce consistently useful results (Dallal, 2012; Bailar, 1997). If you want to learn about meta-analysis, the text by Borenstein and colleagues (2009) is a good place to start.

CHALLENGE	SOLUTION
Investigators may use inconsistent criteria to define a disease and so may include different groups of patients in different studies.	Define which patient groups are included.
Studies asking the same clinical question may use different outcome variables. For example, studies of cardiovascular disease often tabulate the number of patients who have experienced new cardiac events. Some studies may only count people who have documented myocardial infarctions (heart attacks) or who have died. Others may include patients with new angina or chest pain.	Define which clinical outcomes are included.
Some studies, especially those that don't show the desired effect, are not published. Omitting these studies from the meta-analysis causes publication bias (discussed later in this chapter).	Seek unpublished studies; review registries of study protocols.
Some studies are of higher quality than others.	Define what makes a study of high enough quality to be included.
Some relevant studies may be published in a language other than English.	Arrange for translation.
Published studies may not include enough information to perform a proper meta-analysis.	Estimate data from figures; obtain unpublished details from the investigators or Web archives.
Published data may be internally inconsistent. For example, Garcia-Berthou and Alcaraz (2004) found that 11% of reported P values were not consistent with the reported statistical ratios (t, F, etc.) and dfs.	Resolve inconsistencies without bias; request details from original investigators.
Data from some patients may be included in multiple publications. Including redundant analyses in a meta-analysis would be misleading.	When multiple studies are published by the same investigators, ask them to clarify which patients were included in more than one publication.

Table 43.1. The challenges of meta-analysis.

PUBLICATION BIAS

The problem

Scientists like to publish persuasive findings. When the data show a large and statistically significant effect, it is easy to keep up the enthusiasm necessary to write up a paper, create the figures, and submit the resulting manuscript to a journal. If the effect is large, the study is likely to be accepted for publication, as editors prefer to publish papers that report results that are statistically significant. If the study shows a small effect, especially if it is not statistically significant, it is much harder to stay enthusiastic through this process and stick it out until the work is published.

Many people have documented that this really happens. Studies that find large, statistically significant effects are much more likely to get published, while studies showing small effects tend to be abandoned by the scientists or rejected by journals. This tendency is called *publication bias*.

An example of publication bias

Turner and colleagues (2008) demonstrated this kind of selectivity in industry-sponsored investigations of the efficacy of antidepressant drugs. Between 1987 and 2004, the FDA in the US reviewed 74 such studies and categorized them as positive, negative, or questionable. The FDA reviewers found that 38 studies showed a positive result (i.e., the antidepressant worked). All but one of these studies was published. The FDA reviewers found that the remaining 36 studies had negative or questionable results. Of these, 22 were not published, 11 were published with a "spin" that made the results seem somewhat positive, and only 3 were published with clearly presented negative findings.

Think about that. The investigators were forced to dig through the FDA files to find all the studies. Only about half of these studies (38 of 74) showed a positive result. Almost all (49 of 52) of the published studies, however, concluded that the drugs worked. What a difference! Selective publication leads to misleading conclusions and reduces the value of a meta-analysis. Goldacre (2013) gives more examples of this bias in the first chapter ("Missing Data") of his compelling book *Bad Pharma*.

How a meta-analysis can deal with publication bias

Meta-analyses really can only be interpreted at face value when they summarize all the studies that have been performed. Good meta-analyses try hard to include all relevant studies. This may means including articles written in a language the analyst doesn't understand (so needs to arrange for translation) and articles published in obscure journals that are not available online. It is also important to try to include unpublished studies. One approach is to search a registry of study protocols (e.g., https://clinicaltrials.gov) and then track down the investigators and obtain unpublished results.

This is a tough problem. By definition, unpublished studies are hard to learn about, and you can never be sure that you have found them all. And even if the meta-analyst tracks down the investigators of an unpublished study, those investigators may not have completed the study, may not have kept good records, or may choose not to share the data.

Meta-analyses often try to test for publication bias by looking at the pattern of data. These tests are based on a simple idea. The effect sizes reported by smaller studies will usually end up with more variation than large studies. If the analysts found all the studies, the variation among these effect sizes ought to be symmetrical around the overall effect size computed by the meta-analysis. If this distribution is not symmetrical, one possibility is that small studies with an effect that went in the "wrong" direction were never published – publication bias. While that idea is simple, the details of such an analysis are tricky, and beyond the scope of this text. If you want to learn more, read about *funnel plots* and Egger's test (Sedgwick, 2015; Cochrane Handbook, 2017).

RESULTS FROM A META-ANALYSIS

Part of a meta-analysis is a narrative report that summarizes the studies that have been done and evaluates their strengths and weaknesses.

The rest of a meta-analysis is quantitative. The results of each study are summarized by one value, called the *effect size,* along with its CI. The effect size is usually a relative risk or odds ratio, but it could also be some other measure of treatment effect. In some cases, different studies use different designs, and so the meta-analyst has to do some conversions so that the effect sizes are comparable.Combining all the studies, the meta-analysis computes the pooled treatment effect, its CI, and a pooled P value.

The results for the individual studies and the pooled results are plotted on a graph known as a *forest plot* or *blobbogram.* Figure 43.1 is an example. It shows part of a meta-analysis performed by Eyding et al. (2010) on the effectiveness of the drug reboxetine, a selective inhibitor of norepinephrine re-uptake, marketed in Europe as an antidepressant. The researchers tabulated two different effects for this drug: the odds ratio for remission and the odds ratio for any response. They computed these odds ratios comparing reboxetine against either placebo or other antidepressants.

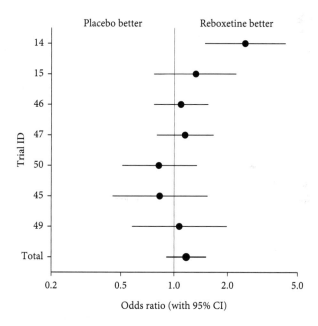

Figure 43.1. Forest plot.
This forest plot, adapted from Eyding et al. (2010), shows the results from seven studies of the effectiveness of reboxetine as a treatment for severe depression. The horizontal axis is the odds ratio. A value of 1.0 means no effect; a value greater than 1.0 means the drug works better than placebo; and a value less than 1.0 means the drug works worse than placebo. The top 7 lines are results of individual studies, identified by labels in the Y-axis (with a key in the meta-analysis publication). For each study, the plot shows the odds ratio (circle) and 95% confidence interval (line). At the very bottom, the graph shows the total (pooled) effect computed from all seven studies, also shown as an odds ratio with its 95% confidence interval.

Figure 43.1 shows the odds ratio for remission comparing reboxetine to placebo. An odds ratio of 1.0 means no effect, an odds ratio greater than 1.0 means that reboxetine was more effective than placebo, and an odds ratio less than 1.0 means that reboxetine was less effective than placebo. The graph shows 95% CIs for seven studies identified with numbers (which can be used to look up details presented in the published meta-analysis). The bottom symbol shows the overall odds ratio and its 95% CI.

In six of the seven studies, the 95% CI includes 1.0. With 95% confidence, you cannot conclude from these six studies that reboxetine works better than placebo. Accordingly, a P value would be greater than 0.05. In only one of the studies (the one shown on top of the graph and identified as Study 14) does the 95% confidence not include 1.0. The data from this study, but not the others, would lead you to conclude with 95% confidence that reboxetine worked better than placebo and that the effect is statistically significant (with $P < 0.05$). As explained in Chapter 22, even if the drug were entirely ineffective, it is not too surprising that one of six studies would find a statistically significant effect just by chance.

The bottom part of the graph shows the total, or pooled, effect, as computed by the meta-analysis. The 95% CI ranges from 0.98 to 1.56. The CI contains 1.0, so the P value must be greater than 0.05. From this CI alone, it might be hard to know what to conclude. The overall odds ratio is 1.24. That is a small effect, but not a tiny one. The 95% CI includes 1.0 (no effect), but just barely. The paper also compares reboxetine to other antidepressants and concludes that reboxetine is less effective than the others.

META-ANALYSIS OF INDIVIDUAL PARTICIPANT DATA

Meta-analyses have been traditionally done by pooling the results from a set of studies. A newer approach is for the meta-analysts to obtain the original data for each participant of each study, and reanalyze all the data together. This approach is called *meta-analysis of individual participant data,* or *meta-analysis of individual patient data*, abbreviated *IPD meta-analysis*. This approach is considered the gold standard of meta-analysis (Thomas, Radji and Benedetti, 2014) and has determined definitive answers to clinical questions, which might not have been obtained from meta-analysis of summary data (Cochrane Methods, 2017).

Performing an IPD meta-analysis requires obtaining all the data from each study, reviewing these data, rearranging and transforming so all the data from all the studies are in a consistent format, and then running the analyses. Often, this requires that the meta-analyst contact the investigators of each study to resolve any ambiguities. This is a huge amount of work compared to a meta-analysis of summarized data.

Why is it worth doing an IPD meta-analysis if it is so much more work? Because IPD meta-analysis has numerous advantages (Riley, 2015; Thomas, 2014), including:

- Ensure that the outcomes are defined in a consistent way.
- In some cases, the investigators may have continued to follow the participants (for a future paper). The IPD meta-analysis can obtain these data, and so include more data than did the original paper.

- Answer questions not posed by the original investigators.
- Analyze all the data using the same methods, even if the original investigators used differing methods.
- Run subgroup analyses that were not run by the original investigators.
- Account for confounding variables not considered by the original investigators.

ASSUMPTIONS OF META-ANALYSIS

The calculations used in a meta-analysis are complicated and beyond the scope of this book. But you should know that there are two general methods that are used, each of which is based on different assumptions:

- Fixed effects. This model assumes that all the subjects in all the studies were really sampled from one large population. Thus, all the studies are estimating the same effect, and the only difference between studies is due to random selection of subjects.
- Random effects. This model assumes that each study population is unique. The difference among study results is due both to differences between the populations and to random selection of subjects. This model is more realistic and is used more frequently.

COMMON MISTAKES: META-ANALYSIS

Mistake: Using the fixed effects model without thinking

Don't use the fixed effects model without considering the possibility that each study population is unique, in which case the random effects model should be used.

Mistake: Thinking that meta-analyses are objective

The meta-analysis calculations are very objective and use standard methods. But before the calculations can be performed, the investigator must decide which studies to include and which to reject. These decisions include some subjective criteria. A different investigator might make different decisions.

Q & A

How does the meta-analysis combine the individual P values from each study?
> It doesn't! A meta-analysis pools weighted effect sizes, not P values. It calculates a pooled P value from the pooled effect size.

Why not just average the P values?
> A meta-analysis accounts for the relative sizes of each study. This is done by computing a weighted average of the effect size. Averaging P values would be misleading.

How is a meta-analysis different from a narrative review article?

In a narrative review, the author discusses all the studies and explains their strengths and weaknesses. This is a very subjective process. In contrast, a meta-analysis is quantitative. It combines the evidence and calculates a pooled effect size (and its CI).

What is the minimum number of studies that can be combined in a meta-analysis?

Two. Meta-analyses don't have to be huge. It can be helpful to combine the results from as few as two studies.

If every study in the meta-analysis had a P value greater than 0.05, is it possible for the meta-analysis to conclude that there is a statistically significant effect?

Yes. Pooling data in a meta-analysis can do this. Figure 43.2 shows an example. The 95% CI for relative risk of stroke for those taking statins crossed 1.0 (no change in risk) for all six of the studies in the meta-analysis. But look at the bottom line. It does not cross the line that marks a relative risk of 1.0. Combined, the data provide evidence (with 95% confidence) that taking statins reduces the risk of stroke.

If every study in the meta-analysis had a P value less than 0.05, is it possible for the meta-analysis to conclude that there is not a statistically significant effect?

Surprisingly, the answer is yes. If you analyze using the random effects model and the effect sizes in the studies are not very consistent, the pooled P value can be greater than any of the individual P values, so it is possible for the pooled P value to be greater than 0.05 even if all the individual P values are all less than 0.05.

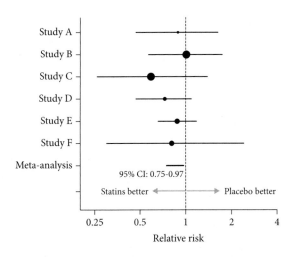

Figure 43.2. Meta-analysis for the risk of strokes after taking statins.

Each line shows the 95% confidence interval for the relative risk of stroke comparing those who take statins with controls. The vertical line is at a relative risk of 1.0, which represents equal risk in both groups. The six lines on top, one for each study, cross that line so show no significant association of statins and stroke. The meta-analysis computed an overall 95% confidence interval, shown at the bottom. It does not cross 1.0, so the meta-analysis shows a statistically significant association of taking statins and fewer strokes (P = 0.02), even though none of the individual studies do. From Thavendiranathan and Bagai (2006).

CHAPTER SUMMARY

- A meta-analysis pools multiple studies and reports an overall effect size (perhaps a relative risk or odds ratio), an overall CI, and an overall P value.
- Studies that show positive results are far more likely to be published than studies that reach negative or ambiguous conclusions. Selective publication (publication bias) makes it impossible to properly interpret the published literature and can make the results of meta-analyses suspect. A good meta-analyst has to work hard to find relevant unpublished data.
- The main results of a meta-analysis are shown in a forest plot, also known as a blobbogram. This kind of graph shows the effect size of each study with its CI, as well as the pooled (overall) effect with its CI.

TERMS INTRODUCED IN THIS CHAPTER

- Effect size (p. 455)
- Fixed effects (p. 457)
- Individual participant data (IPD) meta-analysis (p. 456)
- Forest plot (or blobbogram) (p. 455)
- Meta-analysis (p. 452)
- Meta-analysis of individual participant data (p. 456)
- Meta-analysis of individual patient data (p. 456)
- Publication bias (p. 453)
- Random effects (p. 457)

PART I

Putting It All Together

CHAPTER 44

The Key Concepts of Statistics

*If you know twelve concepts about a given topic you will look
like an expert to people who only know two or three.*

SCOTT ADAMS

When learning statistics, it is easy to get bogged down in the details and lose track of the big picture. This short chapter reviews the most important concepts in statistical inference that every scientist should know.

Many statistical terms are also ordinary words

A large part of the challenge of learning any new field is mastering its vocabulary. This is a special challenge with statistics, because so many of the terms are ordinary words given special meanings. It is easy to think that a term has an ordinary meaning when it actually is being used in a very special way. This causes lots of confusion and miscommunication about statistics.

Statistics helps you make general conclusions from limited data

The whole point of statistics is to extrapolate from limited data to make a general conclusion. *Descriptive statistics* simply describe data without reaching any general conclusions. But the challenging and difficult aspects of statistics are all about making inferences—reaching general conclusions from limited data.

Statistical conclusions are always presented in terms of probability

Myles Hollander said, "Statistics means never having to say you are certain" (quoted in Samaniego, 2008). Every statistical conclusion must include words like "probable," "most likely," or "almost certainly." Be wary if you ever encounter statistical conclusions that seem 100% definitive—you are probably misunderstanding something.

All statistical tests are based on assumptions

Every statistical inference is based on a list of assumptions. Review that list before interpreting any statistical results.

The SD and SEM are quite different

As explained in Chapter 14, the SD and SEM are quite distinct. SD quantifies the average variation among a set of values, while the SEM estimates how close the sample mean is likely to be to the population mean. The SEM is computed by dividing the SD by the square root of n (the sample size), so it is always smaller than the SD. The SEM is used to compute the CI and is sometimes used to plot error bars on graphs.

CIs quantify precision

Almost all results—proportions, relative risks, odds ratios, means, differences between means, slopes, rate constants, and so on—can and should be reported with a CI. Whenever you see a value computed from data, ask for its CI.

Every P value tests a null hypothesis

The logic of P values seems strange at first, and they are commonly misinterpreted. The key to interpreting a P value is to first identify the null hypothesis that the P value refers to. Every P value is associated with a null hypothesis, so you cannot begin to understand a P value until you can precisely state which null hypothesis is being tested. If you can't confidently state the null hypothesis, you have no idea what that P value means.

The sample size has a huge effect on the P value

Table 44.1 demonstrates that sample size has a huge impact on the P value. Each P value in the table was calculated using an unpaired t test (see Chapter 30) comparing two groups. One group has a mean of 10 and an SD of 5, and the other group has a mean of 15 and an SD of 5. The only thing that differs between the difference comparisons is sample size. When n (in each group) is 3, the P value is 0.29. When n equals 50, the P value is 0.000003. This huge spread is due only to differences in sample size, as the samples have exactly the same means and SD in every case.

SAMPLE SIZE (PER GROUP)	P VALUE (UNPAIRED T TEST)
3	0.287864
5	0.152502
7	0.085946
10	0.038250
15	0.010607
20	0.003073
25	0.000912
50	0.000003

Table 44.1. The impact of sample size on P value.

The P value in each row was computed by an unpaired t test comparing one group with mean = 10 and SD = 5 with another group with mean = 15 and SD = 5. The first column shows the sample size for each of the two groups.

"Statistically significant" does not mean the effect is large or scientifically important

The term *significant* has a special meaning in statistics. It means that, by chance alone, a difference (or association, or correlation, etc.) as large (or larger than) the one you observed would happen less than 5% of the time (or some other stated value). That's it! A tiny effect can be statistically significant and yet scientifically or clinically trivial. Spector and Vesell (2006b) give an example. Use of the drug montelukast for treatment of allergic rhinitis reduces symptoms, and the results are statistically significant. However, the drug reduces allergy symptoms by only 7%, so it may not be clinically useful.

The conclusion that a finding is statistically significant does not tell you that the effect is large, does not tell you that the effect is important, and does not tell you that it is worth following up. Chapter 18 lists many explanations for a statistically significant result.

"Not significantly different" does not mean the effect is absent, small, or scientifically irrelevant

If a difference is not statistically significant, you can conclude that the observed results are not inconsistent with the null hypothesis. Note the double negative. You *cannot* conclude that the null hypothesis is true. It is quite possible that the null hypothesis is false and that there really is a small difference between the populations. All you can say is that the data are not strong or consistent enough to persuade you to reject the null hypothesis.

Chapter 19 lists many reasons for obtaining a result that is not statistically significant.

The concept of statistical significance is designed to help you make a decision based on one result

It is helpful to define a result as being either statistically significant or not statistically significant when you plan to make a crisp decision based on a single result. If you don't plan to use this one result to make a clear decision, the concept of statistical significance is not necessary and is likely to be confusing or misleading.

If you decide to classify all results as "statistically significant" and "not statistically significant," then some of your results will inevitably be very close to the deciding line. There really is not much difference between $P = 0.045$ and $P = 0.055$! If you are using the concepts of statistical hypothesis testing and have set the significance level to 5% (as is conventional), then the first result is statistically significant and the second one is not.

Setting the dividing line at 0.05 is just a convention. Figure 44.1 is a cartoon that shows how some scientists seem to mistakenly think that $P = 0.05$ has magic qualities.

Multiple comparisons make it hard to interpret statistical results

When many hypotheses are tested at once, the problem of multiple comparisons makes it very easy to be fooled. If 5% of tests will be statistically significant by chance, you will expect lots of statistically significant results if you test hundreds or thousands of hypotheses. Special methods can be used to reduce the problem

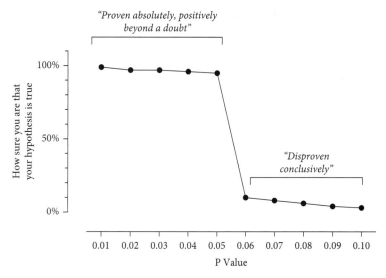

"Proven absolutely, positively beyond a doubt"

"Disproven conclusively"

Figure 44.1. "P = 0.05" does not have magical properties.

This figure exaggerates what many researchers seem to believe—that P values less than 0.05 definitely prove that the experimental hypothesis is true and P values greater than 0.05 definitely prove that the experimental hypothesis is wrong. Of course, this belief is wrong. There is very little difference between 0.04 and 0.06. Adapted from Goldstein (2006).

of finding false but statistically significant results, but these methods also make it harder to find true effects.

Multiple comparisons can be insidious. To correctly interpret statistical analyses, all analyses must be planned (before collecting data), and all planned analyses must be conducted and reported. However, these simple rules are widely broken.

Correlation does not prove causation

If two variables are correlated (or associated) and that difference is statistically significant, you should not conclude that changes in the variable labeled X causes changes in the variable labeled Y. It is possible that changes in Y cause changes in X, that changes in X cause some other change that indirectly changes Y (or vice versa), that both X and Y are influenced by a third variable, or that the correlation is spurious and due to a coincidence of random sampling. Correlation does not prove causation.

Published statistics tend to be optimistic

By the time you read a paper, a great deal of selection has occurred. When experiments are successful, scientists tend to continue the project. Lots of other projects get abandoned. When the project is done, scientists are more likely to write up projects that lead to remarkable results or keep analyzing the data in various ways to extract a statistically significant conclusion. Finally, journals are more likely to publish "positive" studies. If the null hypothesis were true, you would expect a statistically significant result in 5% of experiments. But those 5% are more likely to get published than the other 95%. This is called *publication bias* and was discussed in Chapter 43.

What about the published findings that are not false positives? As pointed out in Chapter 26, these studies tend to inflate the size of the difference or effect. The explanation is simple. If many studies were performed, you'd expect the average of the effects detected in these studies to be close to the true effect. By chance, some studies will happen to find larger effects and some studies will happen to find smaller effects. However, studies with small effects tend not to get published. On average, therefore, the studies that do get published tend to have effect sizes that overestimate the true effect (Ioannidis, 2008). This is called the *winner's curse* (Zollner and Pritchard, 2007). This term was coined by economists to describe why the winner of an auction tends to overpay.

Decisions about how to analyze data should be made in advance

Analyzing data requires many decisions: Collect how many data points? Run a parametric or nonparametric test? Eliminate outliers or not? Transform the data first? Normalize to external control values? Adjust for covariates? Use weighting factors in regression? All these decisions (and more) should be made as part of the experimental design. When decisions about statistical analysis are made only after inspecting the data, it is too easy for statistical analysis to become a high-tech Ouija board—a method to produce preordained results rather than an objective method of analyzing data.

Statistics is only part of interpreting data

Statistical calculations provide limited help in overcoming the following common issues in interpreting data:

- The population you really care about is more diverse than the population from which your data were sampled.
- The subjects in the study were not randomly sampled from a larger population.
- The measured variable is a proxy for another variable you really care about.
- The measurements may be made or recorded incorrectly, and assays may not always measure exactly the right thing.
- Scientific (or clinical) conclusions require examining multiple outcomes, not just one.

Understanding the statistical calculations is only part of evaluating scientific data. Interpreting data also requires a great deal of judgment.

Statistical calculations are helpful, but should augment (not replace) looking at graphs of the data

Always look at graphs (or tables) first to understand the data. Then look at statistical calculations to sharpen your understanding.

TERM INTRODUCED IN THIS CHAPTER

- Winner's curse (p. 467)

CHAPTER 45

Statistical Traps to Avoid

When the data don't make sense, it's usually because you have
an erroneous preconception about how the system works.

ERNEST BEUTLER

To avoid being misled by statistical results when reading scientific
papers or using statistical software, beware of the traps listed in this
chapter. Be sure to review Chapter 44 before reading this one.

TRAP #1: FOCUSING ON P VALUES AND STATISTICAL SIGNIFICANCE RATHER THAN EFFECT SIZE

P values and conclusions about statistical significance can be useful, but there
is more to statistics than P values and asterisks. Focus instead on the effect size
(the size of the difference or association, etc.). Is the difference or association
or correlation (effect) big enough to care about? Is that effect determined with
sufficient precision (as assessed by the CI)? With huge sample sizes, tiny effects might be statistically significant yet irrelevant. With small sample sizes,
you may not have determined the effect with sufficient precision to make any
useful conclusion.

Gelman (2013) suggests a clever approach to avoid getting distracted by
statements of statistical significance. He suggests imagining that the sample size
is huge ("one zillion"). In that case, essentially every effect will be statistically
significant, but most will be trivial, so "you'll need to ignore conclusions about
statistical significance and instead think about what you really care about."

TRAP #2: TESTING HYPOTHESES SUGGESTED BY THE DATA

When studies are designed to answer a focused question, the results are often
straightforward to interpret. But some studies analyze data for many variables,
in many groups of subjects, and have no real hypothesis when they begin. Such
studies can be a useful way to generate hypotheses as part of *exploratory data
analysis*. New data, however, are needed to test those hypotheses.

The trap occurs when a scientist analyzes the data many ways (which might
include many variables and many subgroups), discovers an intriguing relationship,

and then publishes the results so it appears that the hypothesis was stated before the data collection began As Chapter 23 noted, Kerr (1998) coined the acronym *HARK*: hypothesizing after the results are known. Kriegeskorte and colleagues (2009) call this *double dipping*.

It is impossible to evaluate the results from such studies unless you know exactly how many hypotheses were actually tested. You will be misled if the results are published as if only one hypothesis was tested. A XKCD cartoon points out the folly of this approach (Figure 45.1).

Also beware of studies that don't truly test a hypothesis. Some investigators believe their hypothesis so strongly (and may have stated the hypothesis so vaguely) that no conceivable data would lead them to reject that hypothesis. No matter what the data showed, the investigator would find a way to conclude that the hypothesis is correct. Such hypotheses "are more 'vampirical' than 'empirical'—unable to be killed by mere evidence" (Gelman & Weakliem, 2009 citing Freese, 2008).

TRAP #3: ANALYZING WITHOUT A PLAN—"P-HACKING"

Analyzing data requires many decisions. How large should the samples be? Which statistical test should you use? What should you do with outliers? Should you transform the data first? Should the data be normalized to external control values? Should you run a multivariable regression to control for differences in confounding variables? All of these decisions (and more) should be made at the time you design an experiment.

If you make analysis decisions only after you see the data, there is a danger that you will pick and choose among analyses to get the results you want and thus fool yourself. We already discussed the problem of ad hoc sample size selection in Chapter 26, and the folly of p-hacking in Chapter 23.

Gotzsche (2006) presented evidence of p-hacking (without calling it that). He tabulated the actual distribution of P values in journals and found that "P values were much more common immediately below .05 than would be expected based on the number of P values occurring in other ranges." Masicampo and Lalande (2012) found similar results (see Figure 45.2). In addition to being caused by p-hacking, this excess of P values just below 0.05 could be due to ad hoc sample size selection (see Chapter 26) and publication bias (see Chapter 43).

TRAP #4: MAKING A CONCLUSION ABOUT CAUSATION WHEN THE DATA ONLY SHOW CORRELATION

Messerli (2012) wondered why some countries produce many more Nobel Prize winners than others, and to answer this question, he plotted the data shown in Figure 45.3. The Y-axis plots the total number of Nobel Prizes ever won by citizens of each country. The X-axis plots chocolate consumption in a recent year (different years for different countries, based on the availability of data). Both X and Y values are standardized to the country's current population. The correlation

Figure 45.1. Nonsense conclusions via HARKing (hypothesizing after the results are known).

Source: https://xkcd.com.

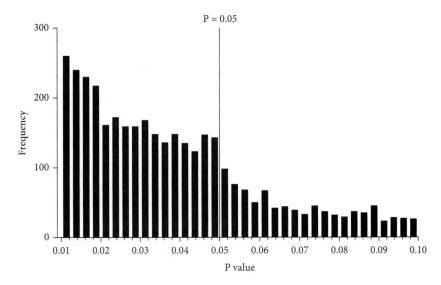

Figure 45.2. Too many published P values are just a tiny bit less than 0.05.

Masicampo and Lalande (2012) collected P values from three respected and peer-reviewed journals in psychology and then tabulated their distribution. The figure shows a "peculiar prevalence" of P values just below 0.05. This figure was made from the raw list of 3,627 P values kindly sent to me by Masicampo, matching a graph published by Wasserman (2012).

is amazingly strong, with r = 0.79. A P value testing the null hypothesis of no real correlation is tiny, less than 0.0001.

Of course, these data don't prove that eating chocolate helps people win a Nobel Prize. Nor does it prove that increasing chocolate imports into a country will increase the number of Nobel Prizes that residents of that country will win.

When two variables are correlated, or associated, it is possible that changes in one of the variables causes the other to change. But it is also likely that both variables are related to a third variable that influences both. There are many variables that differ among the countries shown in the graph, and some of those probably correlate with both chocolate consumption and number of Nobel Prizes.

This point is often summarized as "Correlation does not *imply* causation," but it is more accurate to say that correlation does not *prove* causation.

When each data point represents a different year, it is even easier to find silly correlations. For example, Figure 45.4 shows a very strong negative correlation between the total number of pirates in the world and one measure of global average temperature. But correlation (pirates would say *carrrelation*) does not prove causation. It is quite unlikely that the lack of pirates caused global warming or that global warming caused the number of pirates to decrease. More likely, this graph simply shows that both temperature and the number of pirates have changed over time. The variable time is said to be *confounded* by the other two variables.

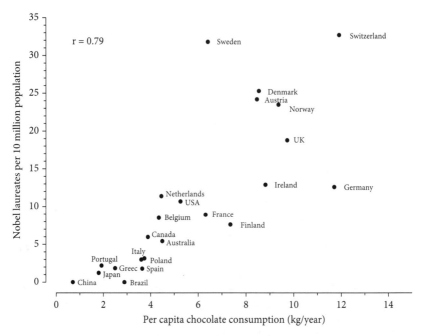

Figure 45.3. Correlation between average chocolate consumption by country and the number of Nobel Prize winners from that country.

Data are from Messerili (2012). The Y-axis plots the total number of Nobel Prizes won by citizens of each country. The X-axis plots chocolate consumption in a recent year (different years for different countries, based on the availability of data). Both X and Y values are normalized to the country's current population. The correlation is amazingly strong, but this doesn't prove that eating chocolate will help you win a Nobel Prize.

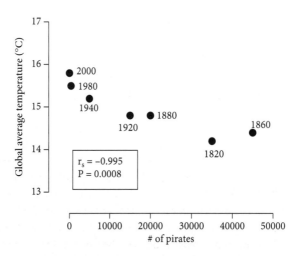

Figure 45.4. Correlation between the number of pirates worldwide and average world temperature.

Adapted from Henderson (2005). Correlation does not prove causation!

Figure 45.5. Correlation does not imply causation.

Sources: Top: https://xkcd.com. Bottom: DILBERT © 2011 Scott Adams. Used By permission of UNIVERSAL UCLICK. All rights reserved.

The cartoons in Figure 45.5 drive home the point that correlation does not prove causation. But note a problem with the XKCD cartoon. Since the outcome (understanding or not) is binary, this cartoon really points out that association does not imply causation.

TRAP #5: OVERINTERPRETING STUDIES THAT MEASURE A PROXY OR SURROGATE OUTCOME

Measuring important outcomes (e.g., survival) can be time consuming and expensive. It is often far more practical to measure *proxy*, or *surrogate*, *variables*. However, an intervention that improves the proxy variable may not improve the more important variable.

Chapter 15 already provided an example. For many years, people who had myocardial infarctions (heart attacks) were treated with anti-arrhythmic drugs. These drugs prevented extra heartbeats (premature ventricular contractions) and thus were thought to reduce the incidence of sudden death due to arrhythmia. The logic was clear. Abnormal results (extra beats) on an electrocardiogram (ECG) were known to be associated with sudden death, and treatment with

TREATMENT	SURROGATE OR PROXY VARIABLE	IMPORTANT VARIABLE(S)
Two anti-arrhythmic drugs	• Fewer premature heart beats • Conclusion: Good treatment	• More deaths • Conclusion: Deadly treatment
Torcetrapib	• Increase in HDL ("good" cholesterol) • Decrease in LDL ("bad" cholesterol) • Conclusion: Good treatment	• More deaths • More heart attacks • Conclusion: Deadly treatment

Table 45.1. Results using proxy or surrogate variables can produce incorrect conclusions.

anti-arrhythmic drugs was known to reduce the number of extra beats. So it made sense that taking those drugs would extend life. The evidence was compelling enough that the FDA approved use of these drugs for this purpose. But a randomized study to directly test the hypothesis that anti-arrhythmic drugs would reduce sudden death showed just the opposite. Patients taking two specific anti-arrhythmic drugs had (fewer extra beats, the proxy variable) but were more likely to die (CAST Investigators, 1989). Fisher and VanBelle (1993) summarize the background and results of this trial.

Another example is the attempt to prevent heart attacks by using drugs to raise HDL levels. Low levels of HDL ("good cholesterol") are associated with an increased risk of atherosclerosis and heart disease. Pfizer Corporation developed torcetrapib, a drug that elevates HDL, with great hope that it would prevent heart disease. Barter and colleagues (2007) gave the drug to thousands of patients with a high risk of cardiovascular disease. LDL ("bad cholesterol") decreased 25% and HDL ("good cholesterol") increased 72%. The CIs were narrow, and the P values were tiny (<0.001). If the goal was to improve cholesterol levels, the drug was a huge success. Unfortunately, however, treatment with torcetrapib also increased the number of heart attacks by 21% and increased the number of deaths by 58%. Similarly, niacin increases HDL but doesn't decrease heart attacks (The HPS2-THRIVE Collaborative Group, 2014).

The take-home message is clear: treatments that improve results of lab tests may not improve health or survival (see Table 45.1). Svennson (2013) lists 14 additional examples.

TRAP #6: OVERINTERPRETING DATA FROM AN OBSERVATIONAL STUDY

Munger and colleagues (2013) wondered whether deficiency of vitamin D (25-hydroxyvitamin D, abbreviated 25(OH)D) predisposed people to developing type 1 diabetes. The researchers compared a group of diabetics with a group of

| | | RESULTS FROM . . . | |
| | | OBSERVATIONAL | |
INTERVENTION	INCIDENCE OF	STUDIES	EXPERIMENT
Hormone replacement therapy after menopause	Cardiovascular events	Decrease	Increase
Megadose vitamin E	Cardiovascular events	Decrease	No change
Low-fat diet	Cardiovascular events and cancer	Decrease	No change
Calcium supplementation	Fractures and cancer	Decrease	No change
Vitamins to reduce homocysteine	Cardiovascular events	Decrease	No change
Vitamin A	Lung Cancer	Decrease	Increase

Table 45.2. Six hypotheses suggested by observational studies proven not to be true by experiment.

people without diabetes who were similar in other ways and then compared the vitamin D levels in blood samples drawn before disease onset. They found that people with average 25(OH)D levels greater than 100 nmol/L had a much lower risk of developing diabetes than those whose average 25(OH)D levels were less than 75 nmol/L. The risk ratio was 0.56, with a 95% CI ranging from 0.35 to 0.90 (P = 0.03).

Intriguing data! Do these findings mean that taking vitamin D supplements will prevent diabetes? No. The association of low vitamin D levels and onset of diabetes could be explained many ways. Sun exposure increases vitamin D levels. Perhaps sunlight exposure also creates other hormones (which we may not have yet identified) that decrease the risk of diabetes. Perhaps the people who are exposed more to the sun (and thus have higher vitamin D levels) also exercise more, and it is the exercise that helps prevent diabetes. Perhaps the people with higher vitamin D levels drink more fortified milk, and it is the calcium in the milk that helps prevent diabetes. The only way to find out for sure whether ingestion of vitamin D prevents diabetes is to conduct an experiment comparing people who are given vitamin D supplements with those who are not.

Table 45.2 lists six examples where the results of experiments were opposite to the results from observational studies. Although observational studies are much easier to conduct than experiments, data from experiments are more definitive. With observational studies, it is difficult to deal with confounding variables, and so it is nearly impossible to convincingly untangle cause and effect. The cartoon of Figure 45.6 adds two more examples.

The first five rows are adapted from Spector and Vesell (2006a). "Cardiovascular events" include myocardial infarction, sudden death, and stroke. The bottom row is from Omenn and colleagues (1996).

Figure 45.6. Multiple explanations for the results of observational studies.

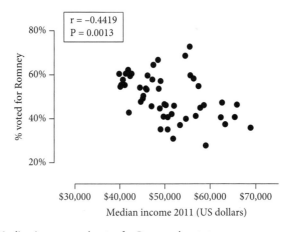

Figure 45.7. Median income and votes for Romney by state.
Each symbol represents one of the 50 states (excluding the District of Columbia). The X-axis shows median income in 2011 as determined by the U.S. Census (2011), and the Y-axis shows actual voting data from the Federal Election Commission (2012).

TRAP #7: MAKING CONCLUSIONS ABOUT INDIVIDUALS WHEN THE DATA WERE ONLY COLLECTED FOR GROUPS

Figure 45.7 shows the relationship between the median income in 2011 and the fraction of voters who voted for Romney, the Republican candidate in the 2012 US presidential election (this example was adapted from Statwing, 2012). There are 50 dots on the graph, one for each state. The relationship is striking. In states with higher income, a smaller percentage of voters tended to vote for Romney.

Does that mean that people with higher income were less likely to vote for Romney? No! The data in Figure 45.7 only let you make conclusions about states. If you want to make a conclusion about individuals, you must analyze data from individuals. Figure 45.8 shows results from polling individuals. There is a strong relationship between income and voting, but it goes in a direction *opposite* to the

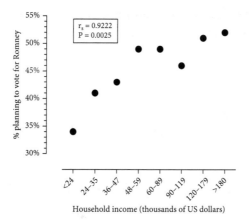

Figure 45.8. Income and votes for Romney by individual.
These are national polling data collected about two months before the election (Gallup, 2012).

state data. People with higher income were *more* likely to vote for Romney, even though states with higher average income tended to have fewer Romney voters.

What accounts for the discrepancy? There are lots of differences between states. A correlation of statewide data does not tell you about the individuals in those states (Gelman & Feller, 2012). Mistakenly using relationships among groups to make inferences about individuals is called the *ecological fallacy*. Another example is the data on Nobel Prizes and chocolate presented earlier in this chapter.

TRAP #8: FOCUSING ONLY ON MEANS WITHOUT ASKING ABOUT VARIABILITY OR UNUSUAL VALUES

"Math Scores Show No Gap for Girls, Study Finds," reported the *New York Times* several years ago (Lewin, 2008). Not quite (Briggs, 2008a). In addition to showing essentially no difference in the average math scores of boys and girls, the study showed that there was a difference in variation (Hyde et al., 2008). The scores of boys were more variable, and twice as many boys as girls were in the top 1%. To fully understand these data, you must consider more than just the averages. Kane and Mertz (2012) continued this line of inquiry using data sets from many countries and showed that boys' math scores are not universally more varied than girls' math scores.

Here is another example. Stephen Jay Gould wondered about changes in professional baseball from 1870 to 1970 (Gould, 1997). Why have there been no players with batting averages over 0.400 since 1941, even though the mean batting average has been consistent (at about 0.260)? It's not enough to look at the mean. Gould discovered that the SD of players' batting averages has dropped considerably (by almost 50%), especially in the first half of that period. This decrease in variability, he concludes, occurred because players, coaching, umpiring, and equipment have all become more consistent. Since the mean hasn't changed and the SD is much smaller than it was in earlier times, a batting average greater than 0.400 is now incredibly rare. Gould wasn't able to understand the changes in baseball until he investigated changes in variability (SD).

Variability in biological or clinical studies often reflects real biological diversity (rather than experimental error). Appreciate this diversity! Don't get mesmerized by comparing averages. Pay attention to variation and to the extreme values. Nobel Prizes have been won from studies of people whose values were far from the mean.

TRAP #9: COMPARING STATISTICALLY SIGNIFICANT WITH NOT STATISTICALLY SIGNIFICANT

Figure 45.9 is similar to many you'll see in scientific journals. It compares a response in wild-type and mutant animals. The results might be summarized in a journal as, "The drug-induced increase in uptake in the wild-type mice was not statistically significant (P = 0.068), but the increase in the mutant mice was statistically significant (P = 0.004)."

The data are simulated. The bars show the mean and SEM of triplicate values. Unpaired t tests were used to compare baseline and stimulated results, with separate t tests done for the wild-type and mutant animals. A seemingly obvious conclusion—that there is a statistically significant difference between the drug-stimulated transmitter uptake in mutant and wild-type mice—is incorrect. If you want to compare one change with another change, you need to run statistical tests with that purpose in mind. Here, the difference between wild-type and mutant drug-induced transmitter uptake is not statistically significant.

What can you conclude from that statement? A seemingly obvious conclusion is that there is a statistically significant difference between the drug-stimulated uptake in mutant mice compared to wild-type mice. But this would be an incorrect

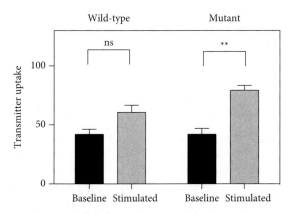

Figure 45.9. The difference between statistically significant and not statistically significant is not itself statistically significant.

conclusion (Nieuwenhuis, Forstmann, & Wagenmakers, 2011). As summarized by Gelman and Stern (2006), "The difference between 'significant' and 'not significant' is not itself statistically significant."

If you want to make a conclusion about the difference between two differences, you need to use a statistical method designed to test that exact question. Since the baseline values are very similar for the two groups of animals, the stimulated values in the wild-type animals can be directly compared to the stimulated values in the mutant animals with an unpaired t test. The difference between means is 18.7, with a 95% CI ranging from −1.8 to 39.1 (P = 0.064). A fancier approach (beyond what is covered in this book) is to use two-way ANOVA to compute a P value that tests the null hypothesis of no interaction (the null hypothesis is that the difference between baseline and stimulated is identical in wild-type and mutant animals). The conclusion would be the same. The difference in drug-stimulated transmitter uptake between mutant and wild-type animals is not significantly significant.

This is not a trivial issue that only applies, say, when one P value is 0.04 and the other is 0.06. Bland and Altman (2011) have done simulations showing how easy it is to be led astray by basing one's conclusion on the difference between a statistically significant change in one group and a not statistically significant change in the other. In their simulations, there was no difference in the actual mean increase in the two populations. Yet in 38% of their simulations, the increase was statistically significant in one group and not in the other.

TRAP #10: MISSING IMPORTANT FINDINGS BECAUSE DATA COMBINE POPULATIONS

Admission was offered to 44.3% of the men but only 34.6% of the women who applied to a graduate school (see Table 45.3; Bickel, Hammel, & O'Connell, 1975). The ratio is 1.28, with a 95% CI ranging from 1.22 to 1.34. The P value is less than 0.0001.

While these data seem to be convincing proof of sexism, in fact the data are very misleading. The problem is that the data pool results from many graduate programs. That would make sense if students applied to many graduate programs and if admissions decisions were centralized. But that is not the way graduate admissions work. Students choose the programs to which they apply, and each

	ACCEPTED	DENIED	% ACCEPTED
Men	3,738	4,704	44.3
Women	1,494	2,827	34.6

Table 45.3. Admissions to the graduate programs in Berkeley, 1973. Pooled data.
At first glance, the data seem to provide evidence of sexism, but that is only because the data pool admission rates from multiple graduate programs, each of which makes its own admissions decisions.

EVIDENCE FOR DISCRIMINATION?	NUMBER OF PROGRAMS
No statistically significant difference in admission decisions	75
Males preferred (P < 0.05)	4
Females preferred (P < 0.05)	6
Total number of programs	85

Table 45.4. Admissions to the graduate programs at Berkeley, 1973, by program.

program makes its own admissions decisions. Therefore, it is essential to analyze the data for each program individually. When the investigators did this, they found that the admission rates of men and women were about the same. The difference was not statistically significant in 75 of the 85 programs (see Table 45.4). In four programs, the difference was statistically significant, and women were less likely to be admitted. In six programs, the difference was statistically significant, and women were more likely to be admitted. Overall, the authors found no evidence of sexism. So what's going on? Why do the pooled data suggest sexism?

Some graduate programs accept a high fraction of applicants, and others accept a low fraction of applicants. The two departments most popular among women admitted only 34% and 24% of the applicants, respectively, whereas the two departments most popular among men admitted 62% and 63% of the applicants, respectively (Freedman, 2007). A smaller fraction of women were admitted overall for a simple reason: women tended to apply to more selective programs than did men.

This is a classic example of *Simpson's paradox*. Analysis of pooled data can lead to misleading results.

The same problem probably happens in some medical studies. Imagine how little progress you'd make if you tested whether a new drug cured cancer by pooling patients with all kinds of cancers. Cancers are a group of many diseases that respond to different drugs. Combining all those diagnoses into one study would lead to frustration and inconclusive results. Most likely, many medical conditions (including, perhaps, septic shock, breast cancers, and autism) really are combinations of distinct disorders. Until we figure out how to identify the individual diseases, studies of therapies are likely to be ambiguous.

TRAP #11: INVALID MULTIPLE REGRESSION ANALYSES AS A RESULT OF AN OMITTED VARIABLE

This example (extended from one presented by Freedman, 2007) is a bit silly but makes an important point. Pretend that your goal is to find a model that predicts the area of a rectangle from its perimeter.

Figure 45.10 shows that rectangles with larger perimeters also tend to have larger areas, with two outliers. Figure 45.11 fits the remaining points (after removing the "outliers") to possible models. The straight-line model (Figure 45.11, left side) might be adequate, but the sigmoid-shaped model (Figure 45.11, right side) fits the data better.

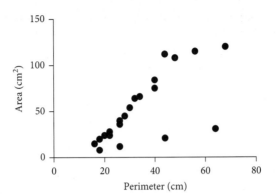

Figure 45.10. The area of rectangles as a function of their perimeter.

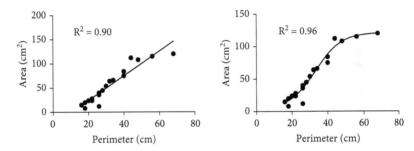

Figure 45.11. Can we fit a model to predict area from perimeter?

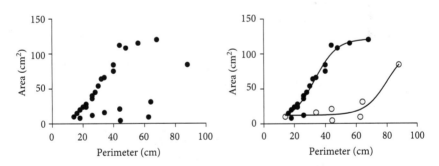

Figure 45.12. Two kinds of rectangles, maybe?

Figure 45.12 adds data from more rectangles. Now it seems that those two outliers were really not so unusual. Instead, it seems that there are two distinct categories of rectangles. The right side of Figure 45.11 tentatively identifies the two types of rectangles with open and closed circles and fits each to a different model.

While this process may seem like real science, it is not. Two rectangles with the same perimeter can have vastly different areas depending on their shapes. It simply is not possible to predict the area of a rectangle from its perimeter. The area must be computed from both height and width (or, equivalently, from perimeter and either height or width, or from perimeter and the ratio of height/width). An important variable (either height, width, or their ratio) encoding the shape of the rectangle was missing from our analysis. Understanding these data required simple thinking to identify the missing variable, rather than fancy statistical analyses. A missing variable that influences both the dependent and independent variables is called a *lurking variable*.

TRAP #12: OVERFITTING

As pointed out in Chapter 37, overfitting happens when you ask more questions than the data can answer—when you have too many independent variables in the model, compared to the number of subjects (Babyak, 2004). The problem with overfitting is that the modeling process ends up fitting aspects of the data caused by random scatter or experimental quirks. Such a model won't be reproduced when tested on a new set of data, so doesn't help you understand what is actually going on in the overall population.

TRAP #13: MIXING UP THE SIGNIFICANCE LEVEL WITH THE FPRP

The FPRP is not the same as the significance level. Mixing up the two is a common mistake.

The significance level α is the fraction of experiments performed when the null hypothesis is true that result in a conclusion that the effect is statistically significant. In Table 45.5, the significance level is defined to equal $A/(A + B)$.

The FPRP is the fraction of experiments leading to the conclusion that the effect is statistically significant where in fact the null hypothesis is really true. In Table 45.5, the FPRP is defined to equal $A/(A + C)$.

	DECISION: REJECT NULL HYPOTHESIS	DECISION: DO NOT REJECT NULL HYPOTHESIS	TOTAL
Null hypothesis is true	A (Type I error)	B	A +B
Null hypothesis is false	C	D (Type II error)	C + D
Total	A + C	B + D	A + B + C + D

Table 45.5. The results of many hypothetical statistical analyses to reach a decision to reject or not reject the null hypothesis.

A, B, C, and D are integers (not proportions) that count numbers of analyses (number of P values). The total number of analyses equals A + B + C + D. This is identical to Table 18.1.

TRAP #14: NOT RECOGNIZING HOW COMMON FALSE POSITIVE FINDINGS ARE

Chapter 18 showed you how to compute the FPRP from three values:

- Definition of statistical significance, α
- Power
- The scientific context, expressed as the prior probability (before collecting data) that the hypothesis is true

Example 4 in Chapter 18 presented a common situation where the sample size is large enough so the power is 80%, the significance level is set at the conventional 5%, and the prior probability is 50%. For these experiments, the FPRP is 5.9% if computed for all P values less than 0.05. But the FPRP is 27% if computed just for P values barely less than 0.05. Think about that. In this common situation, if you get a P value of 0.047 and conclude that the result is "statistically significant," there is a 27% chance that the finding is a false positive.

TRAP #15: NOT REALIZING HOW LIKELY IT IS THAT A "SIGNIFICANT" CONCLUSION FROM A SPECULATIVE EXPERIMENT IS A FALSE POSITIVE

The FPRP is higher when the prior probability is low and when the power is low. It can be really high when both are low as would happen when testing a wild hypothesis with small samples. Some would call this risky research, in the sense that there is a very high risk that the effect you are looking for doesn't exist (but it is not risky in the usual sense of possibly harming someone). You think that there is only a 1% chance your hypothesis is true but figure it is worth testing because if your hypothesis turns out to correct, it would be an important contribution to science (and your career).

Testing the hypothesis involves comparing the means of treated and control groups, which you expect to be sampled from Gaussian populations. The smallest effect (difference between means) you are looking for is equal to the SDs of the groups. If you want a significance level of 5% and power of 80%, you need n = 16 in each group. You don't have the time or budget to run that many samples for this speculative experiment. Instead you reduce the power to 50% so only need 8 subjects in each group. You are already running the controls as part of another experiment, so only have to add 8 more samples to test your hypothesis. You think: What can I lose? You run the experiment and the P value is less than 0.05, so the finding is "statistically significant." But before getting too excited, let's calculate the chance that you have a false positive finding.

Table 45.6 shows the results of 10,000 experiments run in this scenario. Of 10,000 experiments, there is a real effect (given our assumptions) in 1% of them, which is 100 experiments. Of those 100, you expect to detect a statistically significant effect in 50, since the power is only 50%. In the other 99% of experiments, the null hypothesis is true. Given the definition of statistical significance,

	STATISTICALLY SIGNIFICANT: REJECT NULL HYPOTHESIS	NOT STATISTICALLY SIGNIFICANT: DO NOT REJECT NULL HYPOTHESIS	TOTAL
No real effect (null hypothesis true)	495	9,405	9,900
Effect is real	50	50	100
Total	545	9,455	10,000

Table 45.6. Results of 10,000 comparisons with 50% power, a 5% significance level, and a prior probability of 1%.

random sampling will lead to a statistically significant conclusion in 5% of these experiments, which is 495 experiments. Thus, 495/(495 + 50) = 91% of the statistically significant results in this scenario will be false positives. In other words, the FPRP with this kind of experiment is 91%. This is the fraction of experiments with P < 0.05 that are actually false positives (false discoveries). If one were to calculate (via simulations) the fraction of experiments where the P value is just a tiny bit less than 0.05 (say between 0.045 and 0.050) that are false positives, the answer would be even higher than 91%.

To avoid super-high FPRP values in exploratory experiments like this, you need to set the significance level to a tiny value and choose a large enough sample size so the power is very high. Although it is tempting, it really makes no sense to run experiments of unlikely hypotheses with the standard definition of statistical significance (0.05) because a conclusion that the results are statistically significant is almost certainly a false positive.

TRAP #16: NOT REALIZING THAT MANY PUBLISHED STUDIES HAVE LITTLE STATISTICAL POWER

The tradition in many fields is to compute sample size to prove 80% statistical power. But power to detect what effect size? Dumas-Mallet and colleagues (2017) analyzed many studies published in several medical fields and quantified what their power was to detect the pooled effect size found in a meta-analysis of several studies. Ignoring the studies where the meta-analysis found a small effect, they found that the power of published studies varied between medical specialties, ranging from 22% to 77%, with a median of about 50%.

TRAP #17: TRYING TO DETECT SMALL SIGNALS WHEN THERE IS LOTS OF NOISE

Much of statistics is trying to detect a "signal" (difference, association, etc.) from data with "noise" (biological variation and technical variation). Statistical methods can help you to separate signal from noise. But not always (Figure 45.13). In criticizing one study, Andrew Gelman (2015) wrote,

Figure 45.13. Weighing a feather in the pouch of a jumping kangaroo.

Too much noise, too little signal. It is not possible to weigh a feather when it is in the pouch of a jumping kangaroo. Idea by EJ Wagenmakers; art work by Viktor Beekman (with permission; viktor.beekman@gmail.com); protected by Creative Commons license (CC BY 2.0).

Their effect size is tiny and their measurement error is huge. At some point, a set of measurements is so noisy that biases in selection and interpretation overwhelm any signal and, indeed, nothing useful can be learned from them. My best analogy is that they are trying to use a bathroom scale to weigh a feather—and the feather is resting loosely in the pouch of a kangaroo that is vigorously jumping up and down.

TRAP #18: UNNECESSARY DICHOTOMIZING

Many outcomes are binary, and Chapters 4, 27, and 28 discuss how to analyze such data. A common mistake is to convert continuous data into binary data. Three examples:

- Converting a measured weight into a binary variable that encodes if someone is obese or not obese
- Converting blood pressure into a binary variable that encodes if someone is hypertensive (high blood pressure) or not
- Converting a result (on numerical scale) of a personality test into a binary variable that encodes if someone is depressed or not

Senn (2003) pointed out several problems with studies that dichotomize the outcome variable. The biggest problem is that dichotomizing essentially throws away information, so the study loses power. If you want to design a study that uses dichotomizing to have the same power as a study that analyzes the original continuous data, you would need a much larger (about 40% larger) sample size.

There are only a few situations where dichotomization is actually helpful, but it is done quite often.

TRAP #19: INFLATING SAMPLE SIZE BY PSEUDOREPLICATION

When analyzing data with nearly any statistical test, each observation must be observed independently. If you do an experiment on 5 animals and repeat the measurement 3 times for each animal, you don't have 15 independent values. Entering the data into a statistics program as if you had 15 replicates (*pseudoreplication)* would lead to invalid statistical results (Lazic, 2010).

CHAPTER SUMMARY

- Don't get distracted by P values and conclusions about statistical significance without also thinking about the size of difference or correlation and its precision (as assessed by its CI).
- Beware of HARKing (hypothesizing after results are known). Testing a hypothesis proposed as part of the experimental design is quite distinct from "testing" a hypothesis suggested by the data.
- Beware of p-hacking. You cannot interpret P values at face value if the investigator analyzed the data many different ways.
- Correlation does not prove causation.
- Beware of surrogate or proxy variables. A treatment can improve laboratory results while harming patients.
- If you want to know if an intervention makes a difference, you need to intervene. Observational studies can be useful but rarely are definitive.
- Don't fall for the ecological fallacy. If the data were collected for groups, the conclusion can only be applied to the set of groups, not to the individuals within the groups.
- Ask about more than averages. Be on the lookout for differences in variability.
- The difference between "significant" and "not significant" is not itself statistically significant.
- If the study ignores a crucial variable, it doesn't really matter how fancy the analyses are.
- False positive conclusions are much more common than most people realize.
- Avoid dichotomizing outcome variables.

TERMS INTRODUCED IN THIS CHAPTER

- Double dipping (p. 469)
- Ecological fallacy (p. 477)
- Exploratory data analysis (p. 468)
- HARKing (Hypothesizing after the results are known) (p. 469)
- Lurking variable (p. 482)
- Proxy variable (or surrogate variable) (p. 473)
- Simpson's paradox (p. 480)
- Vampirical hypotheses (p. 469)

CHAPTER 46

Capstone Example

I must confess to always having viewed studying statistics as similar to a screening colonoscopy; I knew that it was important and good for me, but there was little that was pleasant or fun about it.

ANTHONY N. DeMARIA

This chapter was inspired by Bill Greco, who is its narrator and coauthor. One definition of capstone is "a final touch," and this chapter reviews many of the statistical principles presented throughout this book. It also demonstrates the usefulness and versatility of simulations. Although based on a true situation, many of the elements of the storyline, people's names, and drug names have been changed to protect the innocent.

THE CASE OF THE EIGHT NAKED IC_{50}S

Then

It was Friday, May 2, 1975. It was a chilly spring day in Buffalo, New York. Gerald Ford was the president of the United States. There were no PCs on the desks of scientists. The internet hadn't been invented yet. I was a new graduate student who had just received an A in a beginning biostatistics course.

An interim chairman, Dr. Jeremy Bentham, asked me for help. Two months ago, he had submitted a manuscript that compared the potency of two drugs to block the proliferation of cultured cancer cells. The key results are shown in Table 46.1, which tabulates the amount of drug needed to inhibit cell growth by 50% (the IC_{50}).

The average IC_{50} for trimetrexate (TMQ) was 21 nM and that for methotrexate (MTX) was 2.4 nM. A smaller IC_{50} corresponds to a more potent drug, because less drug is needed to inhibit growth. MTX was 8.8-fold (21 nM/2.4 nM) more potent than TMQ. MTX was the standard dihydrofolate reductase (DHFR) inhibitor used in the clinic; TMQ was a new lipophilic DHFR inhibitor with "better" cellular uptake properties.

The investigator was sure of the conclusion based on his experience with these types of agents and the specific cell growth inhibition assay. He was very confident that TMQ was about tenfold less potent than MTX in this KB cell line.

TMQ (nM)	MTX (nM)
5.5	0.83
12	1.1
19	1.9
47	5.8

Table 46.1. Sample IC$_{50}$ data for this example.
Human KB cells were grown in monolayer culture in glass T-25 flasks with RPMI 1640 medium and 10% fetal calf serum for seven days. Each IC$_{50}$ was calculated from the concentration-effect curve from a separate experiment, consisting of five control flasks (no added drug) and three flasks at each of seven different logarithmically spaced concentrations of TMQ or MTX. Cell growth was assessed by measuring total protein content with the Lowry assay.

Each of the eight experiments had taken one week to conduct. All eight experiments were conducted with the same passage of the KB cell line over the six-month period preceding the submission of the manuscript. The IC$_{50}$s for TMQ and MTX of 19 and 1.9 nM, respectively, were determined from a joint experiment with shared controls. The other six IC$_{50}$ determinations were made at different times and preceded this last paired experiment. The overall theme of the manuscript was a comparison of the two DHFR inhibitors against the single KB human cell line regarding antiproliferative potency, cellular uptake, metabolism, and inhibition of the target enzyme DHFR.

As part of the usual process of peer review, the journal's editor sent the manuscript to several anonymous reviewers. The paper was rejected largely because of the following comment by one reviewer: "This reviewer performed an unpaired Student's t test on the data [see Table 46.1], and found that there is no significant difference between the mean IC$_{50}$ of TMQ and MTX (P = 0.092) against KB cells. The data do not support your conclusion."

Not believing the reviewer, Dr. Bentham asked me to check the calculations. Using the formulas and tables in my biostatistics text and a scientific calculator, I performed a standard unpaired two-tailed Student's t test (with the usual assumption of equal error variance for the two populations). I found that t = 2.01 and P = 0.092, in total agreement with the reviewer. Because P > 0.05, I informed the investigator that the reviewer was correct: the difference in potency was not statistically significant.

Because the P value was 0.092, the investigator asked, doesn't that mean there is a (1.000 − 0.092) = 0.908, or 90.8%, chance that the findings are real? I explained to him that a P value cannot be interpreted that way. The proper interpretation of the P value is that even if the two drugs really have the same potency (the null hypothesis), there is a 9.2% chance of observing this large a discrepancy (or larger still) by chance. If I had had a crystal ball, perhaps I could have shown him Chapter 19 of this text!

I took the analysis one step further and computed the CI of the difference. The difference in observed IC$_{50}$s was −18.5 nM (the order of subtraction was arbitrary, so the difference could have been +18.5 nM; it is important only to be consistent). The 95% CI for the difference was from −41.0 to +4.06 nM. I pointed out that the lower confidence limit was negative (MTX was more potent) and the upper confidence limit was positive (TMQ was more potent), so the 95% CI included the possibility of no difference in potency (equal IC_{50} values; the null hypothesis). This was consistent with the P value (see Chapter 17). When a P value is greater than 0.05, the 95% CI must include the value specified by the null hypothesis (zero, in this case).

The investigator became enraged. He thought the data in the table would convince any competent pharmacologist that TMQ is, beyond any reasonable doubt, less potent than MTX. Statistics weren't needed, he thought, spouting off half-remembered famous quotes: "When the data speaks for itself, don't interrupt." "There are lies, damned lies and statistics." "If you need statistics, you've done the wrong experiment."

The investigator sent the manuscript unaltered to another journal, which accepted it.

Now

It is Friday, January 23, 2009. It is a chilly winter day in Buffalo, New York. Barack Obama has just become the president of the United States. There are small, powerful PCs on the desks of virtually all scientists. Powerful statistical PC software is available, at reasonable or no cost, to all scientists. The Internet is a major part of professional and personal life. I am the instructor of an applied biostatistics course. I use *Intuitive Biostatistics* as the main textbook in the course. I just retrieved the published paper on TMQ and MTX. As I stare at the IC_{50} table reproduced here as Table 46.1, it stares back at me.

The mismatch of statistical reasoning and common sense gnawed at me. I knew the statistical calculations were correct. But I also felt the senior and experienced investigator's scientific intuition and conclusion were also probably correct. Something was wrong, and this chapter is the culmination of decades of angst. I must resolve this 34-year-old quandary. How should I have analyzed these data back in 1975? How could I analyze these data now in 2009?

Before reading on, think about how you would analyze these data.

LOOK BEHIND THE DATA

If I were asked to consult on this project today, I would not begin with Table 46.1. Instead, I would go back to the raw data. Each of the values in Table 46.1 came from analyzing an experiment that measured cell number at one time point in the presence of several different concentrations of drug. The IC_{50} is defined as the concentration of drug that inhibits cell growth to a value halfway between the maximum and minimum values. This raises many questions:

- What are the details of the cell proliferation assay? Which details provide the opportunity for experimental artifacts? Were cells actually counted? Or did they measure some proxy variable (perhaps protein content) that is usually proportional to cell count? Over what period of time was cell growth measured? Has the relationship between cell growth and time been well characterized in previous experiments? Does cell growth slow down because of cell-to-cell contact inhibition (or depletion of nutrients) as the flasks fill with cells? Are the different drug treatments randomly assigned to the flasks? Is it possible that cell growth depends on where each flask is placed in the incubator?

- How was 100% cell growth defined—as growth with no drug? What if growth was actually a bit faster in the presence of a tiny concentration of drug? Should the 100% point be defined by averaging the controls or by fitting a curve and letting the curve-fitting program extrapolate to define the top plateau?
- How was 0% cell growth defined? Is 0% defined to be no cell growth at all or cell growth with the largest concentration of drug used? What if an even larger concentration had been used? Or was 0% defined by a curve-fitting program that extrapolated the curve to infinite drug concentration? The answer to this seemingly simple technical question can have a large influence on the estimation of drug potency. The determination of the IC_{50} can be no more accurate than the definitions of 0% and 100%.
- How was the IC_{50} determined? By hand with a ruler? By hand with a French curve? By linearizing the raw data and then using linear regression? By fitting the four-parameter concentration-effect model (see Chapter 36) to the raw data using nonlinear regression? If so, with what weighting scheme?

Here, we will skip all those questions and assume that the values in Table 46.1 are reliable. But do note that many problems in data analysis actually occur while planning or performing the experiment or while wrangling the data into a form that can be entered into a statistics program.

STATISTICAL SIGNIFICANCE BY CHEATING

One way to solve the problem is to report a one-tailed P value, rather than a two-tailed P value. It isn't too hard to justify that decision. All I have to do is state that the experimental hypothesis was to confirm that TMQ is less potent than MTX. The one-tailed P value is half the two-tailed P value, so it is 0.046. That is less than 0.05, so now the difference is statistically significant. Mission accomplished. That was easy!

Of course, this is cheating. To justify a one-tailed P value, the decision to report a one-tailed P value would need to have been made before the data were collected. It wasn't. At that time, it would also be necessary to decide that any data showing that TMQ is more potent than MTX would be attributed to chance and not worth pursuing. In reality, if the results had gone the other way, with MTX being substantially less potent, the investigator probably would have found an explanation and reported the findings.

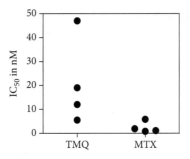

Figure 46.1. The data shown in Table 46.1.

USING A t TEST THAT DOESN'T ASSUME EQUAL SDs

One of the assumptions for the unpaired t test—that the data are sampled from populations with equal SDs—seems dubious with these data. Look at Figure 46.1. The TMQ data are way more scattered. This can be quantified by comparing the two SDs. The sample SD for the TMQ data was 18.3 nM, whereas the sample SD for the MTX data was 2.31 nM.

Chapter 30 provides a formal way to test the assumption of equal variances Calculate F as the square of the ratio of SDs, $(18.3/2.31)^2 = 62.7$. If the null hypothesis that the two populations have the same SD were true (so the observed discrepancy is the result of chance) what is the chance that F would be that high or higher? The answer depends on the df of the numerator and denominator, which are both are $n - 1$, or 3. Calculate P using the free GraphPad QuickCalc (Appendix A) or the Excel formula = FDIST(62.7,3,3). The P value is 0.0033. That is the probability that the first group would have a SD so much larger than the second group by chance alone. Doubling that value to also allow for the chance that the second group would have had the larger SD, produces a P value of 0.0066. With such a low P value, it makes sense to reject the null hypothesis of equal SDs.

Now what? A modified t test allows for unequal SDs. With this modified test, the t value is still 2.01, but the P value is now 0.136, with df = 3. The P value is higher, but perhaps more appropriate, because it does not make an assumption that is probably false. However, according to Moser and Stevens (1992), the decision to use this test should be made as part of the experimental design and should not be based on the results of the F test to compare SDs.

This, perhaps, is a more appropriate way to analyze the data, but it doesn't really help. It doesn't bridge the gap between the result that is scientifically obvious and the result computed by formal statistics. In addition, the P value moved in the "wrong" direction! I know that it is cheating to just run a bunch of rival statistical tests and then pick the one that yields the smallest P value, but the increase in P value with the modified t test surely suggests that something is fundamentally wrong with my thinking!

UNPAIRED t TEST AS LINEAR OR
NONLINEAR REGRESSION

Hidden in Chapters 35 and 36 is an intriguing regression approach that might be appropriate for these data. I defined X to equal zero for the TMQ group and 1 for the MTX group and defined Y to be the individual IC_{50} values. I then used linear regression to fit a straight line to the data, recast as the X,Y pairs shown in Table 46.2. The slope reported by linear regression equals the difference in Y values divided by the difference in X values. The X values for the two drugs differ by 1.0, so the slope quantifies the difference in mean IC_{50}s between TMQ and MTX.

Linear regression calculated that the best-fit slope equals −18.5 nM, with a standard error of 9.21 nM. The 95% CI for the true slope (difference in mean IC_{50}) is −41.0 to 4.06 nM. The associated P value (testing the null hypothesis that the population slope equals zero) is 0.092. These are exactly the same as the results calculated by the simple unpaired t test. I expected this, because Chapter 35 emphasized that the two approaches are equivalent. However, the regression approach is more versatile, because it lets me compensate for nonuniform variance by appropriately weighting the data. Data with more scatter can be given less weight, and data with less scatter can be given more weight. Unfortunately, most linear regression software does not permit this type of weighting, so I switched to nonlinear regression software (which can also fit straight lines).

To apply this weighted regression approach, I fit a straight line to the X,Y data using weighted nonlinear regression (see Chapter 36). I assumed that both sets of data were drawn from Gaussian populations but that the SDs of these populations are not equal but instead are proportional to their mean. To fit this model, I asked the regression program to minimize the square of the relative distance of each point from the fitted line (weighting with the reciprocal of the square of the predicted Y). The slope of this line is the difference between IC_{50} values, which equals −18.5 nM, with a 95% CI ranging from −42.1 to 5.12 nM. One end of the CI is negative, which corresponds to the IC_{50} of MTX being

X: DRUG	Y: IC_{50} (nM)
0	5.5
0	12
0	19
0	47
1	0.83
1	1.1
1	1.9
1	5.8

Table 46.2. Data as entered into a regression program.
The X value was defined to be zero for the drug TMQ and 1 for the drug MTX.

smaller. But the other end of the CI is positive, which means the IC_{50} of TMQ is smaller. Like before, this method concludes that the data are consistent with no difference in IC_{50}. Because the 95% CI includes zero, the P value must be higher than 0.05. In fact, it is 0.104.

In summary, this alternative logical approach lengthens the CI and increases the P value! I don't seem to be making any real progress.

NONPARAMETRIC MANN–WHITNEY TEST

The t test assumes that the data were sampled from populations with Gaussian distributions. With only four values in each group, it is impossible to seriously assess this assumption. Normality tests (see Chapter 24) don't work with such tiny samples.

In fact, I find that normality tests are rarely helpful when used as part of an analysis of a particular experiment, but they can be quite useful when characterizing an assay. I prefer to collect many values (dozens to hundreds) while characterizing (developing) an experimental assay. I then examine the distribution of these values using various tests (including normality tests) to decide how these kinds of data should be analyzed. That decision then applies to all experiments that use the assay. In this example, we don't have the large number of measurements we'd need to tell us whether the Gaussian assumption is reasonable for this kind of data.

Rather than assuming the data were sampled from Gaussian populations, we can use the nonparametric Mann–Whitney test instead. With this test, P = 0.057. Although the P value is closer to 0.05, it is still greater than 0.05. Sure, using 0.05 as a threshold for defining statistical significance is arbitrary, but it is widely accepted. It is also tempting to just erase that last digit and call it 0.05—but that would be cheating. Rounded to two decimal places, the P value is 0.06.

Although you could argue that 0.057 is close enough to "tentatively suggest" that the potencies are different for the two drugs, this conclusion is somehow unsatisfying. In my gut, I really know that TMQ is less potent than MTX.

JUST REPORT THE LAST CONFIRMATORY EXPERIMENT?

I considered just presenting the data from the last paired experiment, in which TMQ and MTX were studied simultaneously, with a set of shared control flasks. The individual IC_{50}s for TMQ and MTX were 19 and 1.9, respectively, with a potency ratio of 10. Perhaps I might consider the first six individual experiments exploratory and the last paired experiment confirmatory.

This last paired experiment's potency ratio is close to the overall potency ratio of the mean IC_{50}s of 8.7. Perhaps it would be best to eliminate the table and just list in the text the individual IC_{50}s for TMQ and MTX, as well as the potency ratio for this last confirmatory experiment, with a statement that this experiment was representative of six previous individual exploratory experiments with the two agents.

This seems to be a scientifically ethical approach to the dilemma. However, because I like to think of myself as a stubborn, honest, and thorough scientist, I suspect that there is a better solution—one that involves calculating, rather than avoiding, statistics. I want to honestly showcase and analyze all of the experimental work.

Studying TMQ and MTX in paired experiments with common controls to assess their potency ratio is a good idea. Pairing (see Chapter 31) is usually recommended because it reduces variability and so increases power. This example combines data from paired experiments with data from unpaired experiments, which makes the analysis awkward.

INCREASE SAMPLE SIZE?

Of course, we'd like to get the results we want without collecting more data. But maybe more data are required. Chapter 26 showed how to use a sample size program to compute the approximate sample size per group (n) needed to detect a specified difference (w) given an anticipated SD (s) with specified power and significance level.

For s, we must enter the anticipated SD. In many situations, we may have lots of prior data to use to estimate this. Here, we have only the data at hand. Our two samples have different SDs. The sample SD for the TMQ data was 18.3 nM, whereas the sample SD for MTX was 2.31 nM. One thought might be to average the two and enter s = 10.3. Actually, with the same sample size in both groups, it is better to use the square root of the average of the variances (the square of the SDs). With this procedure, s is approximately 12 nM.

For w, we must enter the smallest difference between mean potencies that we'd find scientifically interesting. The choice of this number is a bit arbitrary, but to keep the calculations easy, let's set it to 20 nM, which is a bit larger than the observed difference of 18.6 nM.

Using G*Power (http://www.gpower.hhu.de/en.html), we request sample size for unpaired t test with a two-tailed alpha of 0.05, an effect size of $20/12=1.66$, and a desired power of 80%. G*Power reports we need to run seven experiments per drug, and each of those experiments requires 26 flasks of cells measured at multiple time points. This would be an outrageously large amount of work to answer a relatively trivial scientific question for which I am quite sure that I already know the answer.

COMPARING THE LOGARITHMS OF IC_{50} VALUES

Lognormal data

IC_{50} values tend to follow lognormal distributions (see Chapter 11) rather than Gaussian distributions. This leads to larger SDs when the mean is larger, as we saw. It can also lead to false identification of outliers (see Chapter 25).

Analyzing lognormal data is simple. I first calculated the (base 10) logarithms of the eight IC_{50}s. The results are shown in Table 46.3 and Figure 46.2.

The mean $\log(IC_{50})$ for TMQ is 1.19 and that for MTX is 0.251. The variability of the two sets of $\log(IC_{50})$ values is very similar, with nearly identical sample
SDs (0.389 and 0.373, respectively). Thus, there is no problem accepting the assumption of the standard two-sample unpaired t test that the two samples come
from populations with the same SDs.

Reversing the logarithmic transform (i.e., taking the antilog; see Appendix E)
converts the values back to their original units. Taking the antilog is the same as
taking 10 to that power, and taking the antilog of a mean of logarithms computes
the geometric mean (see Chapter 11). The corresponding geometric mean for the
IC_{50} for TMQ is 15.6 nM and that for MTX is 1.78 nM.

Unpaired t test of log-transformed data

The calculated t ratio is 3.49 and the corresponding P value is 0.013, df = 6. MTX
is more potent than TMQ and that discrepancy is statistically significant (according to the conventional definition).

The difference between the $\log(IC_{50})$s is 0.942. Transform this to its antilog to
turn it into a potency ratio (remember that the difference between the logarithms

TMQ	MTX
0.740	−0.081
1.079	0.041
1.279	0.279
1.672	0.763

Table 46.3. The logarithms of the IC50 values of TMQ and MTX.
These are the logarithms of the values (in nM) shown in Table 46.1.

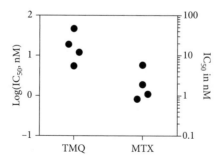

Figure 46.2. The logarithms of the data shown in Figure 46.1.
The left axis shows the logarithms, and the right axis shows the antilogarithms. Comparing
the two lets you review the meaning of logarithms. The logarithm of 100 is 2, because
$10^2 = 100$. The logarithm of 1.0 is 0.0, because $10^0 = 1.0$. The logarithm of any value
between zero and 1 is negative.

of two values equals the logarithm of the ratio of those two values). The antilog is 10 to that power, or $10^{0.942}$, which equals 8.75. In other words, MTX is almost nine times more potent than TMQ. It requires one-ninth as much MTX to have the same effect as TMQ.

95% CI

The 95% CI for the difference in mean $\log(IC_{50})$ ranges from 0.282 to 1.60. Transform each limit to its antilog (10 to that power) to determine the 95% confidence limits of the population potency ratio, 1.91 to 39.8. This interval does not encompass 1.0 (the value of the null hypothesis of equal drug potency), so it is consistent with a P value of less than 0.05. With 95% confidence, we can say that MTX is somewhere between twice as potent as TMQ and 40 times as potent. Analyzed this way, the intuition of the investigator is proven to be correct.

Why this is the optimal method

This is probably the best simple way to analyze the data. The t test of the untransformed IC_{50} values is inferior to the t test of the log-transformed IC_{50} values for three important reasons.

- The log transform equalized the two SDs and thus eliminated the assumption violation of unequal variances.
- As suggested in Chapter 11, the log transformation changed the distribution of IC_{50}s from a lognormal distribution to a Gaussian distribution, another assumption of the t test. With only two sets of data, each with only four values, we really don't know much about the distribution of the population of IC_{50}s. But extensive past experience with concentration-effect experiments tells us to expect a lognormal distribution.
- The t test asks about differences between means. The t test on log-transformed data essentially asks about ratios. Pharmacologists universally ask about ratios of drug potencies, not differences. Transforming to logarithms helps match the statistical method to the scientific question. The t test of the log-transformed IC_{50} values is a nice statistical translation of the scientific question, What is the relative potency of TMQ and MTX?

If I had access to the original raw concentration-effect data, I would likely try to fit a variant of the four-parameter concentration-effect model showcased in Chapter 36 to all of the data at once using nonlinear regression. I would take into account differences among the experiments run at different times and would estimate the overall potency ratio with an accompanying 95% CI. I would take advantage of the true sample size of 203 flasks from eight experiments in my analysis. However, in this case, I don't have the original raw data; I just have the eight naked IC_{50}s shown in Table 46.1.

SAMPLE SIZE CALCULATIONS REVISITED

Now that we realize the data follow a lognormal distribution, let's go back and redo the sample size calculations. On a log scale, we found that the SD of both groups was 0.38, so we enter into the sample size formula a conservative 0.40 as s. How large a difference in $\log(IC_{50})$ are we looking for? Let's say we are looking for an eightfold difference in potency, so the $\log(IC_{50})$ values differ by $\log(8)$, which equals about 0.9. Set w = 0.9. The effect size is 0.9/0.4 = 2.25. Change the effect size in G*Power to 2.25 (while keeping the settings for two-tailed significance threshold of 0.05 and a power of 80%). The required sample size is four in each group.

This might be stated in a paper as follows: a sample size of four in each group was computed to have 80% power to detect as statistically significant (P < 0.05) an eightfold difference in potency, assuming that $\log(IC_{50})$ values follow a Gaussian distribution with a SD of 0.40.

IS IT OK TO SWITCH ANALYSIS METHODS?

This whole exercise to search for a better statistical analysis than was suggested by the reviewer has left me frustrated.

Data analysis should not be a process of fishing for statistical significance. It is not ethical to conduct multiple statistical procedures on the same data set for the purpose of finding the statistical approach that yields the lowest P value. This whole procedure of analyzing the data this way, then that way, and then another way is called P-hacking and leads to misleading results.

At the same time, the logarithm method makes scientific sense. As explained in the previous section, this is the optimal method for comparing IC_{50} values. The method matches the scientific context and goals, whereas the other methods do not. Of course, the decision to analyze the data in this way should have been planned before the data were collected. But it wasn't. Is it OK to switch methods after the fact?

Here's a reality check: this happens all the time in basic research. Many scientists never even think about how to analyze the data until after the experiments are done. But just because something is done commonly doesn't mean it is correct. Chapter 23 pointed out how easy it is to be fooled by multiple analyses. Certainly, clinical trials that are reviewed by the FDA require that the analysis method be precisely specified in advance.

In this case, I have a strong scientific justification for switching to the proper method. Also, I am responding, albeit indirectly, to comments made by an expert reviewer of the manuscript. Does this rationale make it OK? I think so, but I will let you, the reader, decide for yourself. It certainly would be OK if these were preliminary data and if this whole exercise was done as a way to make decisions about how I will analyze data from future experiments.

THE USEFULNESS OF SIMULATIONS

Analyzing $\log(IC_{50})$ values has more power

For this example, analyzing the $\log(IC_{50})$ values produced a smaller P value than analyzing the IC_{50} values and seems to better match the scientific goals of the study. If the data really do follow a lognormal distribution, how much is lost by analyzing the IC_{50} values rather than the $\log(IC_{50})$ values? Our one example gives a hint of the answer, but only a hint. To find out a fuller answer, I performed Monte Carlo simulations.

Table 46.4 shows the simulated results. The first row in Table 46.4 is the result of 1 million simulated experiments. For each, four values were randomly generated to simulate an experimental measurement of potency for one drug, and four other values were randomly generated to simulate the potency of the other drug. All in all, 8 million values were randomly generated to create the first row of Table 46.4.

The data for one drug were simulated with an IC_{50} of 2 nM, so the logarithm of 2 nM, 0.301, was entered as the mean into the simulation program. The data for the other drug were simulated using an IC_{50} of 20 nM, so the logarithm of 20 nM, 1.301, was entered into the simulation program. Each $\log(IC_{50})$—all 8 million of them—was chosen using a random number method that simulates sampling from a Gaussian distribution, assuming a mean of 0.301 or 1.301 and a SD of 0.40.

These simulations mimic the data presented at the beginning of this chapter. The potency of the two drugs differs by a factor of 10, the $\log(IC_{50})$ values vary randomly according to a Gaussian distribution, the SD of the logarithms is 0.4, and n = 4 for each group. The $\log(IC_{50})$ values from each simulated experiment were compared with an unpaired t test, and the P value was tabulated.

The last column of Table 46.4 shows the percentage of the simulated experiments in which the P value is less than 0.05, the usual definition of statistical significance. In other words, that last column shows the power of the experimental design (see Chapter 20). For this design, the power is 83.6%. In the remaining simulated experiments (16.4% of them), the P values are greater than 0.05 and so result in the conclusion that the results were not statistically significant. This happens when the means happen to be close together or the scatter among the values in that simulated experiment happens to be large.

N PER GROUP	RELATIVE POTENCIES OF TWO DRUGS	DATA ANALYZED	HOW OFTEN IS P < 0.05?
4	Differ by factor of 10	$\log(IC_{50})$	83.6%
4	Differ by factor of 10	IC_{50}	45.6%
7	Differ by factor of 10	$\log(IC_{50})$	99.0%
7	Differ by factor of 10	IC_{50}	79.2%

Table 46.4. Simulations in which the drugs have different potencies.

Each row is the result of 1 million simulations. The $\log(IC_{50})$ values were simulated using a random number method that generates a Gaussian distribution, and the IC_{50} values were computed by taking the antilog (10 to that power). Therefore, the $\log(IC_{50})$ values are sampled from a Gaussian distribution, and the IC_{50} values are sampled from a lognormal distribution. The true population mean IC_{50}s of the two groups in each row differ by a factor of 10 (2 and 20 nM), and the samples included either four or seven IC_{50}s per group. The differences in mean $\log(IC_{50})$ (or IC_{50}) were compared using the unpaired t test, with the assumption of uniform variances.

The second row of Table 46.4 shows analyses of IC_{50} values. Each IC_{50} was computed by taking the antilog (10 to that power) of a $log(IC_{50})$ generated as explained for the first row. The first row shows the results of a simulation of sampling $log(IC_{50})$ values from a Gaussian distribution, and the second row shows the results of a simulation of sampling IC_{50} values from a lognormal distribution. The IC_{50} values from each simulated experiment were compared with an unpaired t test, and the P value was tabulated.

In only 45.6% of these experiments was the P value less than 0.05. Because the violation of all three of its assumptions (lognormal distribution, unequal variance, differences examined rather than ratios) discourages the use of the standard t test on untransformed IC_{50} values, this approach has much less statistical power (45.6% vs. 83.6%).

Analyzing $log(IC_{50})$ better controls the Type I error rate

Table 46.5 shows simulations identical to those shown in Table 46.4, except that the simulations were done using the same mean IC_{50} for both drugs. In other words, for these simulations, the null hypothesis was true—the drugs had equal potency.

The first row shows the analyses of the $log(IC_{50})$ values. For both drugs, these values were drawn from the same Gaussian distribution, so the assumptions of a t test are met. If the simulations work as they should, 5% of these P values should be less than 0.05, and that is precisely what I observed. No surprise here, but this is a good way to check that the simulations worked properly.

The second row shows the analyses of the IC_{50} values. Only 3.23% of these simulated experiments resulted in a P value less than 0.05, although the null hypothesis is true. The data violate one of the assumptions of a t test (Gaussian distribution) and so the definition of statistical significance is off. When P < 0.05 is defined to be statistically significant, you can expect that 5% of experiments under the null hypothesis will end up being statistically significant. The Type I error is defined to be 5%. Here, only 3.24% of the experiments under the null hypothesis resulted in a statistically significant result. The Type I error is only 3.24%. Less Type I error might seem to be a

N PER GROUP	RELATIVE POTENCIES OF TWO DRUGS	DATA ANALYZED	HOW OFTEN IS P < 0.05?
4	Equal	$log(IC_{50})$	5.00%
4	Equal	IC_{50}	3.24%
7	Equal	$log(IC_{50})$	4.99%
7	Equal	IC_{50}	3.42%

Table 46.5. Simulations in which the null hypothesis is true (drugs have equal potencies).

The simulations were done as explained in Table 46.4, except that these simulations used equal potencies for the two drugs (here, 20 nM, but nearly identical results were obtained when using 2 nM).

good thing, but it is accompanied by a decrease in power, which is definitely not desirable.

Increasing sample size increases power

The third and fourth rows of Table 46.4 show the results of increasing the sample size to seven per group. The power rose to 99.0% for the t test of the $\log(IC_{50})$s. Is this extra effort worth it? It depends on the context of the work, the cost of the experiments (in money and effort), and the consequences of missing a real difference. For this experiment, I think 80% power is plenty and that it is not necessary to spend the time and money it would take to increase the sample size to seven experiments.

Increasing the sample size to n = 7 increases the power to 79.2% for the t test of the IC_{50}s. This simulation result is similar to our previous sample size calculation with the simple formula.

The third and fourth rows of Table 46.5 show that an increase to n = 7 barely changes the Type I error rates.

Simulating experiments is not hard

These simulations took me an hour or two to set up and less than 20 minutes to compute using a 2008 computer. In contrast, it would have taken many hours to complete such simulations on many "mainframe" computers in 1975.

Many software packages can perform these simulations (e.g., GraphPad Prism, Minitab, Excel, R, etc.). I used 1 million simulations per experiment to be compulsive, but I would have gotten virtually the same results with 10,000 or even 1,000 simulated experiments. If you plan to design experiments, it is well worth your time to learn how to run these kinds of simulations. You can save a lot of time in the lab by using simulations to test the design of your experiments.

CHAPTER SUMMARY

- Lognormal distributions are common in biology. Transforming data to logarithms may seem arcane, but it is actually an important first step for analyzing many kinds of data.
- This chapter was designed to review as many topics as possible, not to be as realistic as possible. It rambles a bit. However, it realistically shows that the search for the best method of analyzing data can be complicated. Ideally, this search should occur before collecting the data or after collecting preliminary data but before doing the experiments that will be reported. If you analyze final data in many different ways in a quest for statistical significance, you are likely to be misled.
- Statistical methods are all based on assumptions. The holy grail of biostatistics is to find methods that match the scientific problem and for which the

assumptions are likely to be true (or at least not badly violated). The solution to this chapter's problem (using logarithms) was not optimal because it yielded the smallest P value but rather because it melded the statistical analysis with the scientific background and goal. It was a nice translation of the scientific question to a statistical model.

- Simulating data is a powerful tool that can reveal deep insights into data analysis approaches. Learning to simulate data is not hard, and it is a skill well worth acquiring if you plan to design experiments and analyze scientific data.

- Statistical textbooks and software aren't always sufficient to help you analyze data properly. Data analysis is an art form that requires experience as well as courses, textbooks, and software. Until you have that experience, collaborate or consult with others who do. Learning how to analyze data is a journey, not a destination.

CHAPTER 47

Statistics and Reproducibility

Truthiness is a quality characterizing a "truth" that a person making an argument or assertion claims to know intuitively "from the gut" or because it "feels right" without regard to evidence, logic, intellectual examination or facts.

STEPHEN COLBERT

In recent years, it has become apparent that much published research is not reproducible. This chapter reviews statistical concepts that explain some of the problems of irreproducibility.

THE REPODUCIBILITY CRISIS

Many results cannot be reproduced

Scientific papers only get submitted after a scientist (or more often, a group of scientists) do a lot of careful work. With most journals, a different group of scientists (called peer reviewers) evaluate the paper and recommend whether it should be published, usually suggesting many improvements before they recommend publication. With that much work and review, you'd expect the quality of published peer-reviewed science to be very high.

For many years, most scientists expected that any experienced investigator with the right tools should be able to reproduce a finding published in a quality peer-reviewed biomedical science journal. Of course, the results won't reproduce exactly. But the expectation was that the main results would be essentially reproduced in most cases, so the data from the repeat experiment would lead to the same general conclusions as the first one, and new studies could be designed and performed on the foundation of the published work.

In fact, a large percentage of published findings cannot be reproduced. Table 49.1 summarizes three big studies showing that the general findings of less than half of published work were substantially reproduced in a repeat experiment. Of course, there is lots of subjectivity in defining what it means for a result to be "substantially" reproduced, but the results shocked many scientists.

Dumas-Mallet and colleagues (2016) looked at it a different way, comparing initial reports of association between a disease and a marker or risk factor with later meta-analyses that combined evidence from at least seven studies. Looking only at

Citation	Field	# Attempts	% Reproduced
Prinz, 2011	Pharmacology	67	24%
Begley and Ellis, 2012	Cancer biology	53	11%
Open Science Consortium, 2015	Psychology	100	39%

Table 49.1. Attempts to reproduce published findings.
The last column gives the percentage of reproduced studies that gave substantially the same results as the original.

the presence or absence of a statistically significant association, the agreement between initial studies and their corresponding meta-analyses ranged from 6% to 80% depending on the field, with most fields having about a 40% reproducibility rate.

Why?

Why can so few findings be reproduced? Undoubtedly, there are many reasons. Part of the problem is that many experiments are complicated, and it is difficult to repeat all the steps exactly. Different labs that think they are working on the same cell line may actually be working with somewhat different cell lines. Reagents or antibodies thought to be identical may differ in subtle, but important, ways. Differences in experimental method thought to be inconsequential (such as choosing plastic or glass test tubes) may actually explain discrepancies between labs. In some cases, exploring reasons for inconsistent results can lead to new scientific insights.

Reality check

The real goal, of course, is not to have multiple studies that give similar reproducible results. The real goal is to obtain results that are correct. When the results are incorrect, it doesn't really matter much whether another study can reproduce the findings. To be sure the results are correct, it is not enough to repeat the experiment exactly. It is also necessary to run similar experiments to make sure the general conclusions can be confirmed, not just that the results of that exact experiment can be reproduced.

MANY ANALYSES ARE BIASED TO INFLATE THE EFFECT SIZE

Many published findings in some fields of science follow a system that inflates the effect size and exaggerates (makes smaller) the P value. If the findings in the initial study are inflated, one wouldn't expect good reproducibility. Ioannidis (2005) argues that *most* published statistically significant research findings are false.

Let's review some reasons for this.

Publication bias inflates published effect sizes

If a study ends up with a small effect size and/or large P value, the investigators may not put in the effort to write the paper and submit it. And even if they do, reviewers and editors tend to reject such papers. So the papers that end up being published are in part selected because they have large effect sizes and small P values. In other words, published papers tend to exaggerate differences between groups or effects of treatments. Publication bias was discussed in more detail in Chapter 43.

Selective presentation of results within a paper inflates published effect sizes

It is tempting to only report results "that worked" and to not show the results of experiments that didn't work. That's ok if you define "not working" to mean experiments with built-in positive and/or negative controls that didn't yield the expected results, and you use consistent preset criteria for rejecting these results. But it is not ok to reject experiments when the effect size is smaller than you'd expect and only show the ones that gave a larger effect. If you reject experiments that show a small effect and show (or pool) experiments that show a large effect, you are biasing the results toward showing a large effect.

Ad hoc sample size selection inflates published effect sizes

You can only evaluate statistical results at face value when the sample size was decided in advance or when a specialized analysis method was used that accounts for the experimental design. However, in some fields of science it is common for investigators to increase the sample size when the result seems promising. What happens if the investigator added more samples when the effect size wasn't as large as desired (or the P value is not as small as desired) and stops adding samples when the investigator gets the result she wants? In these cases, the published findings, on average, show an effect size larger than it really is, and a P value that is smaller than it ought to be. Ad hoc sample size selection was discussed in Chapter 26.

Ad hoc analyses of subsets inflate published effect sizes

If the first analysis does not result in the conclusion the investigator wanted, it is tempting to then analyze subsets of the data. Try only analyzing the results from men. If that doesn't "work," try only analyzing results from a certain age group. If you slice and dice the data in many ways, you are likely to find a large effect size and small P value just by chance. Such results are not likely to be reproducible. Wallach (2017) showed this by studying about a hundred articles that reported statistically significant results by analyzing subgroups. Few of these subgroup analyses held up to scrutiny. Only five were ever repeated. All of those repeat studies found a smaller effect than did the original study, and none were statistically significant.

Trying multiple data analysis methods inflates published effect sizes

Some investigators try multiple approaches to analyzing data to try to find a method that produces a large effect size and/or a small P value. As a result, the published effect size is inflated, and the published P value is too small. Chapter 23 reviewed this problem.

Deleting "outliers" to get the desired result inflates published effect sizes

Deleting outliers that are almost certainly due to experimental mistakes can make sense. But it must be done in a consistent way using criteria set in advance. Results are misleading if scientists delete outliers when it helps them reach the desired conclusion but leave them in when that choice "improves" the results. More on outliers in Chapter 25.

The garden of forking paths inflates published effect sizes

Even if you only analyze the data one way, you might be P-hacking. The *garden of forking paths* (Gelman & Loken, 2014) refers to options available to investigators who didn't finalize their decision about how to analyze the data until after inspecting the data. If the data had come out differently, they would have chosen a different analysis method. Even if the choices are prespecified in a formal plan, this is a form of P-hacking and multiple comparisons that can lead to results that are not reproducible.

Changing the choice of primary outcome can inflate published effect sizes

Some studies collect data from multiple outcomes. Ideally, the investigator should state which outcome is the primary outcome in the study protocol and focus on this outcome in the published paper. Unfortunately, it is common for investigators to report the outcome that gives the strongest results even if another outcome was defined as primary in the study protocol. This was discussed in Chapter 23. If the author picks and chooses which of many outcomes to report, the reported effect size is likely to be exaggerated.

Overfitting inflates published effect sizes

Many clinical studies and studies in social sciences involve analyzing multiple variables analyzed by multiple linear regression (when the outcome is continuous), multiple logistic regression (when there are only two possible outcomes), or Cox proportional hazards regression (when the outcome is survival time). As explained in Chapter 37, overfitting is an issue with these analyses. It happens when you ask more questions than the data can answer. This happens when you have too many independent variables in the model compared to the number of subjects. The problem with overfitting is that the modeling process ends up fitting aspects of the data caused by random scatter or experimental quirks. Such a model will report large regression coefficients and small P values that won't be reproduced when tested on a new set of data.

How many independent variables is too many? For multiple regression, one rule of thumb is to have *at least* 10 to 20 subjects per independent variable. Fitting a model with five independent variables thus requires at least 50 to 100 subjects. For logistic regression that rule applies for the number of subjects with the least common outcome. So if 10% of the subjects in a study have the outcome, then a study with five independent variables needs at least 50 to 100 of that outcome, so needs 500 to 1000 subjects.

Note that the number that matters is the number of independent variables you begin the analysis with. You can't reduce the problem of overfitting by reducing the number of variables through stepwise regression or by preselecting only variables that appear to be correlated with the outcome.

Variable selection in multiple regression inflates published effect sizes

As explained in Chapter 37, if you choose which independent variables to include in a multiple regression model, the model will fit too well (R^2 will be inflated), the regression coefficients (effect sizes) will be too large (or too far from zero if the coefficient is negative) with CIs that are too narrow, and the corresponding P values will be too small. In essence, variable selection is a form of multiple comparisons. You can't expect to reproduce results obtained with multiple regression that used variable selection.

EVEN PERFECTLY PERFORMED EXPERIMENTS ARE LESS REPRODUCIBLE THAN MOST EXPECT

Many statistically significant results are false positive

Chapter 18 emphasized that the P value is not the same as the FPRP. The FPRP answers this question: Of all studies like this one that reach a P value as small as (or smaller than) this one, what fraction are actually false positives?

The FPRP depends on the context of the experiment. Table 49.2 repeats Table 18.2. Focus on the third row of the table, where the scientific context is such that the chance the effect you are looking for actually exists (the prior probability) is only 1%. Even in this context, you may want to run a speculative experiment in hopes of finding something. But what does it mean if you find a P value less than 0.05?

Of the 99% of the potential experiments were there is really no effect, you'll expect to see a statistically significant result in 5% of them. Of the 1% of the potential experiments where there really is an effect, you'll expect to obtain a statistically significant result in 80% of them (because we are assuming the sample

	False Positive Report Rate	
PRIOR PROBABILITY	DISCOVERY: P < 0.05	DISCOVERY: P BETWEEN 0.045 AND 0.050
0%	100%	100%
1%	86%	97%
10%	36%	78%
50%	5.9%	27%
100%	0%	0%

Table 49.2. The FPRP depends on the prior probability and P value.

The calculations assume sample size was sufficient to have a power of 80% and the FPRP values would be lower if the power were lower. The results in the last column were determined by simulation using methods similar to Colquhoun (2014).

size was chosen to give a power of 80%). Chapter 18 explained how to combine these values to compute the FPRP. The table shows that there is an 86% chance that a finding with any P value less than 0.05 is a false positive, which leaves only a 14% chance that it is real. And if the P value is just barely smaller than 0.05, there is a 95% chance that the finding is a false positive, leaving only a 3% chance that it is a real finding. With the FPRP so high, there really is no point in running a speculative experiment with the conventional significance level of 5%. And if you do that kind of experiment and find a "significant" result, you shouldn't expect a repeat experiment to reproduce that conclusion.

Now let's focus on experiments where the prior probability is 50%. If you look at all experiments with P < 0.05, the FPRP is only 5.9%. But if you look only at experiments with P values just barely less than 0.05 (I used the range 0.045 to 0.050 in the simulations), the FPRP is 27%. In other words, in this very common situation, more than a quarter of results with a P value just a bit smaller than 0.05 will be a false positive.

In situations where false positive results are likely, it doesn't make a lot of sense to worry about whether a particular finding can be replicated.

Many experiments have low power

The concept of statistical power was introduced in Chapter 20. Power answers this question: If the true difference or effect is of a certain defined size, what is the chance that a study with a specified sample size will reach a conclusion that the difference or effect is statistically significant. The value of the power depends on how large a difference you are looking for, how much variation you expect to see, how you define statistical significance, and sample size.

If an experiment has low power to find the difference (or effect) you are looking for, you are likely to end up with a result that is not statistically significant, even if the effect is real. And if the low power study does reach a conclusion that the effect is statistically significant, the results are hard to interpret. One problem is that the FPRP will be high. A second problem is the effect size observed in that experiment is likely to be larger than the actual effect size (if there is any effect), because only large observed effects (even if due to chance) will be yield a P value less than 0.05. It is even possible (but quite unlikely) that the observed effect is in the opposite direction of the actual effect.

Experiments designed with low power cannot be very informative, and it should not be surprising when they cannot be reproduced.

P values are not very reproducible

Let's assume that the experimental methods are described in extensive detail, as are the steps used to analyze the data. You have a large prior probability, use a large sample size, repeat the experiment exactly, and analyze the data exactly as the original investigators did. You therefore avoid the problems mentioned in the earlier section. You might think that you'd get a similar P value. But Figure 15.2 showed you this isn't the case. Leaving out the 2.5% highest and lowest P values, the middle 95% of the P values computed from simulated data covered a range spanning more than three orders of magnitude! Boos and Stefanski (2011) have shown this much variability in many situations.

Lazzeroni1, Lu, and Belitskaya-Lévy (2014) have derived an equation for computing the 95% prediction interval for a P value that answers this question: If you obtain a certain P value in one experiment, what is the range of P values you would expect to see in 95% of repeated experiments using the same sample size? It turns out that this interval does not depend on sample size (but does depend on whether the repeat experiment uses the same sample size). If the P value in your first experiment equals 0.05, you can expect 95% of P values in repeat experiments (with the same sample size) to be between 0.000005 and 0.87. If the first P value is 0.001, you expect 95% of P values in repeat experiments to be between 0.000000002 and 0.38.

P values are nowhere near as reproducible as most people expect!

Conclusions about statistical significance are not very reproducible

Conclusions about statistical significance are also less reproducible than most people would expect. Imagine that one experiment has a P value exactly equal to 0.05 (which you use as your threshold value for statistical significance). If you repeat the experiment, what is the chance that the repeat experiment will result in a P value less than 0.05? Your best guess is that the P value will be the same as last time, 0.05. Half the time, random sampling of data will lead the P value to be larger than 0.05, and half the time random sampling will make the P value smaller. So the chance of statistical significance in a repeat experiment if the first one is right at the border of significance is 50%. Cumming (2008, 2011) extended this analysis to show that if the first P value equals 0.05, there is a 20% chance that the P value from a repeat experiment will be greater than 0.38, and a 5% chance it will be greater than 0.82.

Goodman (1992) generalized this with other P values in the first experiment. So did Curran-Everett (2016), and both sets of results are shown in Table 49.3. Even if the P value from the first experiment is 0.001, you can be far from sure

P VALUE IN FIRST EXPERIMENT	CHANCE THAT THE P VALUE IN THE SECOND EXPERIMENT WILL BE LESS THAN 0.05	
Author	Goodman (1992)	Curran-Everett (2016)
0.10	41%	38%
0.05	50%	50%
0.01	66%	73%
0.001	78%	91%
0.00031		95%

Table 49.3. Probability of statistical significance in a second experiment given that you know the P value in the first experiment as computed in two papers by different methods.

The calculations assume that both experiments are performed perfectly and both comply with the assumptions of the analysis.

that the repeat experiment will have a P value less than 0.05. It turns out that that probability is only 78% or 91% (the two values differ because two methods were used to compute).

Conclusions about statistical significance don't have to match for results to be considered reproducible

Schmidt and Rothman (2014) reviewed two case-control studies asking whether there is an association between taking selective COX-2 inhibitors (a class of drug used to reduce pain and inflammation) and the onset of atrial fibrillation (irregular heart rate). Table 49.4 summarizes the results. A 2011 study found a statistically significant association. A study in 2013 found that the association was not significant and pointed out the discrepancy. But look at the odds ratio, the effect size reported by analyses of case-control studies (see Chapter 28). As explained in Chapter 28, you can interpret the odds ratio as a relative risk. Both studies found a relative risk of 1.20, meaning that people taking the drug have a 20% higher risk of going into atrial fibrillation. What differs is the 95% CI. In the 2011 study, the CI does not include 1.0 (the value of the odds ratio that means no association, and so corresponds to the null hypothesis). Therefore, the P value must be less than 0.05, and the authors concluded that the association was statistically significant. The CI for the 2013 study includes 1.0. Therefore, the P value is greater than 0.05, and the authors concluded that the association is not statistically significant. But the two studies found the same effect size, an odds ratio of 1.20. Even though one study found a statistically significant result and the other did not, the two studies don't contradict each other but rather found exactly the same effect size. The last row in the table uses meta-analysis methods (see Chapter 43) to present the combined results.

This example makes two important points:

- It is essential to look at effect sizes and not just conclusions about statistical significance.
- While meta-analysis methods are usually used to pool evidence from many studies, these methods can be used to combine evidence from just two studies.

STUDY	ODDS RATIO	95% CI	STATISTICALLY SIGNIFICANT?
Schmidt (2011)	1.20	1.09 to 1.33	Yes
Chao (2013)	1.20	0.97 to 1.48	No
Combined by Schmidt and Rothman (2014)	1.20	1.10 to 1.31	Yes

Table 49.4. Look at CIs, not conclusions about statistical significance.
The top two rows summarize two case control studies that looked at the association between taking COX-2 inhibitors (a kind of drug) and the new onset of atrial fibrillation (an arrhythmia with an irregular heart rhythm that can lead to strokes). The bottom row shows the results combined by meta-analysis.

SUMMARY

There are many reasons why the results of many experiments or studies cannot be reproduced. This chapter reviewed statistical reasons:

- Many studies are designed in a way to exaggerate effect sizes and so report P values that are small. When the study results are exaggerated, you can't expect a repeat experiment to get the same results.
- Even with perfectly performed studies, the FPRP can be high, especially when the prior probability is low or when the sample size is low so the power is low. If the reported result is a false positive, you can't expect a repeat experiment to reproduce the results.
- Many experiments are designed with low power to find the effect the experimenters are looking for. In this case, one may not expect a repeat experiment to find the same results.
- It is important to focus on how large the effects are and not just on P values and conclusions about statistical significance.
- P values are far less reproducible than most people appreciate.
- Meta-analysis is a tool to combine results from several studies. When studies that appear to be done properly reach conflicting results, it can make sense to combine the results with meta-analysis.
- Science is hard to do properly.

CHAPTER 48

CHECKLISTS FOR REPORTING STATISTICAL METHODS AND RESULTS

"Surprising, newsworthy, statistically significant, and wrong: it happens all the time."

ANDREW GELMAN

Many people reading this book will never publish statistical results. But for those who do, the following checklists are an opinionated guide about how to present data and analyses. You may also want to consult similar guides by Curtis and colleagues (2015) and by Altman and colleagues (1983).

REPORTING METHODS USED FOR DATA ANALYSIS

Every paper reporting statistical analysis should report all methods (including those used to process and analyze the data) completely enough so someone else could reproduce the work exactly. These can be included either in the paper or an online supplement.

Reporting the analyses before the analyses

- Did you decide to normalize? Transform to logarithms? Smooth? Remove a baseline? Justify these decisions and report enough details so anyone who started with your raw data would get exactly the same results. State whether these calculations were preplanned or only decided upon after seeing the data.
- If outliers were eliminated, say how many there were, what criteria you used to identify them, and whether these criteria were chosen in advance as part of the experimental design.

Reporting sample size

- Report how you chose sample size.
- Explain exactly what was counted when reporting sample size. When you say n = 3, do you mean three different animals, three different assays on tissue from one animal, one assay from tissue pooled from three animals,

three repeat counts in a gamma counter from a preparation made from one run of an experiment . . . ?

- State whether you choose sample size in advance or adjusted sample size as you saw the results accumulate. If the latter, explain whether you followed preplanned rules.
- If the sample sizes of the groups are not equal, explain why.
- If you started with one sample size and ended with another sample size, explain exactly what happened. State whether these decisions were based on a preset protocol or were decided during the course of the experiment.

Reporting use of statistical tests

- State the full name of the test. Don't say "t test," say "paired t test."
- Identify the program of the program that did the calculations including detailed version number, which for GraphPad Prism might be 7.03.
- State all options you selected. Repeated measures? Correcting for unequal variances? Robust regression? Constraining parameters? Sharing parameters? Report enough detail so anyone who started with your raw data would get precisely the same results you got.
- If the analyses used a system like R where all the analysis steps are specified in a command file, post that file to a repository where future investigators can download. Likewise, if the analyses used a program like GraphPad Prism that stores the analysis steps within the overall project file, post that file.

Reporting planning (or lack of planning) of analyses

- For each figure and table, state whether the analysis was planned. Did every data analysis step follow a preplanned protocol exactly? Did you analyze it in the logical way based on the experimental protocol, with no plan, but only that one analysis? Or did you try several methods for processing and analyzing the data until you decided upon the analysis strategy used in the results included in the paper?
- If you only decided to remove outliers after seeing the data, say so. If you only decided to use a nonparametric test after seeing the data (or running a normality test, or disliking the results of a t test), say so. If you only decided to analyze the logarithms of the data after viewing the data, say so.
- If you don't show every analysis and every experiment you did, describe and enumerate the ones you didn't include.

GRAPHING DATA

General principles

- Every figure and table should present the data clearly and not be exaggerated in a way to emphasize your conclusion.
- Every figure should be reported with enough detail about the methods used so that no reader needs to guess at what was actually done.

- Consider posting the raw data files in a standard format, so others can repeat your analyses or analyze the data differently. This is required in some scientific fields.
- State if any data points are off scale so don't show on the graph.

Graphing variability

- When possible, graph the individual data, not a summary of the data. If there are too many values to show in scatter plots, consider box-and-whisker plots, violin plots, or frequency distributions.
- If you choose to plot means with error bars, graph SD error bars, which show variability, rather than SEM error bars, which do not. Be sure to state in the figure legend what the error bars represent.
- Plot confidence intervals of differences or ratios to graphically present the effect size.
- Don't try to hide the variability among your data. That is an important part of your data, and showing expected variability in measurements of physiological or molecular variables is a way to reassure readers the experiment was properly done.

REPORTING STATISTICAL RESULTS

Reporting and graphing effect size

- The most important result of most experiments is an effect size. How big was the difference (or ratio or percentage increase)? Or how strongly were two variables correlated? In almost all cases, you can summarize this effect size with a single value and should report this effect with a CI, usually the 95% interval. This is by far the most important finding to report in a paper and its abstract.
- Consider showing a graph of effect sizes (i.e., differences or ratios) with 95% CIs.

Reporting P values

- Don't report a P value unless you really think it will help the reader interpret the results.
- When possible, report the P value as a number rather than as an inequality. For example, say "the P value was 0.0234" rather than "P < 0.05."
- If there is any possible ambiguity, clearly state the null hypothesis the P value tests. If you don' know the null hypothesis, then you shouldn't report a P value (since every P value tests a null hypothesis)!
- When comparing two groups, state if the P value is one- or two-sided (which is the same as one- or two-tailed). If one-sided, state that you predicted the direction of the effect before collecting data and recorded that decision and prediction. If you didn't make this decision and prediction before collecting data, you should not report a one-sided P value.

Reporting statistical hypothesis testing

- Statistical hypothesis testing is used to make a firm decision based on a single P value. One use is choosing between the fit of two alternative models. If the P value is less than a preset threshold you pick the simpler model, otherwise the other. When doing this, state both models, the method you are using to choose between them, the preset threshold P value, and the model you chose. Perhaps also report the goodness of fit of both models.
- When comparing groups, you don't always make a decision based on the result. If you are making a crisp decision, report the threshold P value, whether the computed P value was greater or less than the threshold, and the accompanying decision. If you are not making a decision, report the effect with its CI and perhaps a P value. If you are not making a decision based on that P value, then it doesn't really matter whether the P value was less than a threshold, and the whole idea of statistical hypothesis testing isn't really useful.
- The word "significant" has two related meanings, so has caused lots of confusion in science. The two previous bullet points demonstrate that the results of statistical hypothesis testing can (and in my opinion, should) be reported without using the word "significant." If you do choose to use the word *significant* in this context, always use the phrase *statistically significant*, so there is no confusion.
- Never use the word *significant* when discussing the clinical or physiological impact of a result. Instead use words like *large*, *substantial*, and *clinically relevant*. Using *significant* in this context just leads to confusion.
- When using the word *hypothesis*, be sure to specify whether you are referring to a scientific hypothesis or a null statistical hypothesis.

Reporting regression results with one independent variable

- Be sure to state what model you fit to the data. Be complete. Which parameters were fixed to constant values and which were fit? If you fit a family of data sets at once, which parameters were shared?
- If you fit a model with multiple parameters, state how you arrived at this model. Preplanned? Or did you try multiple models before deciding to report this one?
- What assumption did you make about variation? Uniform Gaussian scatter? Nonuniform scatter, accounted for by differential weighting?
- If you report the best-fit values of the parameters, also report their CIs. Consider reporting the covariation matrix to show how independent or intertwined the parameters are.
- When you graph the best-fit curve, consider also graphing the confidence bands.
- Report the goodness of fit, as the sum-of-squares, the SD of the residuals, R^2, or perhaps all three.

Reporting regression results with multiple independent variables

- Be sure to state what model you fit to the data. Be complete. Did you include interactions?
- State how you arrived at this model. Preplanned? Or did you try multiple models before deciding to report this one?
- If you report the best-fit values of the parameters, also report their CIs. Consider reporting the correlation matrix to show how independent/intertwined the parameters are.
- Show a graph where one axis shows the actual Y values and the other axis shows the Y values predicted by the model. This gives a visual sense of goodness of fit.

Reporting multiple comparisons

- Multiple comparisons must be handled thoughtfully, and all steps must be documented.
- Explain how you handled the issue of multiple comparisons in all situations. Follow-up testing after ANOVA is only one of many situations where multiple comparisons must be accounted for.
- State whether all comparisons were planned. If you report unplanned comparisons, the results should be identified as preliminary.
- State whether all planned comparisons were reported. If some planned comparisons aren't reported, at least describe them.
- If you used any correction for multiple comparisons, explain the details.
- If you report multiplicity adjusted P values, point out clearly that these P values were adjusted.

PART J

❖❖❖

Appendices

APPENDIX A

Statistics with GraphPad

GRAPHPAD PRISM

All the figures in this book were created with GraphPad Prism, and most of the analyses mentioned in this book can be performed with Prism (although not logistic regression).

GraphPad Prism, available for both Windows and Macintosh computers, combines scientific graphing, curve fitting, and basic biostatistics. It differs from other statistics programs in many ways:

- Statistical guidance. Prism helps you make the right choices and make sense of the results. If you like the style of this book, then you'll appreciate the help built into Prism.
- Analysis checklists. After completing any analysis in Prism, click the clipboard icon to see an analysis checklist. Prism shows you a list of questions to ask yourself to make sure you've picked an appropriate analysis.
- Nonlinear regression. Some statistics programs can't fit curves with nonlinear regression and others provide only the basics. Nonlinear regression is one of Prism's strengths, and it provides many options (e.g., remove outliers, compare models, compare curves, interpolate standard curves, etc.).
- Automatic updating. Prism remembers the links among data, analysis choices, results, and graphs. When you edit or replace the data, Prism automatically updates the results and graphs.
- Analysis choices can be reviewed and changed at any time.
- Error bars. You don't have to decide in advance. Enter raw data and then choose whether to graph each value or the mean with SD, SEM, or CI. Try different ways to present the data.

Prism is designed so you can just plunge in and use it without reading instructions. But avoid the temptation to sail past the Welcome dialog so you can go straight to the analyses. Instead, spend a few moments understanding the choices on the Welcome dialog (see Figure A1). This is where you start a new data table and graph, open a Prism file, or clone a graph from one you've made before.

The key to using Prism effectively is to choose the right kind of table for your data, because Prism's data tables are arranged differently than those used by most statistics programs. For example, if you want to compare three means with one-way ANOVA, you enter the three sets of data into three different columns in a data table formatted for entry of column data. With some other programs, you'd enter all the data into one column and also enter a grouping variable into another column to define the group to which each value belongs. Prism does not use grouping variables.

To understand how the various tables are organized, start by choosing some of the sample data sets built into Prism. These sample data sets come with explanations about how the data are organized and how to perform the analysis you want. After experimenting with several sample data sets, you'll be ready to analyze your own data.

You can download a free 30-day trial from www.graphpad.com.

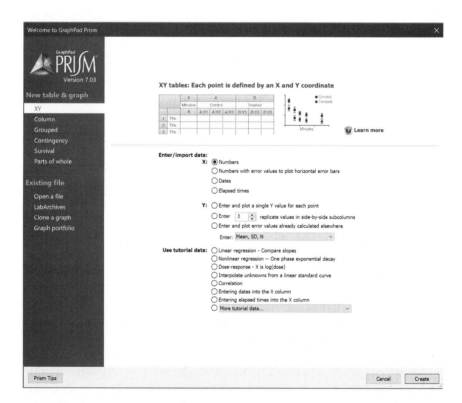

Figure A1. The Welcome dialog of GraphPad Prism.

Before entering data, you must choose which kind of data table is appropriate for your data. You can also choose sample data sets, accompanied by instructions.

FREE GRAPHPAD WEB QUICKCALCS

QuickCalcs are free calculators on www.graphpad.com. We call them calculators because you enter data and instantly get back results, without saving or opening files.

The QuickCalcs system offers a diverse array of calculations (see Figure A2). The most popular calculators include t test, outlier detection, chi-square, CI of a proportion or count, and calculating a P value from a statistical ratio (i.e., t or F). All of the calculators are either self-explanatory or link to longer explanations.

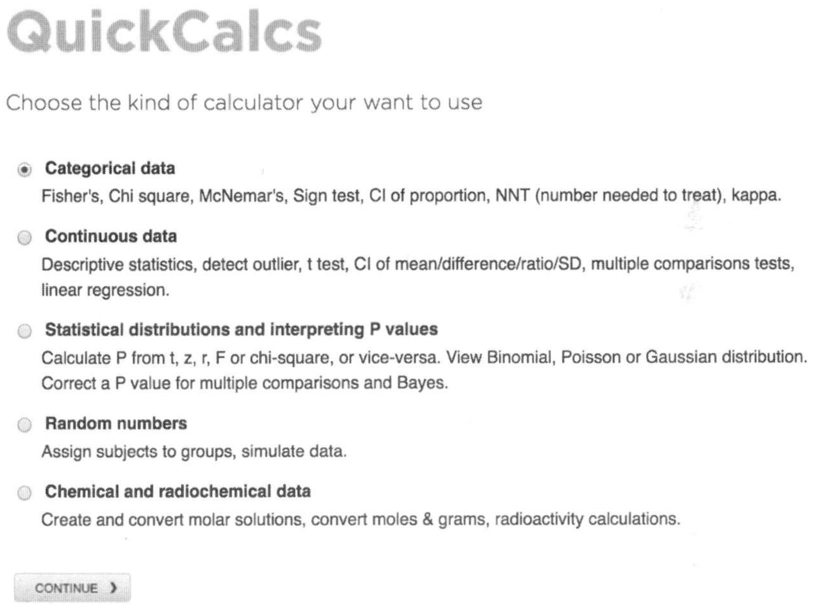

QuickCalcs

Choose the kind of calculator your want to use

- ⦿ **Categorical data**
  Fisher's, Chi square, McNemar's, Sign test, CI of proportion, NNT (number needed to treat), kappa.

- ○ **Continuous data**
  Descriptive statistics, detect outlier, t test, CI of mean/difference/ratio/SD, multiple comparisons tests, linear regression.

- ○ **Statistical distributions and interpreting P values**
  Calculate P from t, z, r, F or chi-square, or vice-versa. View Binomial, Poisson or Gaussian distribution. Correct a P value for multiple comparisons and Bayes.

- ○ **Random numbers**
  Assign subjects to groups, simulate data.

- ○ **Chemical and radiochemical data**
  Create and convert molar solutions, convert moles & grams, radioactivity calculations.

CONTINUE ❯

Figure A2: The main screen of the free QuickCalcs at www.graphpad.com.

APPENDIX B

Statistics with Excel

THE PROS AND CONS OF USING EXCEL
FOR STATISTICAL CALCULATIONS

Microsoft Excel is widely used and is a great program for managing and wrangling data sets. Excel has some statistical capabilities, and many also use it to do some statistical calculations.

The use of Excel for statistics is somewhat controversial, and some recommend that Excel not be used for statistics. One problem is that Excel is far from a complete statistics program. It lacks nonparametric tests, multiple comparisons tests following ANOVA, and many other tests. Another problem is that Excel reports statistical results without including all the supporting details other programs provide.

More seriously, Excel uses some poor algorithms for computing statistics, which can lead to incorrect results (Knusel, 2005; McCullough & Wilson, 2005). Microsoft has responded to these criticisms and fixed many issues in Excel 2003 (Microsoft, 2006). Don't use earlier versions of Excel for statistical work.

Unfortunately, some errors remain in Excel 2007 for Windows and Excel 2008 for Mac. McCullough and Hellser (2008) have pointed out many erroneous results produced by Excel 2007 (especially its Solver) and have concluded, "Microsoft has repeatedly proved itself incapable of providing reliable statistical functionality." Yalta (2008) reached a similar conclusion, stating that "the accuracy of various statistical functions in Excel 2007 range from unacceptably bad to acceptable but inferior." In contrast, Pace (2008) concluded that Microsoft has fixed the important bugs, leaving only statistical bugs that are trivial or obscure. He contended that Excel 2007 is a reasonable choice for analyzing the kinds of data most academics and professionals collect.

Given these problems, you should use another program to check important calculations, especially if your data seem unusual or include missing values.

WHAT YOU NEED TO KNOW BEFORE USING EXCEL FOR STATISTICS

- If you want to compute the mean of a range of numbers, there is no MEAN() function. Use the AVERAGE() function.
- Excel has no function to compute the SEM. You'll need to compute from the SD and n using this equation:

  = STDEV (range) / SQRT (COUNT (range))

- To compute most CIs, you must know the critical value of t (called t* in this book) for specified number of df and confidence (C, in percentage, usually 95%). Do not use Excel's CONFIDENCE() function, which is based on the z (normal) distribution, rather than the t distribution, and so has limited utility. Use this syntax:

  = TINV (1 − 0.01 * C, df)

- The Help within Excel does not always provide useful information about statistical functions. Google for other sources of information.
- Excel provides an Analysis ToolPak, which can perform some statistical tests. You need to use the Add-in Manager to install it before you can use it. Unlike Excel equations, the results of the ToolPak are not linked to the data. If you edit the data, the results will remain fixed until you run the analysis again.
- Beware of Excel's RANK() function. Nonparametric tests assign all tied values the average of the ranks for which they tie. Excel's RANK() function assigns all tied values the lowest of the ranks for which they tie, and the other ranks are not used.
- Beware of the NORMDIST(z) function (Gaussian distribution). It looks like it ought to be similar to the TDIST(t) function (Student t distribution), but the two work very differently. Experiment with sample data before using this function.
- The excellent book by Pace (2008) gives many details about using Excel to do statistical calculations. It can be purchased as either a printed book or as a pdf download.
- Starting with Excel 2010 (Windows) and 2011 (Mac), Excel includes a new set of statistical functions. When you have a choice, use the newer ones, which all are named using several words separated by periods. For example, the old function NORMINV (which still exists) has been replaced with a function called NORM.INV (which is more accurate).

APPENDIX C

Statistics with R

R is a popular software environment (for Windows, Macintosh, and Unix computers) that is used for manipulating, analyzing, and graphing data. Short introductions are provided by Burns (2005) and Briggs (2008c) and a longer one by Robinson (2008). Learn about R and download it from www.r-project.org.

WHAT YOU NEED TO KNOW BEFORE LEARNING R

- R is free and widely available.
- R is a programming environment. You write commands and get back answers. This gives it great power but also makes it hard to get started.
- You don't interact with R via dialogs, menus, and buttons like most Windows and Mac software. Instead, using R feels like having an instant message session with the computer.
- Almost every statistical and bioinformatics analysis has been implemented in R. It is comprehensive.
- Running some analyses requires that you first install add-on software (called a package), as well as R.
- When you run an analysis, the results are stored in an object and R presents only a few key results. You can then enter new commands to query that object to get more details.
- The R language is case sensitive. If you accidentally enter an uppercase letter ("A") when R expects a lowercase letter ("a"), the command won't work.
- You can save R commands in files and then run those R programs. This means you can share R files.
- R is related to S. If you read about a statistical method implemented in S (or S-plus), it will almost certainly also work in R.
- R comes with only a rudimentary spreadsheet-like data editor. You can type values into its session or ask it to read text files. Extensions provide table editors and let it access databases or files created by other statistics programs.
- R includes lots of online help to explain its syntax, but this online help generally assumes you already understand the tests you are using and the results they produce.

APPENDIX D

Values of the t Distribution Needed to Compute Cis

The margin of error of many CIs equals the standard error times a critical value of the t distribution tabulated in Table D1. This value depends on the desired confidence level and the number of df, which equals n minus the number of parameters estimated. For example, when computing a CI of a mean, df = (n – 1), because you

| DESIRED CONFIDENCE LEVEL | | | | | DESIRED CONFIDENCE LEVEL | | | |
df	80%	90%	95%	99%	df	80%	90%	95%	99%
1	3.0777	6.3138	12.7062	63.6567	27	1.3137	1.7033	2.0518	2.7707
2	1.8856	2.9200	4.3027	9.9248	28	1.3125	1.7011	2.0484	2.7633
3	1.6377	2.3534	3.1824	5.8409	29	1.3114	1.6991	2.0452	2.7564
4	1.5332	2.1318	2.7764	4.6041	30	1.3104	1.6973	2.0423	2.7500
5	1.4759	2.0150	2.5706	4.0321	35	1.3062	1.6896	2.0301	2.7238
6	1.4398	1.9432	2.4469	3.7074	40	1.3031	1.6839	2.0211	2.7045
7	1.4149	1.8946	2.3646	3.4995	45	1.3006	1.6794	2.0141	2.6896
8	1.3968	1.8595	2.3060	3.3554	50	1.2987	1.6759	2.0086	2.6778
9	1.3830	1.8331	2.2622	3.2498	55	1.2971	1.6730	2.0040	2.6682
10	1.3722	1.8125	2.2281	3.1693	60	1.2958	1.6706	2.0003	2.6603
11	1.3634	1.7959	2.2010	3.1058	65	1.2947	1.6686	1.9971	2.6536
12	1.3562	1.7823	2.1788	3.0545	70	1.2938	1.6669	1.9944	2.6479
13	1.3502	1.7709	2.1604	3.0123	75	1.2929	1.6654	1.9921	2.6430
14	1.3450	1.7613	2.1448	2.9768	80	1.2922	1.6641	1.9901	2.6387
15	1.3406	1.7531	2.1314	2.9467	85	1.2916	1.6630	1.9883	2.6349
16	1.3368	1.7459	2.1199	2.9208	90	1.2910	1.6620	1.9867	2.6316
17	1.3334	1.7396	2.1098	2.8982	95	1.2905	1.6611	1.9853	2.6286
18	1.3304	1.7341	2.1009	2.8784	100	1.2901	1.6602	1.9840	2.6259
19	1.3277	1.7291	2.0930	2.8609	150	1.2872	1.6551	1.9759	2.6090
20	1.3253	1.7247	2.0860	2.8453	200	1.2858	1.6525	1.9719	2.6006
21	1.3232	1.7207	2.0796	2.8314	250	1.2849	1.6510	1.9695	2.5956
22	1.3212	1.7171	2.0739	2.8188	300	1.2844	1.6499	1.9679	2.5923
23	1.3195	1.7139	2.0687	2.8073	350	1.2840	1.6492	1.9668	2.5899
24	1.3178	1.7109	2.0639	2.7969	400	1.2837	1.6487	1.9659	2.5882
25	1.3163	1.7081	2.0595	2.7874	450	1.2834	1.6482	1.9652	2.5868
26	1.3150	1.7056	2.0555	2.7787	500	1.2832	1.6479	1.9647	2.5857

Table D1

are only estimating one parameter (the mean); when computing a CI for a slope obtained by linear regression, df = (n − 2), because you estimated two parameters (the slope and the intercept).

These values were computed using the following formula from Excel 2011 (C is the desired confidence level in percentage):

$$= \text{T.INV.2T} \ (1 - 0.01 * C, \ df)$$

If you have an older version of Excel (prior to 2010) use the TINV function instead of T.INV. Don't use Excel's CONFIDENCE() function, which is based on the z (normal) distribution rather than the t distribution and so has limited utility.

APPENDIX E

A Review of Logarithms

COMMON (BASE 10) LOGARITHMS

The best way to understand logarithms is through an example. If you take 10 to the third power ($10 \times 10 \times 10$), the result is 1,000. The logarithm is the inverse of that power function. The logarithm (base 10) of 1,000 is the power of 10 that gives the answer 1,000. So the logarithm of 1,000 is 3. If you multiply 10 by itself 3 times, you get 1,000.

You can take 10 to a negative power. For example, taking 10 to the -3 power is the same as taking the reciprocal of 10^3. So 10^{-3} equals $1/10^3$, or 0.001. The logarithm of 0.001 is the power of 10 that equals 0.001, which is -3.

You can take 10 to a fractional power. Ten to the ½ power equals the square root of 10, which is 3.162. So the logarithm of 3.162 is 0.5.

Ten to the zero power equals 1, so the logarithm of 1.0 is 0.0.

You can take the logarithm of any positive number. The logarithms of values between zero and 1 are negative; the logarithms of values greater than 1 are positive. The logarithms of zero and all negative numbers are undefined, because there is no power of 10 that results in a negative number or zero.

OTHER BASES

The logarithms shown in the previous section are called base 10 logarithms, because the computations take 10 to some power. These are also called *common logarithms*.

You can compute logarithms for any power. Mathematicians prefer natural logarithms, using base e (2.7183 . . .). By convention, natural logarithms are used in logistic and proportional hazards regression (see Chapter 37).

Biologists sometimes use base 2 logarithms, often without realizing it. The base 2 logarithm is the number of doublings it takes to reach a value. So the log base 2 of 16 is 4, because if you start with 1 and double it four times (2, 4, 8, and 16), the result is 16. Immunologists often serially dilute antibodies by factors of 2, so they often graph data on a log2 scale. Cell biologists use base 2 logarithms to convert cell counts to number of doublings.

Logarithms using different bases are proportional to each other. Consequently, converting from natural logs to common logs is sort of like changing units. Divide a natural logarithm by 2.303 to compute the common log of the same value. Multiply a common log by 2.303 to obtain the corresponding natural log.

NOTATION

Unfortunately, the notation is used inconsistently.

The notation "log(x)" usually means the common (base 10) logarithm, but some computer languages use it to mean the natural logarithm.

The notation "ln(x)" always means natural logarithm.

The notation "$\log_{10}(x)$" clearly shows that the logarithm uses base 10.

LOGARITHMS CONVERT MULTIPLICATION INTO ADDITION

Logarithms are popular because of this equation:

$$\log(A \cdot B) = \log(A) + \log(B)$$

Similarly, as shown in Chapter 11, logarithms transform a lognormal distribution into a Gaussian distribution.

ANTILOGARITHMS

The antilogarithm (also called an antilog) is the inverse of the logarithm transform. Because the logarithm (base 10) of 1,000 equals 3, the antilogarithm of 3 is 1,000. To compute the antilogarithm of a base 10 logarithm, take 10 to that power.

To compute the antilogarithm of a natural logarithm, take e to that power. The natural logarithm of 1,000 is 6.908. So the antilogarithm of 6.908 is $e^{6.908}$, which is 1,000. Spreadsheets and computer languages use the notation exp(6.908).

APPENDIX F

Choosing a Statistical Test

If you are not sure which statistical test to use, the tables in this appendix may help you decide. Reviewing these tables is also a good way to review your understanding of statistics. Of course, this appendix can't include every possible statistical test (and some kinds of data require developing new statistical tests).

OUTCOME: CONTINUOUS DATA FROM GAUSSIAN DISTRIBUTION

Examples	• Cholesterol plasma level (mg/dL) • Change in systolic blood pressure (mmHg)
Test Gaussian assumption	• Normality tests • Outlier tests
Describe one sample	• Frequency distribution • Sample mean • Minimum and maximum value and range • 25th and 75th percentiles • Sample SD
Make inferences about one population	• One-sample t test
Compare two unmatched (unpaired) groups	• Unpaired t test
Compare two matched (paired) groups	• Paired t test
Compare three or more unmatched (unpaired) groups	• One-way ANOVA, followed by multiple comparisons tests
Compare three or more matched (paired) groups	• Repeated-measures ANOVA followed by multiple comparisons tests
Quantify association between two variables	• Pearson's correlation
Explain/predict one variable from another	• Simple linear regression • Simple nonlinear regression (*simple* means you are predicting the outcome from a single independent variable)
Explain/predict one variable from several others	• Multiple linear regression (*multiple* means you are predicting the outcome from more than one independent variable) • Multiple nonlinear regression

OUTCOME: CONTINUOUS DATA FROM NONGAUSSIAN DISTRIBUTION (OR RANKED DATA)

Examples	• 17β-estradiol levels (pg/mL) • IIEF-5 scoring system (5–25) • Gleason score for prostate cancer (2–10) • Confidence (5 = very high, 4 = high, 3 = moderate, 2 = low, 1 = very low) • Number of headaches in a week
Describe one sample	• Frequency distribution • Sample median • Minimum and maximum value and range • 25th and 75th percentiles
Make inferences about one population	• Wilcoxon's rank-sum test
Compare two unmatched (unpaired) groups	• Mann–Whitney test
Compare two matched (paired) groups	• Wilcoxon's matched pairs test
Compare three or more unmatched (unpaired) groups	• Kruskal–Wallis test • Dunn's posttest
Compare three or more matched (paired) groups	• Friedman's test • Dunn's posttest
Quantify association between two variables	• Spearman's correlation

OUTCOME: SURVIVAL TIMES (OR TIME TO AN EVENT)

Examples	• Months until death for prostate cancer patient • Number of days until cold symptoms go away • Minutes until REM sleep begins
Describe one sample	• Kaplan–Meier survival curve • Median survival time • Five-year survival percentage
Make inferences about one population	• Confidence bands around survival curve • CI of median survival
Compare two unmatched (unpaired) groups	• Log-rank test • Gehan–Breslow test • CI of ratio of median survival times • CI of hazard ratio
Compare two matched (paired) groups	• Conditional proportional hazards regression
Compare three or more unmatched (unpaired) groups	• Log-rank test • Gehan–Breslow test
Compare three or more matched (paired) groups	• Conditional proportional hazards regression
Explain/predict one variable from one or several others	• Cox's proportional hazards regression

OUTCOME: BINOMIAL

Examples	• Cure (yes/no) of acute myeloid leukemia • Within a specific time period • Success (yes/no) of preventing motion sickness • Recurrence of infection (yes/no)
Describe one sample	• Proportion
Make inferences about one population	• CI of proportion • Binomial test to compare observed distribution with a theoretical (expected) distribution
Compare two unmatched (unpaired) groups	• Fisher's exact test
Compare two matched (paired) groups	• McNemar's test
Compare three or more unmatched (unpaired) groups	• Chi-square test • Chi-square test for trend
Compare three or more matched (paired) groups	• Cochran's Q
Explain/predict one variable from one or several others	• Logistic regression

APPENDIX G

Problems and Answers

The first three editions contained a chapter of problems and another chapter with extensive discussion of the answers. These have not been updated for the fourth edition, but the problems and answers for the third edition are available online at www.oup.com/us/motulsky-4e

REFERENCES

Agnelli, G., Buller, H. R., Cohen, A., Curto, M., Gallus, A. S., Johnson, M., Porcari, et al. (2012). Apixaban for extended treatment of venous thromboembolism. *New England Journal of Medicine, 368*, 699–708.

Agresti, A., & Coull, B. A. (1998). Approximate is better than exact for interval estimation of binomial proportions. *American Journal of Statistics, 52*, 119–126.

Alere. (2014). DoubleCheckGold™ HIV 1&2. Retrieved May 14, 2014, http://www.alere.com/ww/en/product-details/doublecheckgold-hiv-1-2.html.

Altman, D. G. (1990). *Practical statistics for medical research.* London: Chapman & Hall/CRC.

Altman, D. G., & Bland, J. M. (1995). Absence of evidence is not evidence of absence. *BMJ, 311*, 485.

———. (1998). Time to event (survival) data. *BMJ, 317*, 468–469.

Altman D. G, Gore, S. M., Gardner, M. J., & Pocock S. J. (1983). Statistical guidelines for contributors to medical journals. *BMJ, 286*, 1489–1493.

Anscombe, F. J. (1973). Graphs in statistical analysis. *American Statistician, 27*, 17–21.

Arad, Y., Spadaro, L. A., Goodman, K., Newstein, D., & Guerci, A. D. (2000). Prediction of coronary events with electron beam computed tomography. *Journal of the American College of Cardiology, 36*, 1253–1260.

Arden, R., Gottfredson, L. S., Miller, G., & Pierce, A. (2008). Intelligence and semen quality are positively correlated. *Intelligence, 37*, 277–282.

Austin, P. C., & Goldwasser, M. A. (2008). Pisces did not have increased heart failure: Data-driven comparisons of binary proportions between levels of a categorical variable can result in incorrect statistical significance levels. *Journal of Clinical Epidemiology, 61*, 295–300.

Austin, P. C., Mamdani, M. M., Juurlink, D. N., & Hux, J. E. (2006). Testing multiple statistical hypotheses resulted in spurious associations: A study of astrological signs and health. *Journal of Clinical Epidemiology, 59*, 964–969.

Babyak, M. A. (2004). What you see may not be what you get: A brief, nontechnical introduction to overfitting in regression-type models. *Psychosomatic Medicine, 66*, 411–421.

Bailar, J. C. (1997). The promise and problems of meta-analysis. *New England Journal of Medicine, 337*, 559–561.

Bakhshi, E., Eshraghian, M. R., Mohammad, K., & Seifi, B. (2008). A comparison of two methods for estimating odds ratios: Results from the National Health Survey. *BMC Medical Research Methodology, 8,* 78.

Barter, P. J., Caulfield, M., Eriksson, M., Grundy, S. M., Kastelein, J. J., Komajda, M., Lopez-Sendon, J., et al. (2007). Effects of torcetrapib in patients at high risk for coronary events. *New England Journal of Medicine, 357,* 2109–2122.

Bausell, R. B. (2007). *Snake oil science: The truth about complementary and alternative medicine.* Oxford: Oxford University Press.

Begley, C. G. C., & Ellis, L. M. L. (2012). Drug development: Raise standards for preclinical cancer research. *Nature, 483,* 531–533.

Benjamin, D. J., Berger, J., Johannesson, M.,et al, (2017). Redefine statistical significance. Retrieved from psyarxiv.com/mky9j

Benjamini, Y., & Hochberg, Y. (1995). Controlling the false discovery rate: A practical and powerful approach to multiple testing. *Journal of Royal Statistical Society, B, 57,* 290–300.

Bennett, C. M., Baird, A. A., Miller, M. B., & Wolford, G. L. (2011). Neural correlates of interspecies perspective taking in the post-mortem Atlantic salmon: An argument for proper multiple comparisons correction. *Journal of Serendipitous and Unexpected Results, 1,* 1–5.

Bernstein, C. N., Nugent, Z., Longobardi, T., & Blanchard, J. F. (2009). Isotretinoin is not associated with inflammatory bowel disease: A population-based case–control study. *American Journal of Gastroenterology, 104,* 2774–2778.

Berry, D. A. (2007). The difficult and ubiquitous problems of multiplicities. *Pharmaceutical Statistics, 6,* 155–160.

Bhatt, D.L., and Mehta, C. (2016). Adaptive Designs for Clinical Trials. *New England Journal of Medicine 375,* 65–74.

Bickel, P. J., Hammel, E. A., & O'Connell, J. W. (1975). Sex bias in graduate admissions: Data from Berkeley. *Science, 187,* 398–404.

Bishop, D. (2013) Interpreting unexpected significant results. *BishopBlog* (blog), June 7. Accessed July 2, 2013, http://www.deevybee.blogspot.co.uk/2013/06/interpreting-unexpected-significant.html.

Bland, J. M. J., & Altman, D. G. D. (2011). Comparisons against baseline within randomized groups are often used and can be highly misleading. *Trials, 12,* 264–264.

Blumberg, M. S. (2004). *Body heat: Temperature and life on earth.* Cambridge, MA: Harvard University Press.

Bleyer, A. & Welch, H. G. (2012). Effect of three decades of screening mammography on breast-cancer incidence. *New England Journal of Medicine, 367,* 1998–2005.

Boos, D. D., & Stefanski, L. A. (2011). P-value precision and reproducibility. *American Statistician, 65,* 213–221.

Borenstein, M., Hedges, L, Higgins, J., & Rothstein, H. (2009). *Introduction to meta-analysis.* Chichester, UK: Wiley.

Borkman, M., Storlien, L. H., Pan, D. A., Jenkins, A. B., Chisholm, D. J., & Campbell, L. V. (1993). The relation between insulin sensitivity and the fatty-acid composition of skeletal-muscle phospholipids. *New England Journal of Medicine, 328,* 238–244.

Briggs, W. M. (2008a). On the difference between mathematical ability between boys and girls. *William M. Briggs* (blog), July 25. Accessed June 21, 2009, http://www.wmbriggs.com/blog/?p=163/.

———. (2008b). Do not calculate correlations after smoothing data. *William M. Briggs* (blog), February 14. Accessed June 21, 2009, wmbriggs.com/blog/?p=86/.

————. (2008c). Stats 101, Chapter 5: R. *William M. Briggs* (blog), May 20. Accessed June 21, 2009, http://www.wmbriggs.com/public/briggs_chap05.pdf.

Brown, L. D., Cai, T. T., & DasGupta, A. (2001). Interval estimation for a binomial proportion. *Statistical Science, 16,* 101–133.

Burnham, K., & Anderson, D. (2003). *Model selection and multi-model inference.* 2d ed. New York: Springer.

Burns, P. (2005). *A guide for the unwilling S user.* Accessed January 26, 2009, http://www.burns-stat.com/pages/Tutor/unwilling_S.pdf.

Bryant, F. B., & Brockway, J. H. (1997). Hindsight bias in reaction to the verdict in the O. J. Simpson criminal trial. *Basic and Applied Social Psychology, 19,* 225–241.

Campbell, M. J. (2006). *Statistics at square two.* 2d ed. London: Blackwell.

Cantor, W. J., Fitchett, D. L., Borgundvagg, B., Ducas, J., Heffernam, M., Cohen, E. A., Morrison, L. J., et al. (2009). Routine early angioplasty after fibrinolysis for acute myocardial infarction. *New England Journal of Medicine, 360,* 2705–2718.

Cardiac Arrhythmia Suppression Trial (CAST) Investigators. (1989). Preliminary report: Effect of encainide and lecainide on mortality in a randomized trial of arrhythmia suppression after myocardial infarction. *New England Journal of Medicine, 3212,* 406–412.

Carroll, L. (1871). *Through the looking glass.* Accessed December 19, 2012, http://www.gutenberg.org/ebooks/12.

Casadevall, A., Fang, F. C. (2014). Diseased science. *Microbe, 9,* 390–392.

Central Intelligence Agency. (2012). *The world factbook.* Accessed August 26, 2012, http://www.cia.gov/library/publications/the-world-factbook/fields/2018.html.

Chan, A. W., Hrobjartsson, A., Haahr, M. T., Gotzsche, P. C., & Altman, D. G. (2004). Empirical evidence for selective reporting of outcomes in randomized trials: Comparison of protocols to published articles. *Journal of the American Medical Association, 291,* 2457–2465.

Chang, M., & Balser, J. (2016). Adaptive design-recent advancement in clinical trials. *Journal of Bioanalysis and Biostatistics, 1,* 1–14.

Chao, T. A., Liu, C. J., Chen, S. J., Wang, K. L., Lin, Y. J., Chang, S. L., Lo, L. W., et al. (2012). The association between the use of non-steroidal anti-inflammatory drugs and atrial fibrillation: a nationwide case-control study. *International Journal of Cardiology, 168,* 312–316.

Clopper, C. J., & Pearson, E. S. (1934). The use of confidence or fiducial limits illustrated in the case of the binomial. *Biometrika, 26,* 404–413.

Cochrane Methods (2017) About IPD meta-analysis. Accessed June 18, 2017, http://methods.cochrane.org/ipdma/about-ipd-meta-analyses.

Cochrane Handbook (2017). Detecting reporting biases. Accessed June 20, 2017, http://handbook.cochrane.org/index.htm#chapter_10/10_4_1_funnel_plots.htm

Cohen, J. (1988). *Statistical power analysis for the behavioral sciences.* 2d ed. Hillsdale, NJ: Erlbaum.

Cohen, T. J., Goldner, B. G., Maccaro, P. C., Ardito, A. P., Trazzera, S., Cohen, M. B., & Dibs, S. R. (1993). A comparison of active compression–decompression cardiopulmonary resuscitation with standard cardiopulmonary resuscitation for cardiac arrests occurring in the hospital. *New England Journal of Medicine, 329,* 1918–1921.

Colquhoun, D. D. (2003). Challenging the tyranny of impact factors. *Nature, 423,* 479–480.

————. (2014). An investigation of the false discovery rate and the misinterpretation of p-values. *Royal Society Open Science, 1,* 140216–140216.

————. (2017). The reproducibility of research and the misinterpretation of P values. bioRxiv preprint 144337; doi: https://doi.org/10.1101/144337. Accessed Aug. 24, 2017.

Cramer, H. (1999). *Mathematical methods of statistics*. Princeton, NJ: Princeton University Press.

Crichton, M. (2003). *Environmentalism* as religion. Speech to the Commonwealth Club, San Francisco, September 13. Accessed November 9, 2008, http://www. michaelcrichton. net/speech-environmentalismaseligion.html.

————. (2005). The case for skepticism on global warming. Accessed November 9, 2008, http://www.michaelcrichton.net/speech-ourenvironmentalfuture.html.

Cumming, G. (2008). Replication and p intervals: P values predict the future only vaguely, but confidence intervals do much better. *Perspectives on Psychological Science, 3,* 286–300.

————. (2011). *Understanding the new statistics: Effect sizes, confidence intervals, and meta-analysis.* New York: Routledge.

Cumming, G., and Maillardet, R. (2006). Confidence intervals and replication: Where will the next mean fall? Psychological Methods 11: 217–227.

Cumming, G., Fidler, F., & Vaux, D. L. (2007). Error bars in experimental biology. *Journal of Cell Biology, 177,* 7–11.

Curran-Everett, D. (2016). Explorations in statistics: statistical facets of reproducibility. Advances in Physiology Education 40: 248–252.

Curtis, M. J., Bond, R. A., Spina, D., Ahluwalia, A., Alexander, S. P. A., Giembycz, M. A., et al. (2015). Experimental design and analysis and their reporting: New guidance for publication in BJP. *British Journal of Pharmacology, 172,* 3461–3471.

Dallal, G. E. (2012). *The little handbook of statistical practice.* Kindle edition. Amazon Digital Services.

Darwin, C. (1876). *The effects of cross and self fertilisation in the vegetable kingdom.* London: Murray.

Davenas, E., Beauvais, F., Amara, J., Oberbaum, M., Robinzon, B., Miadonna, A., Tedeschi, A., Pomeranz, B., Fortner, P., & Belon, P. (1988). Human basophil degranulation triggered by very dilute antiserum against IgE. *Nature, 333,* 816–818.

Delacre, M., Lakens, D.L., and Leys, C. (2017). Why Psychologists Should by Default Use Welch's t-test Instead of Student's t-test. International Review of Social Psychology, 30: 92–10.

Denes-Raj, V., & Epstein, S. (1994). Conflict between intuitive and rational processing: When people behave against their better judgment. *Journal of Personality and Social Psychology, 66,* 819–829.

de Winter, J.C.F. (2013). Using the Student's t-Test with Extremely small samples, *Practical Research & Evaluation. 18,* 1–12.

Dumas-Mallet, E., Button, K. S., Boraud, T., Gonon, F., & Munafò, M. R. (2017). Low statistical power in biomedical science: A review of three human research domains. *Royal Society Open Science, 4.* doi: 10.1098/rsos.160254

Dumas-Mallet, E., Button, K., Boraud, T., Munafo, M., & Gonon, F. (2016). Replication validity of initial association studies: A comparison between psychiatry, neurology and four somatic diseases. *PLoS ONE, 11*(6). doi: 10.1371/journal.pone.0158064

Environmental Protection Agency. (2011). Body weight scales. Chapter 8, *Exposure factors handbook.* Accessed July 7, 2013, https://www.epa.gov/sites/production/files/2015-09/documents/efh-chapter08.pdf.

Evans, S. (2007). When and how can endpoints be changed after initiation of a randomized clinical trial. *PLoS Hub for Clinical Trials, 2*(4), e18. doi: 10.1371/journal.pctr.0020018

Ewigman, B. G., Crane, J. P., Frigoletto, F. D., LeFevre, M. L., Bain, R. P., & McNellis, D. (1993). Effect of prenatal ultrasound screening on perinatal outcome. RADIUS Study Group. *New England Journal of Medicine, 329,* 821–827.

Eyding, D. D., Lelgemann, M., Grouven, U., Härter, M., Kromp, M., Kaiser, T., Kerekes, M. F., Gerken, M., & Wieseler, B. (2010). Reboxetine for acute treatment of major depression: Systematic review and meta-analysis of published and unpublished placebo and selective serotonin reuptake inhibitor controlled trials. *BMJ, 341*. doi: 10.1136/bmj.c4737.

Faul, F., Erdfelder, E., Lang, A.-G., and Buchner, A. (2007). G*Power 3: A flexible statistical power analysis program for the social, behavioral, and biomedical sciences. *Behavior Research Methods 39*: 175–191.

Federal Election Commission. (2012). Official 2012 presidential general election results. Accessed June 2, 2013, http://www.fec.gov/pubrec/fe2012/2012presgeresults.pdf.

Feinstein, A. R., Sosin, D. M., & Wells, C. K. (1985). The Will Rogers phenomenon. Stage migration and new diagnostic techniques as a source of misleading statistics for survival in cancer. *New England Journal of Medicine, 312*, 1604–1608.

Fisher, R. A. (1935). *The design of experiments*. New York: Hafner.

———. (1936). Has Mendel's work been rediscovered? *Annals of Science, 1*, 115–137.

Fisher, L. D, & Van Belle, G. (1993). *Biostatistics. A methodology for the health sciences*. New York: Wiley Interscience.

Fleming, T. R. (2006). Standard versus adaptive monitoring procedures: A commentary. *Statistics in Medicine, 25*, 3305–3512; discussion 3313–3314, 3326–3347.

Flom, P. L., & Cassell, D. L. (2007). Stopping stepwise: Why stepwise and similar selection methods are bad, and what you should use. Paper presented at the NESUG 2007, Baltimore. Accessed January 2, 2017, http://www.lexjansen.com/pnwsug/2008/DavidCassell-StoppingStepwise.pdf.

Frazier, E. P., Schneider, T., & Michel, M. C. (2006). Effects of gender, age and hypertension on beta-adrenergic receptor function in rat urinary bladder. *Naunyn-Schmiedeberg's Archives of Pharmacology, 373*, 300–309.

Freedman, D. (1983). A note on screening regression equations. *American Statistician, 37*, 152–155.

———. (2007). *Statistics*. 4th ed. New York: Norton.

Freese, J. (2008). The problem of predictive promiscuity in deductive applications of evolutionary reasoning to intergenerational transfers: Three cautionary tales. In Intergenerational caregiving, ed. A. Booth, A. C. Crouter, S. M. Bianchi, & J. A. Seltzer. Rowman & Littlefield Publishers, Inc

Fung, K. (2011). The statistics of anti-doping. *Big Data, Plainly Spoken* (blog), Accessed December 9, 2012, at http://www.junkcharts.typepad.com/numbersruleyourworld/2011/03/the-statistics-of-anti-doping.html.

Gabriel, S. E., O'Fallon, W. M., Kurland, L. T., Beard, C. M., Woods, J. E., & Melton, L. J. (1994). Risk of connective-tissue diseases and other disorders after breast implantation. *New England Journal of Medicine, 330*, 1697–1702.

García-Berthou, E., & Alcaraz, C. (2004). Incongruence between test statistics and P values in medical papers. *BMC Medical Research Methodology, 4*, 13–18.

Garnæs, K. K., Mørkved, S., Salvesen, Ø., & Moholdt, T. (2016). Exercise training and weight gain in obese pregnant women: A randomized controlled trial (ETIP Trial). *PLoS Medicine, 13*(7), e1002079–18. doi: 10.1371/journal.pmed.1002079

Gelman, A. (1998). Some class-participation demonstrations for decision theory and Bayesian statistics. *American Statistician, 52*, 167–174.

———. (2009). How does statistical analysis differ when analyzing the entire population rather than a sample? *Statistical Modeling, Causal Inference, and Social Science* (blog), July 3. Accessed December 11, 2012, http://www.andrewgelman.com/2009/07/how_does_statis/.

———. (2010). Instead of "confidence interval," let's say "uncertainty interval." Accessed November 5, 2016, andrewgelman.com/2010/12/21/lets_say_uncert/.

———. (2012). Overfitting. *Statistical Modeling, Causal Inference, and Social Science* (blog), July 27. Accessed November 15, 2012, http://www.andrewgelman.com/2012/07/27/15864/.

———. (2013). Don't let your standard errors drive your research agenda. *Statistical Modeling, Causal Inference, and Social Science* (blog), February 1. Accessed February 8, 2013, http://www.andrewgelman.com/2013/02/dont-let-your-standard-errors-drive-your-research-agenda/.

———. (2015) The feather, the bathroom scale, and the kangaroo. *Statistical Modeling, Causal Inference, and Social Science* (blog). Accessed August 2, 2016, http://andrewgelman.com/2015/04/21/feather-bathroom-scale-kangaroo/.

Gelman, A., & Feller, A. (2012) Red versus blue in a new light. *New York Times,* September 12. Accessed January 2013, http://www.campaignstops.blogs.nytimes.com/2012/11/12/ red-versus-blue-in-a-new-light/.

Gelman, A., & Carlin, J. (2014). Beyond power calculations assessing Type S (sign) and Type M (magnitude) errors. *Perspectives on Psychological Science, 9,* 641–51.

Gelman, A., & Loken, E. (2014). The statistical crisis in science: Data-dependent analysis—a "garden of forking paths"— explains why many statistically significant comparisons don't hold up. *American Scientist, 102,* 460. doi: 10.1511/2014.111.460

Gelman, A., & Stern, H. (2006). The difference between "significant" and "not significant" is not itself statistically significant. *American Statistician, 60,* 328–331.

Gelman, A., & Tuerlinckx, F. (2000). Type S error rates for classical and Bayesian single and multiple comparison procedures. *Computational Statistics, 15,* 373–390.

Gelman, A., & Weakliem, D. (2009). Of beauty, sex and power: Statistical challenges in estimating small effects. *American Scientist, 97,* 310–311.

Gigerenzer, G. (2002). *Calculated risks.* New York: Simon and Schuster.

Glantz, S. A., Slinker, B. K., & Neilands, T. B. (2016). *Primer of applied regression and analysis of variance.* 2d ed. New York: McGraw-Hill.

Glickman, M. E., Rao, S. R., & Schultz, M. R. (2014). False discovery rate control is a recommended alternative to Bonferroni-type adjustments in health studies. *Journal of Clinical Epidemiology, 67,* 850–857.

Goddard, S. (2008). Is the earth getting warmer, or cooler? *The Register,* May 2. Accessed June 13, 2008, http://www.theregister.co.uk/2008/05/02/a_tale_of_two_thermometers/.

Goldacre, B. (2013) *Bad pharma.* London: Faber & Faber.

Goldacre, B., Drysdale, H., Dale, A., Hartley, P., Milosevic, I., Slade, E., Mahtani, K., Heneghan, C., & Powell-Smith, A. (2016). Tracking switched outcomes in clinical trials. Centre for Evidence Based Medicine Outcome Monitoring Project (COMPare). http://www.COMPare-trials.org.

Goldstein, D. (2006). The difference between significant and not significant is not statistically significant. *Decision Science News.* Accessed February 8, 2013, http://www.decisionsciencenews.com/2006/12/06/the-difference-between-significant-and-not-significant-is-not-statistically-significant/.

Good, P. I., & Hardin, J. W. (2006). *Common errors in statistics (and how to avoid them).* Hoboken, NJ: Wiley.

Goodman, S. (2008). A dirty dozen: Twelve p-value misconceptions. *Seminars in Hematology, 45,* 135–140.

Goodman, S. N. (1992). A comment on replication, p-values and evidence. *Statistics in Medicine, 11,* 875–879.

Goodman, S. N. (2016). Aligning statistical and scientific reasoning. *Science, 352,* 1180–1181.

Goodwin, P. (2010). Why hindsight can damage foresight. *Foresight, 17,* 5–7.

Gotzsche, P. C. (2006). Believability of relative risks and odds ratios in abstracts: Cross sectional study. *BMJ, 333.* doi: 10.1136/bmj.38895.410451.79.

Gould, S. J. (1997). *Full house: The spread of excellence from Plato to Darwin.* New York: Three Rivers.

Greenland, S., Senn, S. J., Rothman, K. J., Carlin, J. B., Poole, C., Goodman, S. N., & Altman, D. G. (2016). Statistical tests, P values, confidence intervals, and power: A guide to misinterpretations. *European Journal of Epidemiology, 31,* 337–350.

Greenland, S., Thomas, D. C., & Morgenstern, H. (1986). The rare-disease assumption revisited. A critique of "estimators of relative risk for case-control studies." *American Journal of Epidemiology, 124,* 869–883.

Grobbee, D. E. & Hoes, A. W. (2015). *Clinical epidemiology.* 2d ed. Burlington, MA: Jones and Bartlett.

Hankins, M. (2013). Still not significant. *Psychologically Flawed* (blog). April 21. Accessed June 28, 2013, http://www.mchankins.wordpress.com/2013/04/21/still-not-significant-2/.

Hanley, J. A., & Lippman-Hand, A. (1983). If nothing goes wrong, is everything alright? *Journal of the American Medical Association, 259,* 1743–1745.

Harmonic mean. (2017). Wikipedia. Accessed Jan 20, 2017 at https://en.wikipedia.org/wiki/Harmonic_mean.

Harrell, F. (2015). *Regression modeling strategies: With applications to linear models, logistic and ordinal regression, and survival analysis.* 2d ed. Cham, Switzerland: Springer International.

Harter, H. L. (1984). Another look at plotting positions. *Communications in Statistics: Theory and Methods, 13,* 1613–1633.

Hartung, J. (2005). Statistics: When to suspect a false negative inference. In *American Society of Anesthesiology 56th annual meeting refresher course lectures,* Lecture 377, 1–7. Philadelphia: Lippincott.

Heal, C. F., Buettner, P. G., Cruickshank, R., & Graham, D. (2009). Does single application of topical chloramphenicol to high risk sutured wounds reduce incidence of wound infection after minor surgery? Prospective randomised placebo controlled double blind trial. *BMJ, 338,* 211–214.

Henderson, B. (2005). Open letter to Kansas School Board. *Church of the Flying Spaghetti Monster* (blog). Accessed December 8, 2012, http://www.venganza.org/about/open-letter/.

Hetland, M. L., Haarbo, J., Christiansen, C., & Larsen, T. (1993). Running induces menstrual disturbances but bone mass is unaffected, except in amenorrheic women. *American Journal of Medicine, 95,* 53–60.

Hintze, J. L., & Nelson, R. D. (1998). Violin plots: A box plot-density trace synergism. *American Statistician, 52,* 181–184.

Hoenig, J. M., & Heisey, D. M. (2001). The abuse of power: The pervasive fallacy of power. Calculations for data analysis. *American Statistician, 55,* 1–6.

Hollis, S., & Campbell, F. (1999). What is meant by intention to treat analysis? Survey of published randomized controlled trials. *BMJ, 319,* 670–674.

HPS2-THRIVE Collaborative Group. (2014). Effects of extended-release niacin with laropiprant in high-risk patients. *New England Journal of Medicine, 371,* 203–12.

Hsu, J. (1996). *Multiple comparisons: Theory and methods.* Boca Raton, FL: Chapman & Hall/CRC.

Huber, P. J. (2003). *Robust statistics.* Hoboken, NJ: Wiley Interscience.

Hunter, D. J., Manson, J. E., Colditz, G. A., Stampfer, M. J., Rosner, B., Hennekens, C., Speizer, F. E., & Willet, W. C. (1993). A prospective study of the intake of vitamins C, E, and A and the risk of breast cancer. *New England Journal of Medicine, 329,* 234–240.

Hviid, A., Melbye, M., & Pasternak, B. (2013). Use of selective serotonin reuptake inhibitors during pregnancy and risk of autism. *New England Journal of Medicine, 369,* 2406–2415.

Hyde, J. S., Lindberg, S. M., Linn, M. C., Ellis, A. B., & Williams, C. C. (2008). Diversity: Gender similarities characterize math performance. *Science, 321,* 494–495.

Ioannidis, J. P. (2005). Why most published research findings are false. *PLoS Medicine, 2*(8), e124. doi: 10.1371/journal.pmed.0020124

———. (2008). Why most discovered true associations are inflated. *Epidemiology, 19,* 640–648.

Johnson, V.E. (2013). Revised standards for statistical evidence. Proc Natl Acad Sci USA 110: 19313–19317.

Julious, S. A. (2005). Sample size of 12 per group rule of thumb for a pilot study. *Pharmaceutical Statistics, 4,* 287–291.

Kahan, D. M., Peters, E., Dawson, E. C., & Slovic, P. (2013). Motivated numeracy and enlightened self-government. Yale Law School, Public Law Working Paper No. 307. http://doi.org/10.2139/ssrn.2319992

Kane, J. M., & Mertz, J. E. (2012). Debunking myths about gender and mathematics performance. *Notices of the AMS, 59,* 10–21.

Katz, M. H. (2006). *Multivariable analysis: A practical guide for clinicians.* Cambridge, UK: Cambridge University Press.

Kaul, A., & Diamond, G. (2006). Good enough: A primer on the analysis and interpretation of noninferiority trials. *Annals of Internal Medicine, 145,* 62–69.

Kelley, K., & Maxwell, S. E. (2008). Sample size planning with applications to multiple regression: Power and accuracy for omnibus and targeted effects. In *The SAGE handbook of social research methods,* ed. P. Alasuutari, L. Bickman, & J. Brannen, pp. 166–192. Thousand Oaks, CA: SAGE.

Kerr, N. L. (1998). HARKing: Hypothesizing after the results are known. *Personality and Social Psychology Review, 2,* 196–217.

Kirk, A. P., Jain, S., Pocock, S., Thomas, H. C., & Sherlock, S. (1980). Late results of the Royal Free Hospital prospective controlled trial of prednisolone therapy in hepatitis B surface antigen negative chronic active hepatitis. *Gut, 21,* 78–83.

Kirkwood, T. (1979). Geometric means and measures of dispersion. *Journal of Physical Oceanography, 17,* 1817–1836.

Kline, R. B. (2004). *Beyond significance testing: Reforming data analysis methods in behavioral research.* New York: American Psychological Association.

Knol, M. J., Pestman, W. R., & Grobbee, D. E. (2011). The (mis)use of overlap of confidence intervals to assess effect modification. *European Journal of Epidemiology, 26,* 253–254.

Knottnerus, J. A. J., & Bouter, L. M. L. (2001). The ethics of sample size: Two-sided testing and one-sided thinking. *Journal of Clinical Epidemiology, 54,* 109–110.

Knusel, L. (2005). On the accuracy of statistical distributions in Microsoft Excel 2003. *Computational Statistics and Data Analysis, 48,* 445–449.

Kraemer, H.C. & Blasey, C. (2016), *How many subjects? Statistical power analysis in research.* Los Angeles, Sage.

Kriegeskorte, N., Simmons, W. K., Bellgowan, P. S. F., & Baker, C. I. (2009). Circular analysis in systems neuroscience: The dangers of double dipping. *Nature Neuroscience, 12,* 535–540.

Kruschke, J. K. (2015), *Doing Bayesian data analysis: A tutorial with R and BUGS.* 2d ed. New York: Academic/Elsevier.

Kuehn, B. (2006). Industry, FDA warm to "adaptive" trials. *Journal of the American Medical Association, 296,* 1955–1957.

Lakens, D.L. (2017). Equivalence Tests. Social Psychological and Personality Science 5: in press. Preprint at: http://dx.doi.org/10.1177%2F1948550617697177

Lamb, E. L. (2012). 5 sigma: What's that? *Scientific American* (blog), July 17. Accessed November 16, 2012, http://www.blogs.scientificamerican.com/observations/2012/07/17/five-sigmawhats-that/.

Lanzante, J. R. (2005). A cautionary note on the use of error bars. *Journal of Climate, 18,* 3699–3703.

Larsson, S. C., & Wolk, A. (2006). Meat consumption and risk of colorectal cancer: A meta-analysis of prospective studies. *International Journal of Cancer, 119,* 2657–2664.

Lazzeroni, L. C., Lu, Y., & Belitskaya-Lévy, I. (2014). P-values in genomics: Apparent precision masks high uncertainty. *Molecular Psychiatry, 19,* 1336–40.

Lazic, S. E. (2010). The problem of pseudoreplication in neuroscientific studies: Is it affecting your analysis? *BMC Neuroscience, 11*(5). doi: 10.1186/1471-2202-11-5

Laupacis, A., Sackett, D. L., & Roberts, R. S. (1988). An assessment of clinically useful measures of the consequences of treatment. *New England Journal of Medicine, 318,* 1728–1733.

Lee, K. L., McNeer, J. F., Starmer, C. F., Harris, P. J., & Rosati, R. A. (1980). Clinical judgment and statistics. Lessons from a simulated randomized trial in coronary artery disease. *Circulation, 61,* 508–515.

Lehman, E. (2007). *Nonparametrics: Statistical methods based on ranks.* New York: Springer.

Lenth, R. V. (2001). Some practical guidelines for effective sample size determination. *American Statistician, 55,* 187–193.

Levine, M., & Ensom, M. H. (2001). Post hoc power analysis: An idea whose time has passed? *Pharmacotherapy, 21,* 405–409.

Levins, R. (1966). The strategy of model building in population biology. *American Scientist, 54,* 421–431.

Lewin, T. (2008). Math scores show no gap for girls, study finds. *New York Times,* July 25. Accessed July 26, 2008, http://www.nytimes.com/2008/07/25/education/25math.html.

Limpert, E., Stahel, W. A., & Abbt, M. (2001). Log-normal distributions across the sciences: Keys and clues. *Biosciences, 51,* 341–352.

Limpert, E., and Stahel, W.A. (2011). Problems with Using the Normal Distribution–and Ways to Improve Quality and Efficiency of Data Analysis. PLoS ONE 6: e21403–8.

Lucas, M. E. S., Deen, J. L., von Seidlein, L., Wang, X., Ampuero, J., Puri, M., Ali, M., et al. (2005). Effectiveness of mass oral cholera vaccination in Beira, Mozambique. *New England Journal of Medicine, 352,* 757–767.

Ludbrook, L., & Lew, M. J. (2009). Estimating the risk of rare complications: Is the "rule of three" good enough? *Australian and New Zealand Journal of Surgery, 79,* 565–570.

Mackowiak, P. A., Wasserman, S. S., & Levine, M. M. (1992). A critical appraisal of 98.6 degrees F, the upper limit of the normal body temperature, and other legacies of Carl Reinhold August Wunderlich. *Journal of the American Medical Association, 268,* 1578–1580.

Macrae, D., Grieve, R., Allen, E., Sadique, Z., Morris, K., Pappachan, J., Parslow, R., et al. (2014). A randomized trial of hyperglycemic control in pediatric intensive care. *New England Journal of Medicine, 370*, 107–118.

Manly, B. F. J. (2006). *Randomization, bootstrap and Monte Carlo methods in biology,* 3d ed. London: Chapman & Hall/CRC.

Marsupial, D. (2013). Comment on: Interpretation of p-value in hypothesis testing. *Cross Validated,* January 3. Accessed January 5, 2013, at stats.stackexchange.com/questions/46856/interpretation-of-p-value-in-hypothesis-testing.

Masicampo, E. J., & Lalande, D. R. (2012). A peculiar prevalence of p values just below .05. *Quarterly Journal of Experimental Psychology, 65*, 2271–2279.

Mathews, P. (2010) *Sample size calculations. Practical methods for engineer and scientists.* Mathews, Malnar and Bailey. ISBN 978-0-615-32461-6

McCullough, B. D., & Hellser, D. A. (2008). On the accuracy of statistical procedures in Microsoft Excel 2007. *Computational Statistics and Data Analysis, 52,* 4570–4578.

McCullough, B. D., & Wilson, B. (2005). On the accuracy of statistical procedures in Microsoft Excel 2003. *Computational Statistics and Data Analysis, 49,* 1244–1252.

Mendel, G. J. (1865). Versuche über Pflanzen-Hybriden. *Verhandlungen des naturforschenden Vereines in Brünn, 4,* 3–47.

Messerili, F. H. (2012). Chocolate consumption, cognitive function, and Nobel laureates. *New England Journal of Medicine, 367,* 1562–1564.

Metcalfe, C. (2011). Holding onto power: Why confidence intervals are not (usually) the best basis for sample size calculations. *Trials, 12*(Suppl. 1), A101. doi: 10.1186/1745-6215-12-S1-A101.

Meyers, M. A. (2007). *Happy accidents: Serendipity in modern medical breakthroughs.* New York: Arcade.

Micceri, T. (1989). The unicorn, the normal curve, and other improbable creatures. *Psychological Bulletin, 105,* 156–166.

Microsoft Corporation. (2006). Description of improvements in the statistical functions in Excel 2003 and in Excel 2004 for Mac. Accessed December 16, 2008, http://www.support.microsoft.com/kb/828888.

Mills, J. L. (1993). Data torturing. *New England Journal of Medicine, 329,* 1196–1199.

Montori, V. M., & Guyatt, G. H. (2001). Intention-to-treat principle. *Canadian Medical Association Journal, 165,* 1339–1341.

Montori, V. M., Kleinbart, J., Newman, T. B., Keitz, S., Wyer, P. C., Moyer, V., & Guyatt, G. (2004). Tips for learners of evidence-based medicine: 2. Measures of precision (confidence intervals). *Canadian Medical Association Journal, 171,* 611–615.

Moser, B. K., & Stevens, G. R. (1992). Homogeneity of variance in the two-sample means test. *American Statistician, 46,* 19–21.

Motulsky, H. J. (2015). *Essential biostatistics.* New York: Oxford University Press.

Motulsky, H. J. (2016). GraphPad curve fitting guide. Accessed July 1, 2016, http://www.graphpad.com/guides/prism/7/curve-fitting/.

Motulsky, H. J., & Brown, R. E. (2006). Detecting outliers when fitting data with nonlinear regression: A new method based on robust nonlinear regression and the false discovery rate. *BMC Bioinformatics, 7*(123). doi: 10.1186/1471-2105-7-123

Motulsky, H., & Christopoulos, A. (2004). *Fitting models to biological data using linear and nonlinear regression: A practical guide to curve fitting.* New York: Oxford University Press.

Motulsky, H. J., O'Connor, D. T., & Insel, P. A. (1983). Platelet alpha 2-adrenergic receptors in treated and untreated essential hypertension. *Clinical Science, 64,* 265–272.

Moyé, L. A., & Tita, A. T. N. (2002). Defending the rationale for the two-tailed test in clinical research. *Circulation, 105,* 3062–3065.

Munger, K. L, Levin, L. I., Massa, J., Horst, R., Orban, T., and Ascherio, A. (2013). Preclinical serum 25-hydroxyvitamin D levels and risk of Type 1 diabetes in a cohort of US military personnel. *American Journal of Epidemiology, 177,* 411–419.

NASA blows millions on flawed airline safety survey. (2007). *New Scientist,* November 10. Accessed May 25, 2008, http://www.newscientist.com/channel/opinion/mg19626293.900-nasa-blows-millions-on-flawed-airline-safety-survey.html.

Newport, F. (2012). Romney has support among lowest income voters. Gallup, September 12. http://www.gallup.com/poll/157508/romney-support-among-lowest-income-voters. aspx.

Nieuwenhuis, S., Forstmann, B. U., & Wagenmakers, E.-J. (2011). Erroneous analyses of interactions in neuroscience: A problem of significance. *Nature Neuroscience, 14,* 1105–1107.

Omenn, G. S., Goodman, G. E., Thornquist, M. D., Balmes, J., Cullen, M. R., Glass, A., Keogh, J. P., et al. (1996). Risk factors for lung cancer and for intervention effects in CARET, the Beta-Carotene and Retinol Efficacy Trial. *Journal of the National Cancer Institute, 88,* 1550–1559.

Open Science Collaboration. (2015). Estimating the reproducibility of psychological science. *Science, 349.* doi: 10.1126/science.aac4716.

Oremus, W. (2013). The wedding industry's pricey little secret. *Slate,* June 12. Accessed June 13, 2013, http://www.slate.me/1a4KLuH.

Pace, L. A. (2008). *The Excel 2007 data and statistics cookbook.* 2d ed. Anderson, SC: TwoPaces.

Parker, R. A., & Berman, N. G. (2003). Sample size: More than calculations. *American Statistician, 57,* 166–170.

Paulos, J. A. (2008). *Irreligion: A mathematician explains why the arguments for God just don't add up.* New York: Hill and Wang.

Payton, M. E., Greenstone, M. H., & Schenker, N. (2003). Overlapping confidence intervals or standard error intervals: What do they mean in terms of statistical significance? *Journal of Insect Science, 3,* 34–40.

Pielke, R. (2008). Forecast verification for climate science, part 3. *Prometheus,* January 9. Accessed April 20, 2008, at sciencepolicy.colorado.edu/prometheus/archives/climate_change/001315forecast_verificatio.html.

Pocock, S. J., & Stone, G. W. (2016a). The primary outcome fails: What next? *New England Journal of Medicine, 375,* 861–870.

Prinz, F. F., Schlange, T. T., & Asadullah, K. K. (2011). Believe it or not: How much can we rely on published data on potential drug targets? *Nature Reviews Drug Discovery, 10*(712). doi: 10.1038/nrd3439-c1

Price, A. L., Zaitlen, N. A., Reich, D., & Patterson, N. (2010). New approaches to population stratification in genome-wide association studies. *Nature Reviews Genetics, 11,* 459–463.

Pukelsheim, F. (1990). Robustness of statistical gossip and the Antarctic ozone hole. *IMS Bulletin, 4,* 540–545.

Pullan, R. D., Rhodes, J., Gatesh, S., Mani, V., Morris, J. S., Williams, G. T., Newcombe, R. G., et al. (1994). Transdermal nicotine for active ulcerative colitis. *New England Journal of Medicine, 330,* 811–815.

Ridker, P. M., Danielson, E., Fonseca, F. A. H., Genest, J., Gotto, A. M., Jr., Kastelein, J. J. P., Koening, W., et al. (2008). Rosuvastatin to prevent vascular events in men and women with elevated C-reactive protein. *New England Journal of Medicine, 359,* 2195–2207.

Riley, R.D., Lambert, P.C., and Abo-Zaid, G. (2010). Meta-analysis of individual participant data: rationale, conduct, and reporting. Bmj 340: c221–c228.

Roberts, S. (2004). Self-experimentation as a source of new ideas: Ten examples about sleep, mood, health, and weight. *Behavioral and Brain Sciences, 27,* 227–262; discussion 262–287.

Robinson, A. (2008) *icebreakeR.* Accessed January 26, 2009, from https://cran.r-project.org/doc/contrib/Robinson-icebreaker.pdf.

Rosman, N. P., Colton, T., Labazzo, J., Gilbert, P. L., Gardella, N. B., Kaye, E. M., Van Bennekom, C., & Winter, M. R. (1993). A controlled trial of diazepam administered during febrile illnesses to prevent recurrence of febrile seizures. *New England Journal of Medicine, 329,* 79–84.

Rothman, K. J., Greenland S, & Lash, T. L. (2008). *Modern Epidemiology.* 3d ed. Philadelphia, Wolters Klewer.

Rothman, K. J. (1990). No adjustments are needed for multiple comparisons. *Epidemiology, 1,* 43–46.

Rothman, K. J. (2016). Disengaging from statistical significance. *European Journal of Epidemiology, 31,* 443–444.

Russo, J. E., & Schoemaker, P. J. H. (1989). *Decision traps. The ten barriers to brilliant decision-asking and how to overcome them.* New York: Simon & Schuster.

Ruxton, G.D. (2006). The unequal variance t-test is an underused alternative to Student's t-test and the Mann-Whitney U test. Behavioral Ecology 17: 688–690.

Samaniego, F. J. A (2008). Conversation with Myles Hollander. *Statistical Science, 23,* 420–438.

Sawilowsky, S. S. (2005). Misconceptions leading to choosing the t test over the Wilcoxon Mann–Whitney test for shift in location parameter. *Journal of Modern Applied Statistical Methods, 4,* 598–600.

Schmidt, M., Christiansen, C. F., Mehnert, F., Rothman, K. J., & Sorensen, H. T. (2011). Non-steroidal anti-inflammatory drug use and risk of atrial fibrillation or flutter: population based case-control study. *BMJ, 343.* doi: 10.1136/bmj.d3450

Schmidt, M., & Rothman, K. J. (2014). Mistaken inference caused by reliance on and misinterpretation of a significance test. *International Journal of Cardiology, 177,* 1089–1090.

Schoemaker, A. L. (1996). What's normal? Temperature, gender, and heart rate. *Journal of Statistics Education, 4*(2). Accessed May 5, 2007, http://www.amstat.org/publications/jse/v4n2/datasets.shoemaker.html.

Seaman, M. A., Levin, J. R., & Serlin, R. C. (1991). New developments in pairwise multiple comparisons: Some powerful and practicable procedures. *Psychological Bulletin, 110,* 577–586.

Sedgwick, P. (2015). How to read a funnelplot in a meta-analysis. *BMJ* 351:h4718.

Senn, S. (2003). Disappointing dichotomies. *Pharmaceutical Statistics, 2,* 239–240.

Sheskin, D. J. (2011). *Handbook of parametric and nonparametric statistical procedures,* 5th ed. New York: Chapman & Hall/CRC.

Shettles, L. B. (1996). *How to choose the sex of your baby.* Monroe, WA: Main Street Books.

Simmons, J. P., Nelson, L. D., & Simonsohn, U. (2011). False-positive psychology: Undisclosed flexibility in data collection and analysis allows presenting anything as significant. *Psychological Science, 22,* 1359–1366.

———. (2013). P-curve: A key to the file drawer. *Journal of Experimental Psychology: General, 143*, 534–547

Simon, S. (2005). Stats: Standard deviation versus standard error. *PMean* (blog), May 16. Accessed December 20, 2012, http://www.pmean.com/05/StandardError.html.

Simonsohn, U. (2016) Evaluating replications: 40% full ≠ 60% empty. *Data Colada* (blog), March 3. Accessed May 8, 2016, http://datacolada.org/47.

Snapinn, S. M. (2000). Noninferiority trials. *Current Control Trials in Cardiovascular Medicine, 1,* 19–21.

Sparling, B. (2001). Ozone history. NASA Advanced Supercomputing Division, May 30. Accessed June 13, 2008, http://www.nas.nasa.gov/About/Education/Ozone/history.html

Spector, R., & Vesell, E. S. (2006a). The heart of drug discovery and development: Rational target selection. *Pharmacology, 77,* 85–92.

———. (2006b). Pharmacology and statistics: Recommendations to strengthen a productive partnership. *Pharmacology, 78,* 113–122.

Squire, P. (1988). Why the 1936 *Literary Digest* poll failed. *Public Opinion Quarterly, 52,* 125–133.

Staessen, J. A., Lauwerys, R. R., Buchet, J. P., Bulpitt, C. J., Rondia, D., Vanrenterghem, Y., Amery, A., & the Cadmibel Study Group. (1992). Impairment of renal function with increasing blood lead concentrations in the general population. *New England Journal of Medicine, 327,* 151–156.

Statwing. (2012). The ecological fallacy. *Statwing* (blog), December 20. Accessed February 8, 2013, at blog.statwing.com/the-ecological-fallacy/.

Svensson, S., Menkes, D. B., & Lexchin, J. (2013). Surrogate outcomes in clinical trials: a cautionary tale. *JAMA Internal Medicine,173,* 611–612.

Thavendiranathan, P., & Bagai, A. (2006). Primary prevention of cardiovascular diseases with statin therapy: A meta-analysis of randomized controlled trials. *Archives of Internal Medicine, 166,* 2307–2313.

Taubes, G. (1995). Epidemiology faces its limits. *Science, 269,* 164–169.

Thomas, D., Radji, S., and Benedetti, A. (2014). Systematic review of methods for individual patient data meta- analysis with binary outcomes. BMC Medical Research Methodology 14: 79.

Thun, M. J., & Sinks, T. (2004). Understanding cancer clusters. *CA: A Cancer Journal for Clinicians, 54,* 273–280.

Tierney, J. (2008). A spot check of global warming. *New York Times,* January 10. Accessed April 20, 2008, at https://nyti.ms/2pLmpTT.

Turner, E. H., Matthews, A. M., Linardatos, E., Tell, R. A., & Rosenthal, R. (2008). Selective publication of antidepressant trials and its influence on apparent efficacy. *New England Journal of Medicine, 358,* 252–260.

Twain, M. (1883). *Life on the Mississippi.* London: Chatto & Windus.

U.S. Census (2011). Accessed January 30, 2012, http://www.census.gov/hhes/http://www/income/data/historical/household/2011/H08_2011.xls.

U.S. Food and Drug Administration. (2012). Orange book preface. *Approved drug products with therapeutic equivalence evaluations.* Accessed December 23, 2012, http://www.fda.gov/Drugs/DevelopmentApprovalProcess/ucm079068.htm.

Van Belle, G. (2008). *Statistical rules of thumb.* 2d ed. New York: Wiley Interscience.

Vandenbroucke, J. P., & Pearce, N. (2012). Case-control studies: basic concepts. *International Journal of Epidemiology, 41,* 1480–1489.

Vaux, D. L., Fidler, F., & Cumming, G. (2012). Replicates and repeats: What is the difference and is it significant? *EMBO Reports, 13,* 291–296.

Velleman, P. F., & Wilkinson, L. (1993). Nominal, ordinal, interval, and ratio typologies are misleading. *American Statistician, 47,* 65–72.

Vera-Badillo, F. E., Shapiro, R., Ocana, A., Amir, E., & Tannock, I. F. (2013). Bias in reporting of end points of efficacy and toxicity in randomized, clinical trials for women with breast cancer. *Annals of Oncology, 24,* 1238–1244.

Vickers, A. J. (2006). Shoot first and ask questions later: How to approach statistics like a real clinician. *Medscape Business of Medicine, 7*(2). Accessed June 19, 2009, http://www.medscape.com/viewarticle/540898.

———. (2010). *What is a P-value anyway?* Boston: Addison-Wesley.

Vittinghoff, E., Glidden, D. V., Shiboski, S. C., & McCulloch, C. E. (2007). *Regression methods in biostatistics: Linear, logistic, survival, and repeated measures models.* New York: Springer.

von Hippel, P.T. (2005). Mean, median, and skew: Correcting a textbook rule. *Journal of Statistics Education, 13*(2). ww2.amstat.org/publications/jse/v13n2/vonhippel.html.

Vos Savant, M. (1997). *The power of logical thinking: Easy lessons in the art of reasoning . . . and hard facts about its absence in our lives.* New York: St. Martin's Griffin.

Walker, E., & Nowacki, A. S. (2010). Understanding equivalence and noninferiority testing. *Journal of General Internal Medicine, 26,* 192–196.

Wallach J. D., Sullivan P. G. , Trepanowski J. F. , Sainani K. L. , Steyerberg E. W., Ioannidis J. P. A. (2017). Evaluation of Evidence of Statistical Support and Corroboration of Subgroup Claims in Randomized Clinical Trials. *JAMA Intern Med.*177(4):554–560.

Walsh, M., Srinathan, S. K., McAuley, D. F., Mrkobrada, M., Levine, O., Ribic, C., Molnar, A. O., et al. (2014). The statistical significance of randomized controlled trial results is frequently fragile: A case for a fragility index. *Journal of Clinical Epidemiology, 67,* 622–628.

Wasserman, L. (2012). P values gone wild and multiscale madness. *Normal Deviate* (blog), August 16. Accessed December 10, 2012, http://normaldeviate.wordpress.com/2012/08/16/p-values-gone-wild-and-multiscale-madness/.

Wasserstein, R. L., & Lazar, N. A. (2016). The ASA's statement on p-values: context, process, and purpose. *American Statistician, 70,* 129–133.

Wellek, S. (2002). *Testing statistical hypotheses of equivalence.* Boca Raton, FL: Chapman & Hall/CRC.

Westfall, P. H. (2014). Kurtosis as Peakedness, 1905–2014. R.I.P. *American Statistician, 68,* 191–195.

Westfall, P., Tobias, R., Rom, D., Wolfinger, R., & Hochberg, Y. (1999) *Multiple comparisons and multiple tests using the SAS system.* Cary, NC: SAS.

Wilcox, R. R. (2001). *Fundamentals of modern statistical methods: Substantially improving power and accuracy.* New York: Springer-Verlag.

———. (2010). *Fundamentals of modern statistical methods: Substantially improving power and accuracy.* 2d ed. New York: Springer.

Wilcox, A. J., Weinberg, C. R., & Baird, D. D. (1995). Timing of sexual intercourse in relation to ovulation. Effects on the probability of conception, survival of the pregnancy, and sex of the baby. *New England Journal of Medicine, 333,* 1517–1521.

Winston, W. (2004). Introduction to optimization with the Excel Solver tool. Accessed March 6, 2013, at https://support.office.com/en-us/article/An-introduction-to-optimization-with-the-Excel-Solver-tool-1F178A70-8E8D-41C8-8A16-44A97CE99F60%20#office.

Wolff, A. (2002). That old black magic. *Sports Illustrated,* January 21. Accessed May 25, 2008, http://sportsillustrated.cnn.com/2003/magazine/08/27/jinx/.

Wormuth, D. W. (1999) Actuarial and Kaplan-Meier survival analysis: There is a difference. *Journal of Thoracic and Cardiovascular Surgery, 118*, 973–975.

Wright, S. P. (1992). Adjusted P-values for simultaneous inference. *Biometrics, 48*, 1005–1013.

Xu, F., & Garcia, V. (2008). Intuitive statistics by 8-month-old infants. *Proceedings of the National Academy of Sciences of the United States of America, 105*, 5012–5015.

Yalta, A. T. (2008). The accuracy of statistical distributions in Microsoft Excel 2007. *Computational Statistics and Data Analysis, 52*, 4579–4586.

Zhang, J. H., Chung, T. D., & Oldenburg, K. R. (1999). A simple statistical parameter for use in evaluation and validation of high throughput screening assays. *Journal of Biomolecular Screening, 4*, 67–73.

Zhang, J., & Yu, K. F. (1998). What's the relative risk? A method of correcting the odds ratio in cohort studies of common outcomes. *Journal of the American Medical Association, 280*, 1690–1691.

Ziliak, S., & McCloskey, D. N. (2008). *The cult of statistical significance: How the standard error costs us jobs, justice, and lives.* Ann Arbor: University of Michigan Press.

Zollner, S., & Pritchard, J. K. (2007). Overcoming the winner's curse: Estimating penetrance parameters from case–control data. *American Journal of Human Genetics, 80*, 605–615.

INDEX

Page numbers followed by *f* or *t* indicate figures or tables, respectively.